D1174850

CONSTRUCTION ESTIMATING

CONSTRUCTION ESTIMATING

ROBERT W. PETRI

RESTON PUBLISHING COMPANY, INC.

A Prentice-Hall Company

Reston, Virginia

Illustrations by Thomas M. Williams

Library of Congress Cataloging in Publication Data

Petri, Robert W.
Construction estimating.

Includes index.
1. Building—Estimates. I. Title.
TH435.P44 692'.5 78-4821
ISBN 0-87909-152-5

1 2 3 4 5 6 7 8 9 10

PRINTED IN THE UNITED STATES OF AMERICA

To the late C. Robert McCormick who devoted his life to the development of youth and the advancement of tradespeople of California.

Table of Contents

List of Tables

Preface

This book has been designed to help the student, tradesman, prospective builder, or contractor learn how to estimate materials for room additions or for residential structures. Divided into ten sections, the book deals primarily with the rules and estimating procedures needed to make a material list of the various structural members found in a wood-frame structure. The first section deals with the estimator, his relationship with others he works with, and the background knowledge he must possess to become successful in the construction business.

The book was written for both the student and the experienced tradesman. For the student numerous illustrations are used to help clarify the written text and emphasize construction terminology, plan reading, basic code requirements, and good construction practices. For the experienced tradesman the estimating rules, tables, and procedures are an excellent reference and will help him speed up his material take-off of concrete, rough framing lumber, finish material, and hardware. Although every estimating situation the builder encounters cannot be handled in one book, the example problems do express a variety of estimating situations which will help in solving future problems.

CONSTRUCTION ESTIMATING

INTRODUCTION

1

The Estimator

The estimator's job is to prepare estimates of building project costs. Its importance cannot be overemphasized because the success of a contracting business depends on the accuracy of these estimates. The estimator's success will be based on his previous experience in and knowledge of the construction industry. A contractor or estimator lacking this experience may over- or under-estimate his project costs. That, in either case, could be detrimental to the success of his company: if costs are too high, his jobs will be few and far between; if costs are too low, he will not be able to stay in business.

THE ESTIMATOR'S BACKGROUND

Many successful estimators in the field of residential construction come from the craftsmen's rank and file. They have usually learned their own particular trade well and kept their eyes open to other trades around them. To become successful estimators they have had to acquire both experience and knowledge of the whole building industry.

Estimators working for large commercial firms may also have come from the craftsmen's ranks, but they usually have supplemented their knowledge and experience through formal education, perhaps only with a few night school courses, perhaps with a two-year or four-year degree in business or construction technology.

Some estimators in specialized trades may have acquired their knowledge of construction in the office as estimator trainees. In such cases, they are trained to take off and estimate the particular material or equipment of a specific trade even though they have no construction experience.

Those trades that use area measure as a basis of estimating are usually the easiest and fastest to learn, such as insulating, drywalling, plastering, stuccoing, painting, and roofing. Trades with harder to learn estimating involve the listing of many parts, such as plumbing, heating, electrical, and framing. It must be kept in mind that all trades require the estimator to read and understand building plans to some extent.

WHAT THE ESTIMATOR MUST KNOW

1. *He must have a thorough knowledge of the building trades.* This includes the various trades themselves, types of construction, methods of construction, and trade terminology. Although all the areas mentioned above are important to the estimator, trade terminology will often be the key to his success. For if he is to communicate effectively with subcontractors and material dealers, he must possess the ability to speak their language.

2. *He must be able to read building plans and notes and understand the specifications.* By studying the plans and specifications, the estimator will get a general view of those things that pertain to his phase of the project. If he finds any discrepancies between the plans and specifications, he will jot them down and bring them to the attention of the architect or owner for solution. When all the questions are answered and problems solved, he can then prepare and finish the cost estimate.

3. *He must have a thorough understanding of the building code and ordinances in the area the building project is to be constructed.* The estimator should understand that even though adjoining cities use the same basic code, certain items will vary because of the interpretations of different building departments. For this reason when questions arise about building practices and code requirements, the contractor-estimator should check with the local building department for the answers. Building officials and inspectors are public servants and are there to answer your questions at any time.

4. *He must have a thorough understanding of construction materials.* A contractor-estimator through experience will acquire the knowledge of a vast variety of construction materials. He must understand, therefore, the sizes, strengths, and capabilities of the materials with which he works. He must be able to intelligently substitute equal quality materials when specified materials are not available. By having a good working relationship with his material suppliers, he will know what materials are or are not available in the local area of construction. And last, he must keep in touch with and understand present and future prices of building materials.

5. *He must keep up with the new construction products and materials that are continually being developed.* To keep current, the contractor-estimator should visit trade and home shows and subscribe to construction trade and building magazines. There are some excellent magazines available today, and the advertising in these will keep him aware of what is new on the market.

6. *He has to possess some basic mathematical ability.* The estimator doesn't have to be a genius in math, but he must possess the ability to work accurately with the four basic functions: addition, subtraction, multiplication, and division. Every estimator should have a good basic or trade math book available to handle any special math problems that may arise.

7. *He has to have at his fingertips reference materials, books, tables, and tabulating equipment to speed his job.* In time, he will acquire reference materials in the form of

material catalogs, brochures, and manufacturer's specification sheets for the products he uses on many of his projects. He will acquire various reference books related to building and the construction industry. There are some excellent estimating, building material, and carpentry books on the market today that would add to and fill in the areas not covered in this book. And he will have to acquire time-saving estimating tables, usually printed tables from various sources, that may be used as is or modified for his personal needs. He may even choose to make up his own tables and waste factors.

With the new inexpensive pocket and desk calculators available today the estimator can tabulate and check his estimate with a minimum of time and effort.

8. *Finally, he must project labor cost changes.* The estimator must realize that labor costs may vary in different geographical areas of the state or country. He must also realize and project in his estimates future increases in labor costs because of upcoming union contract negotiations, for instance.

THE ESTIMATOR'S DUTIES

1. ***Quantity Material Take-off.*** This is an itemized list of materials needed to build a project or phase of a project. The items are listed in the order in which they will be used in the construction of the building. In this book we will be concerned with the material take-off of concrete, rough framing lumber, interior finish lumber, and hardware. These are the categories, with the occasional exception of concrete, that are usually handled by the general contractor. The other areas of a project are considered subcontracts; these would be subcontracted out to the specialty trades.

2. ***Labor Take-off.*** This concerns the number of man-hours or approximate hours and the cost for those hours for a particular phase of work on a project. There are estimating books and materials available today on labor costs, but these should only be used for fast or approximate estimates; they take into account a work average and not the work efficiency of one crew of men over another. Many contractor-builders on small jobs break down the project into small units of work (man-hours). They do this by studying the building plans to answer the questions of how many hours it will take for demolition on this job, how many hours to joist and cover the floor, how many hours to frame the walls, how many hours to add for any unforeseeable problems that may arise, and so forth. This time is all then added up and multiplied by the going rate per hour for a total labor cost. This method requires a good deal of field and estimating experience.

An extension of the above method for breaking down labor costs is the unit-labor cost for piecework. In other words, one- to four-men specialized crews do various phases of rough framing or finishing work and get paid by the lineal foot or square foot of material applied or constructed—12¢ a lineal foot for wall framing, or 50¢ a square foot for exterior siding. This method makes it easy for the estimator to plug in labor figures and come up with a total labor cost. Piecework, fortunately, is not used very extensively throughout the country, for it does not produce a very high quality of workmanship.

Other estimators, using previous experience, set up labor costs based on square foot costs. They may have various costs for different types of construction which, when multiplied by the square footage of a project, determines the total labor cost.

Whichever method is used, it is important to keep in mind that the ability of the work crew, the workmanship required, the tools used, and the weather conditions all play a major part in the effective production of a skilled craftsman.

3. ***Subcontractor Bids.*** Those phases of work on a project that the contractor does not do with his own personnel have to be subcontracted. The estimator must acquire from the various subcontractors the cost estimate, known as the *bid,* for labor and/or materials for their phase of work. Over the years a contractor-estimator will compile a list of reliable subcontractors. He should have from three to six "subs" for each phase of work on a project. The estimator contacts these subs by phone or mail to let them know there is work available to bid. Some subs will respond, though others may be too busy to take on any more work at that time. In any case, the estimator should acquire a minimum of three bids on all phases of work to keep the subcontractors competitive.

4. ***Compilation and Presentation of Total Bid Costs.*** The cost of material, labor, subcontracts, and the contractor's overhead-profit have to be added together to determine the total cost of a project. This total bid is then submitted by the contractor to the owner for his approval.

ESTIMATOR'S DUTIES DURING CONSTRUCTION

1. ***Organize Construction Time Schedules.*** The estimator may be called on to organize a tentative time schedule of events from the beginning to the end of each job. This schedule may involve ordering materials and lining up subcontractors to make sure the job runs smoothly. As the job progresses from phase to phase, the estimator may also be called on to schedule the building inspections.

2. ***Check on Fluctuating Prices and Availability of Materials.*** Quite a bit of time may elapse between the contractor's being awarded a project and the start of that project. For this reason, the estimator may be called on periodically to check with his material suppliers for price fluctuations and availability of certain materials. The estimator will also have to check on the time it will take to order and receive certain building products or specially milled materials.

3. ***Award Subcontracts.*** The estimator may be called on to contact and line up in his time schedule the subcontractors with the lowest bids. Some subs may be too busy to be scheduled for a certain time in the general contractor's time schedule. If this is the

case, the general contractor has the choice of holding up the job until the subcontractor can get there or calling the sub with the next lowest bid and scheduling him for the job. The general contractor's cost for doing this will be higher but may be offset by getting the job done on time.

4. ***Record Job Progress.*** The estimator, as previously mentioned, may set up a tentative time schedule before a project starts. This tentative schedule, after construction does start, will be checked by the estimator against the actual job progress. In an ideal situation the job runs smoothly, according to the tentative schedule. But unforeseeable problems may arise that will not allow the work to progress on schedule—problems such as late deliveries of material, subs not reporting on time, inclement weather, labor strikes, etc. A form like the one illustrated on p. 43 of Unit 6 will help the estimator understand where each of his jobs should be, and where the job actually is, so he can schedule the unfinished work accordingly.

5. ***Do an Analysis at Job Completion.*** It is advisable for the beginning or experienced estimator to analyze a few jobs now and then to come up with an accurate or sound basis for doing fast or preliminary estimates in the future. The analysis should also tell the contractor or estimator if he is within the profit margin set by the company. If a contractor did not make the percentage of profit he expected from a job, where did the excessive cost go? Was it in the labor, in the materials, or in an item that was forgotten in the final bid?

If a contractor made a higher percent of profit than expected (there really is no such thing as excessive profit in the opinion of contractors), the estimator should try to find out where it came from. Was it saved on labor? If so, at what stage of construction? Was it saved on material? If so, on what type of material?

When the construction industry is booming, a contractor will usually bid his jobs higher. When the industry slows down, he will have to sharpen his pencil and become more competitive if he is to survive. In slow times the contractor who has analyzed his previous jobs carefully will be more competitive, thus keeping himself and his staff in business.

STEPS IN ESTIMATING

1. ***Building Plans, Notes, and Specifications.*** The estimator must familiarize himself with the building plans and specifications. This may take a few minutes or many hours, depending on the estimator's past experience and the complexity of the project. He will take notes on items in question to be answered later by the architect or owner. After these questions are answered, he is ready to go on to the next step.

2. ***Material Take-off.*** The estimator must make an itemized list of the materials his company will be responsible for on a project. The items on this list, as mentioned before, are usually arranged in the order they are to be used in the construction of the building. After completing this itemized take-off, the estimator sends it to the mate-

rial suppliers for a competitive cost breakdown.

3. *Labor Take-off.* The estimator's method of assessing labor costs will be used to determine the hours and cost for those hours of his phase of work on a building. The various methods of labor take-off mentioned on pp. 5 and 6 all rely on the contractor-estimator's previous estimating and field experience. Without this previous experience, a beginning estimator will find it almost impossible to do an accurate labor take-off.

4. *Subcontractor Bids.* The estimator must make the building plans available to the subcontractor so he can take off the materials and labor for his phase of the building. Ethically, the estimator should award the work to the subcontractor with the lowest bid, but this does not always hold true. Some contractors may favor certain subs even though their bid is not the lowest because they believe that good workmanship, dependability, and fast performance are worth the extra cost.

5. *Equipment Rentals.* Occasionally, certain equipment is needed on a job that is not worth the contractor's investment. This equipment, when needed, may be rented from an equipment rental yard. The estimator must survey the job, decide what rental equipment will be needed, check on the costs of these items, and add them to his estimate. Items usually rented include fencing, power poles, power tools and equipment, machinery, toilet facilities, scaffolding, etc.

6. *Complete Building Costs.* The estimator must add up the building permit fees, material costs, labor costs, subcontract costs, equipment rental costs, and add to this his contingency, overhead, and profit figures. The contingency figure is an amount of money added to the bid for any unforeseeable or forgotten items. The contingency and profit can be lumped together because any money left in the contingency fund becomes profit.

Overhead refers to a contractor's operating costs. It includes such items as office rental and utilities, office equipment, salaries of office and management personnel, equipment (e.g., trucks, power and hand tools), and maintenance. These costs are continuous even though there are no projects presently scheduled. Therefore, overhead costs are spread over each of the contracted jobs.

A contractor often draws a salary for his time and effort on a job. The added responsibilities of supervision, plus the financial risk he assumes, also entitle him to a certain percent of the profit. This percentage is usually based on the availability of work and on competition.

METHOD OF ESTIMATING

The size of a contractor's business and the use of the company personnel determine the estimator's method of estimating his jobs. For example, the contractor-estimator who estimates the materials for a job he is going to supervise or build will be able to order specific lengths of lumber for each phase of framing, and when he starts the job, he will know just where each length should be used. This builder may estimate his job with a minimum of waste.

On the other hand, the contractor who employs an estimator to estimate the material and who has a foreman or superintendent on the job must order enough material so that the foreman and crew do not run short. With a crew of men there is often indiscriminate cutting of certain lengths of material. The estimator must therefore increase his estimates to allow for waste material. It is less expensive for the contractor to have a few extra lengths of lumber at the end of a job, than to delay the job while obtaining the needed material.

PRELIMINARY OR FAST ESTIMATE

A preliminary or fast estimate is one in which the estimator doesn't have to go through the process of completing a material and labor take-off or compiling subcontractor bids to arrive at a cost for a project. He can devise a method of determining an approximate or actual accurate cost. This cost is usually based on an analysis of previous jobs and is often given as a square foot price for a certain type of construction, at so many dollars per square foot. The estimator must also take into account various framing methods and materials used when making up this square foot cost. Many architects also use a preliminary estimate in their first contact with clients. This estimate gives their clients a ballpark figure of what the project will cost.

GENERAL CONTRACTOR'S COMPLETE ESTIMATE

A general contractor's complete estimate includes the building permit fees, the material, the labor, and the machinery and equipment necessary to complete a project. If the general contractor employs an architect or home designer and provides the drawings, he will also include the cost of the building plans.

SUBCONTRACTOR'S ESTIMATE

Subcontractors hired by the general contractor usually furnish the materials, equipment, and labor necessary for their phase of work on a building project. Often, they take care of the costs of building permit fees as well. Occasionally, the general contractor, possibly because of his buying power, may want to furnish the materials and have the subcontractor simply bid on the labor. The subcontractor's bid, in this case, is known as a labor estimate. In such a situation the general contractor is responsible for the safety and security of the materials. In other instances, the general contractor may not want this responsibility and will ask the subcontractor to provide the material and labor for the project. For example, a framing subcontractor will be asked to provide the framing lumber. In this case the framing subcontractor becomes responsible for the security of the materials. This may cost the general contractor more, but it relieves him of many headaches on large jobs.

THE OWNER-BUILDER

Sometimes to cut building costs, the owner may want to act as his own general contractor. When he does, he must secure the building permits himself and contract with various subcontractors for the different phases of work. Before assuming this responsibility, the owner ought to know something about construction. If he doesn't, he and his subcontractors may end up in many disputes before the job is finished. The owner should also keep in mind that the more competitive bids he gets for each phase of work, the lower his building costs may be. And he should make sure that his subcontractors are reliable and that they specify in a written contract the following:

1. What work will be done and when.
2. What work will *not* be done.
3. What materials will be furnished.

2

Building Plans

The estimator has to be able to read and interpret building plans and notes along with the specifications if he is to do an accurate material take-off.

A simple set of building plans consists of the following drawings.

1. ***Plot Plan.*** This shows the size and shape of the lot and the position of the buildings on the lot. It may also show some or all of the following:

A. Building setbacks
B. Gas lines
C. Electrical lines
D. Sewer lines
E. Water lines
F. Driveways
G. Sidewalks
H. Patio slabs
I. Fences
J. Easements
K. Topography (contour of lot)
L. Landscaping.

2. ***Foundation Plan.*** This shows the locations of concrete footings, foundation walls, underpinning, concrete piers, flatwork, concrete steps, masonry fences, and retaining walls. It will also show some or all of the following:

A. Types of steel reinforcing
B. Position of steel
C. Direction and size of girder
D. Direction and size of floor joist
E. Type and size of subflooring material
F. Foundation access holes or basement windows
G. Dropped floors for tile
H. Doubled joist
I. Lavatories, tubs, water closets, showers, and water heater positions
J. Hose bib positions.

3. ***Floor Plan.*** This is the most important drawing in a set of building plans. It contains some information for most of the building trades involved in constructing a project. The floor plan will usually include the following:

A. Building dimensions
B. Location of doors, windows, and wall partitions
C. Equipment and fixtures
D. Lowered ceilings
E. Built-in fixtures
F. Cabinet layout
G. Wall finishes
H. Flooring material
I. Fireplace layout
J. Floor elevation (steps)
K. Access holes
L. Plumbing and heating fixtures.

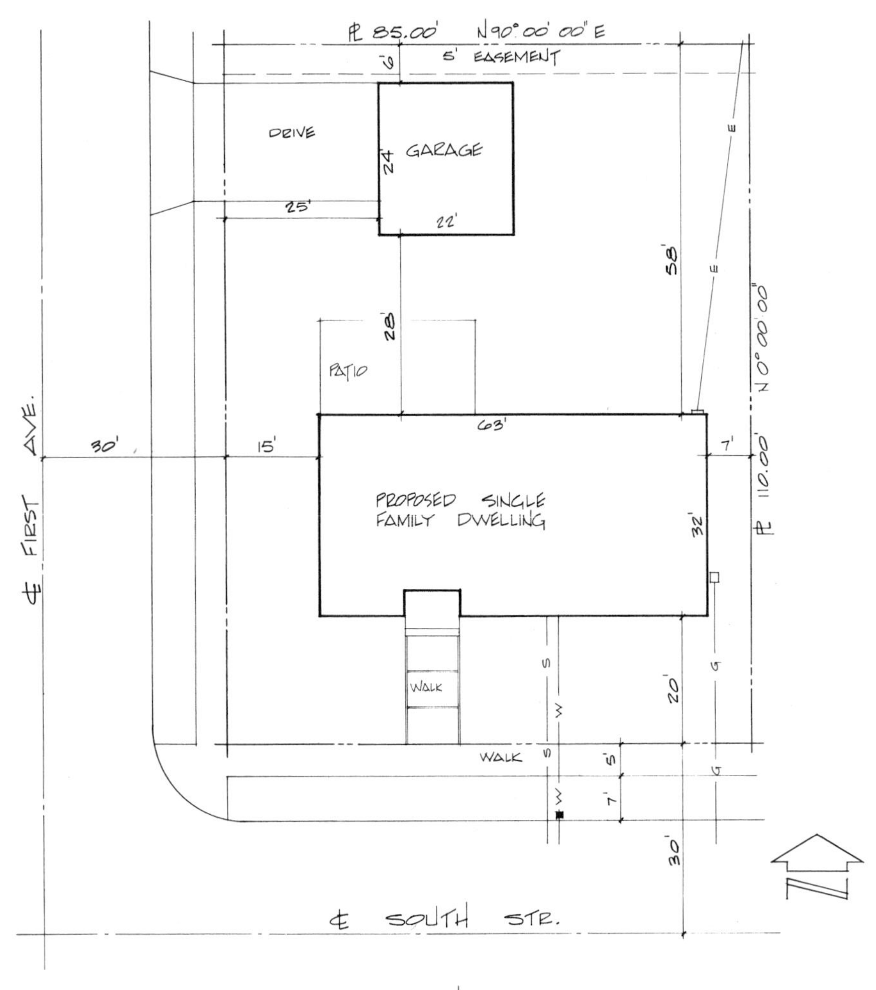

PLOT PLAN SC. 1"=10'

LEGAL DESCRIPTION:

LOT 1242, TRACT 19780
CITY OF CONSTRUCTION

INDEX:

SHEET	TITLE
1	PLOT PLAN
2	FOUNDATION PLAN
3	FLOOR PLAN
4	ELEVATIONS
5	DETAILS

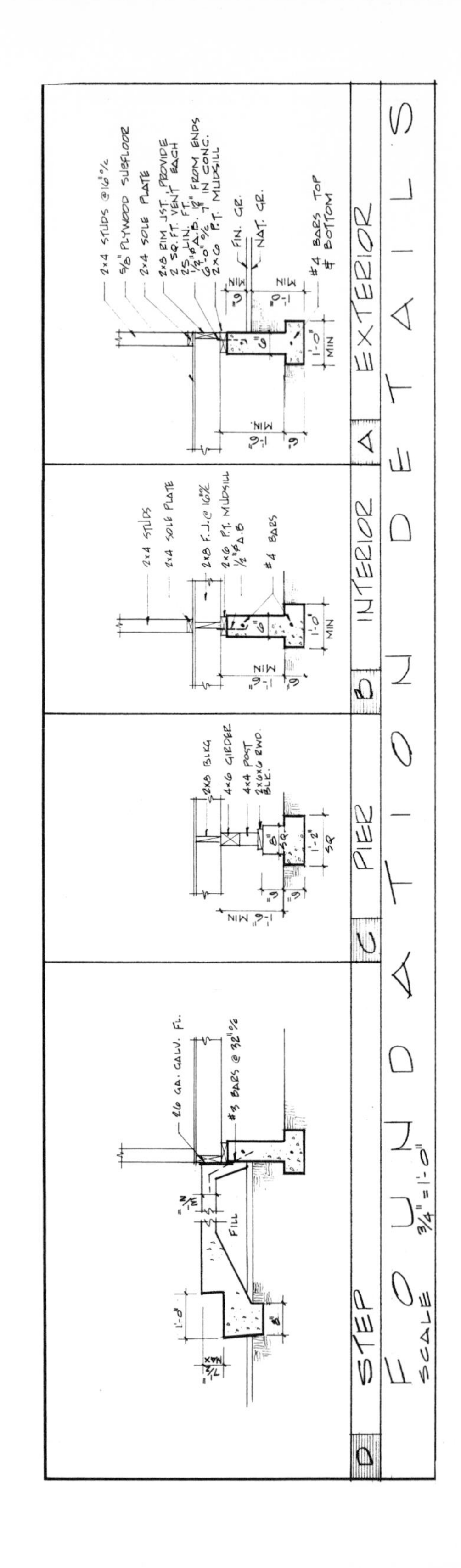

FOUNDATION DETAILS
SCALE 3/4" = 1'-0"
A EXTERIOR
2x4 STUDS @16" o/c
5/8" PLYWOOD SUBFLOOR
2x4 SOLE PLATE
2x8 RIM JST. PROVIDE 2 SQ. FT. VENT EACH 25 LIN. FT.
1/2" Ø A.B. 12" FROM ENDS 6'-0" o/c 7" IN CONC.
2x6 P.T. MUDSILL
FIN. GR.
NAT. GR.
#4 BARS TOP & BOTTOM
1'-6" MIN.
6"
1'-0" MIN
B INTERIOR
2x4 STUDS
2x4 SOLE PLATE
2x8 F.J. @ 16" o/c
2x6 P.T. MUDSILL
1/2" Ø A.B.
#4 BARS
C PIER
2x8 BLKG
4x6 GIRDER
4x4 POST
2x6x6 RWD. BLK.
8" SQ.
1'-2" SQ
D STEP
26 GA. GALV. FL.
#3 BARS @ 32" o/c
FILL
1'-0"
8"
7 1/2" MAX

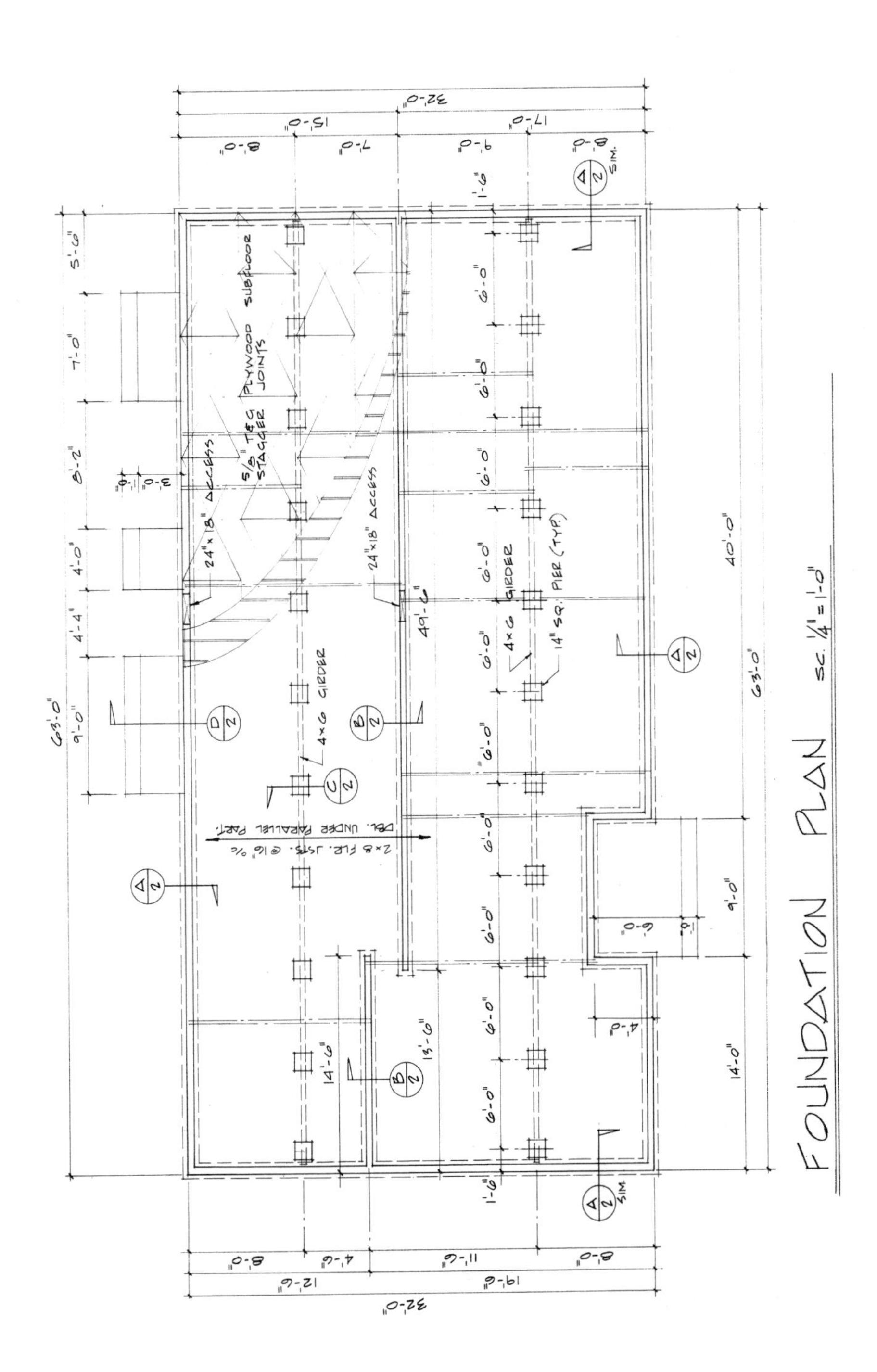
FOUNDATION PLAN
SC. 1/4" = 1'-0"
5/8" T&G PLYWOOD SUBFLOOR
STAGGER JOINTS
24"x18" ACCESS
24"x18" ACCESS
4x6 GIRDER
4x6 GIRDER
14" SQ. PIER (TYP.)
2x8 FLR. JSTS. @ 16" o/c
DBL. UNDER PARALLEL PART.
63'-0"
40'-0"
32'-0"
15'-0"
17'-0"
19'-6"
12'-6"
14'-0"
9'-0"

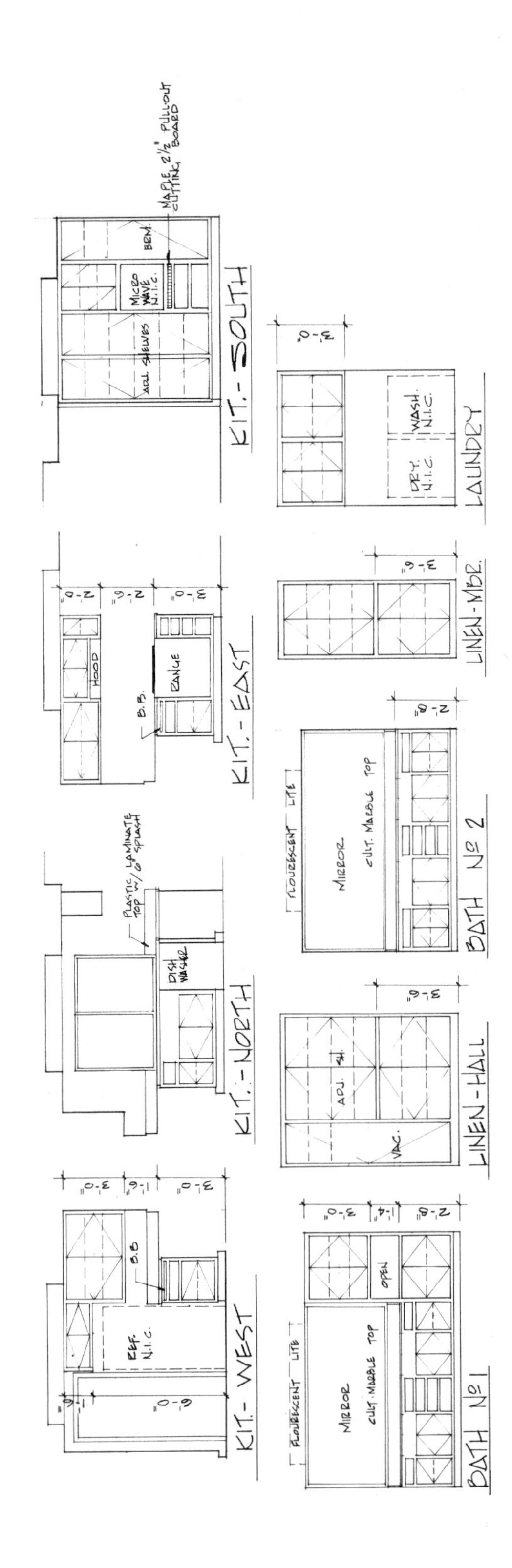
KIT.- SOUTH
MAPLE 2½" PULLOUT CUTTING BOARD
BRM.
MICRO WAVE N.I.C.
ADJ. SHELVES
LAUNDRY
DRY. N.I.C.
WASH. N.I.C.
3'-0"
KIT.- EAST
HOOD
RANGE
B.B.
2'-0"
2'-6"
3'-0"
LINEN-MBR.
3'-6"
KIT.- NORTH
PLASTIC LAMINATE TOP W/ 6" SPLASH
DISH WASHER
BATH № 2
FLOURESCENT LITE
MIRROR
CULT. MARBLE TOP
2'-8"
LINEN-HALL
ADJ. SH.
VAC.
3'-6"
KIT.- WEST
REF. N.I.C.
B.B
3'-0"
1'-6"
3'-0"
6'-0"
BATH №1
FLOURESCENT LITE
MIRROR
CULT. MARBLE TOP
OPEN
3'-0"
1'-4"
2'-8"

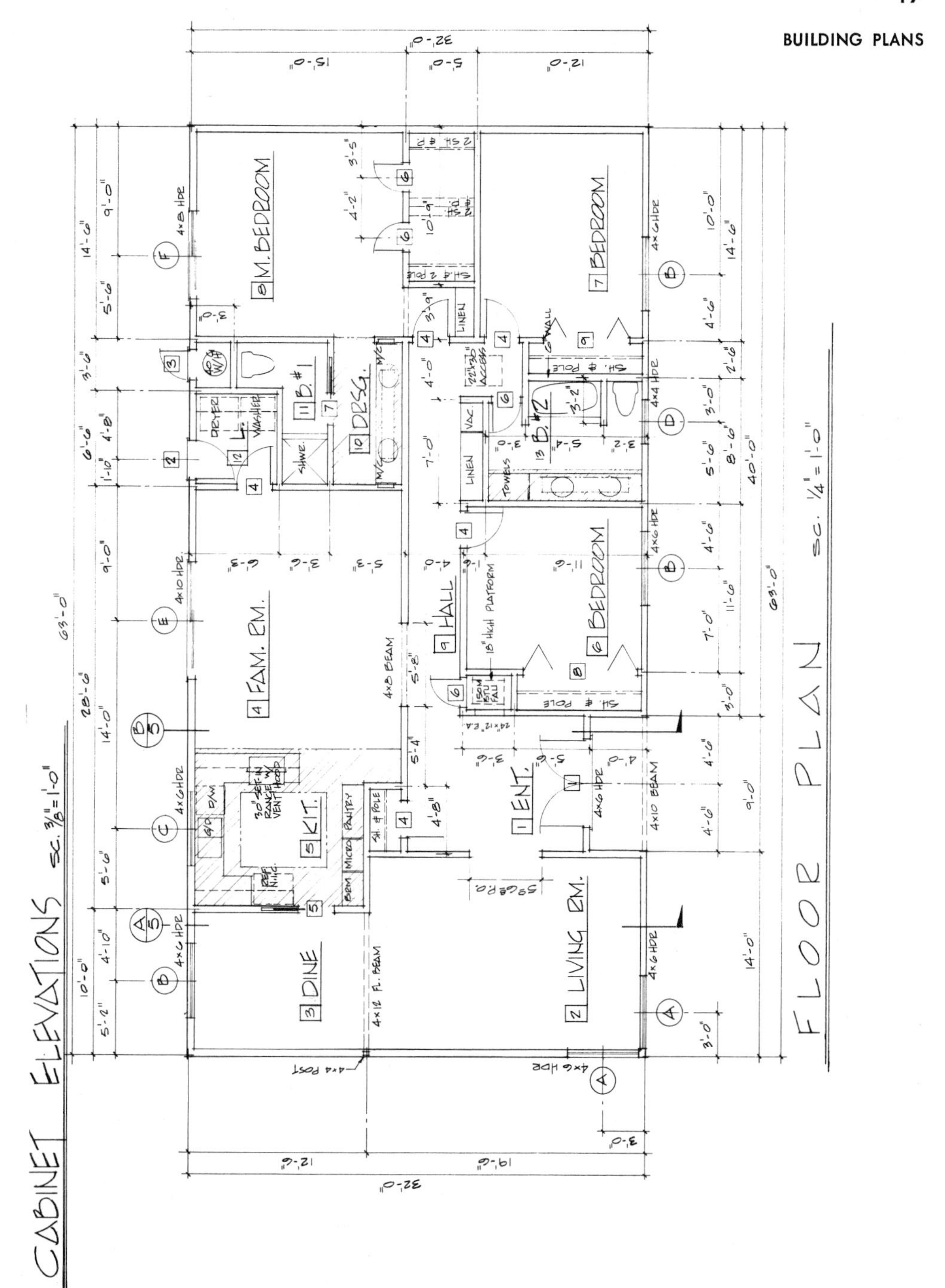
FLOOR PLAN
SC. 1/4" = 1'-0"
CABINET ELEVATIONS
SC. 3/8" = 1'-0"
M. BEDROOM
BEDROOM
FAM. RM.
HALL
KIT.
DINE
LIVING RM.
ENT.
LINEN
32'-0"
63'-0"

4. *Elevation Views.* These will show what the building will look like when finished. It will include information such as:

- **A.** Types of wall coverings
- **B.** Roofing materials
- **C.** Floor and ceiling lines
- **D.** Exterior finish material
- **E.** Roof slopes

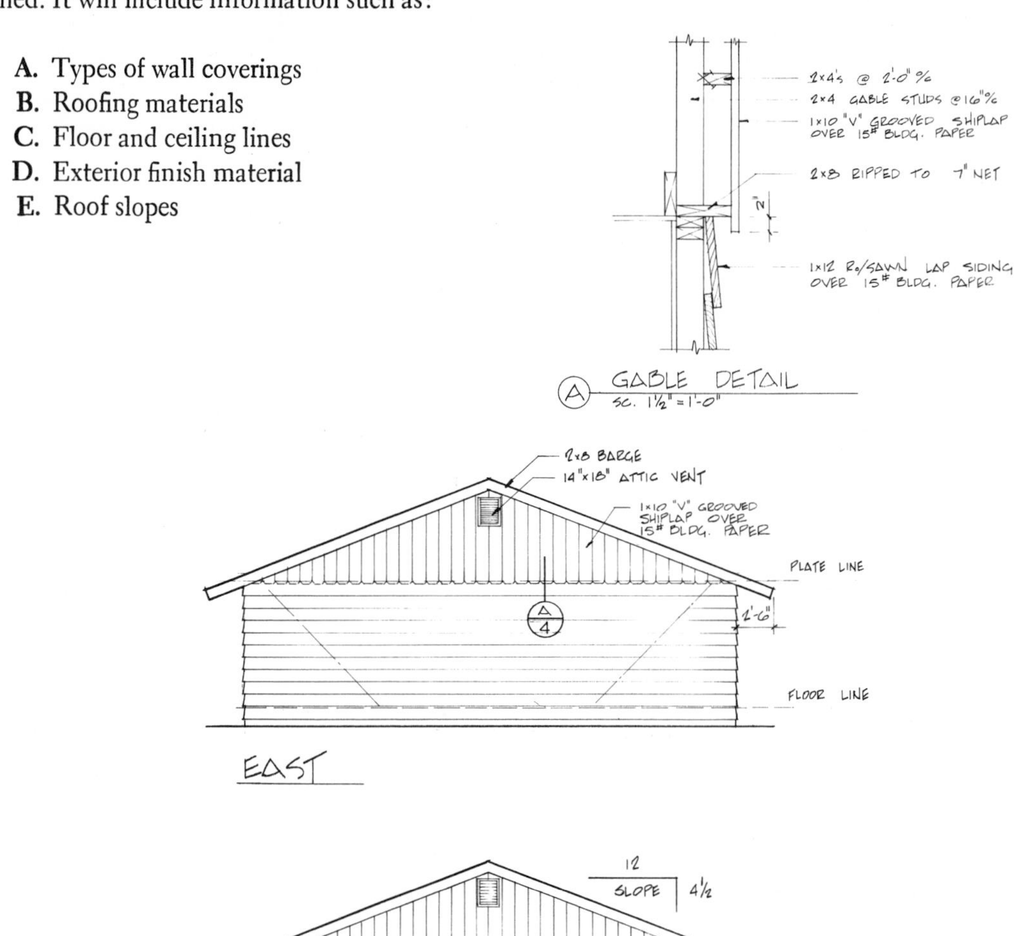

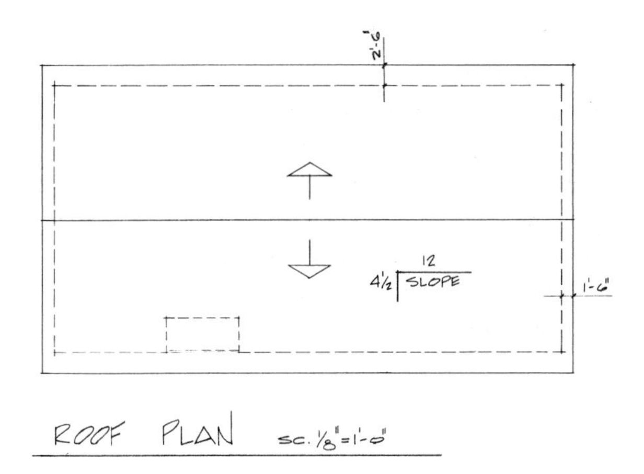

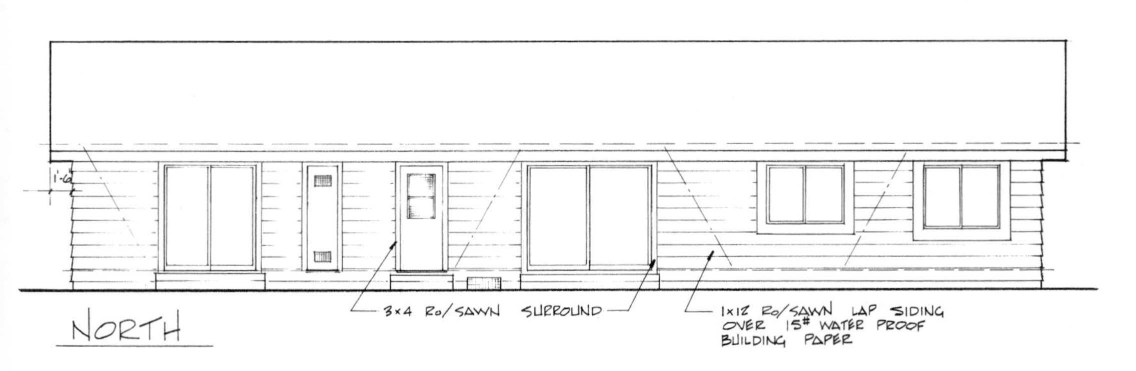

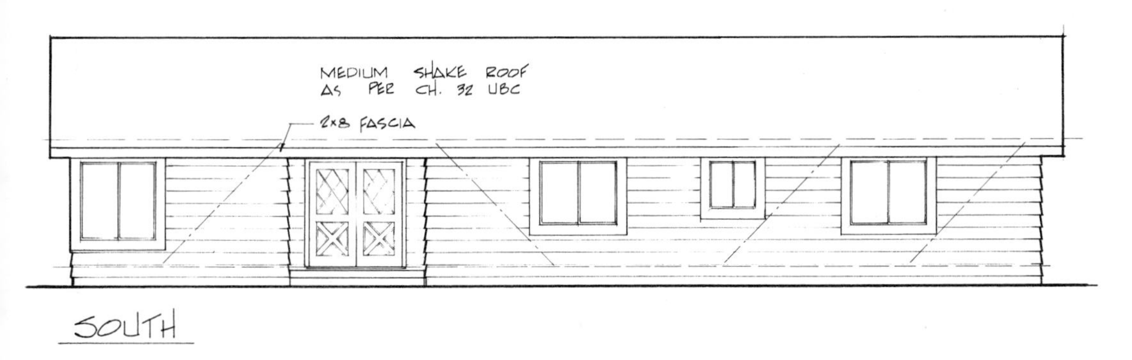

ELEVATIONS sc. 1/4"=1'-0"

5. ***Details and Section Views.*** When doing a material take-off, the estimator must locate details and reference material from one drawing to another. The architect co-ordinates these details from drawing to drawing with a system of reference symbols. The following include examples of typical reference symbols used by many architects.

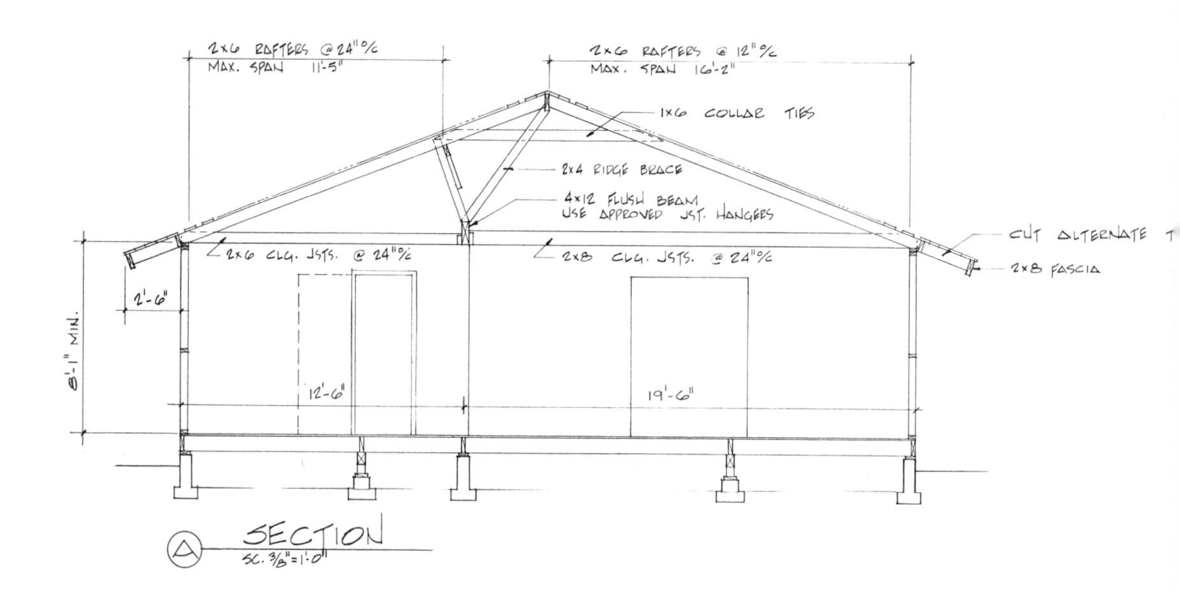

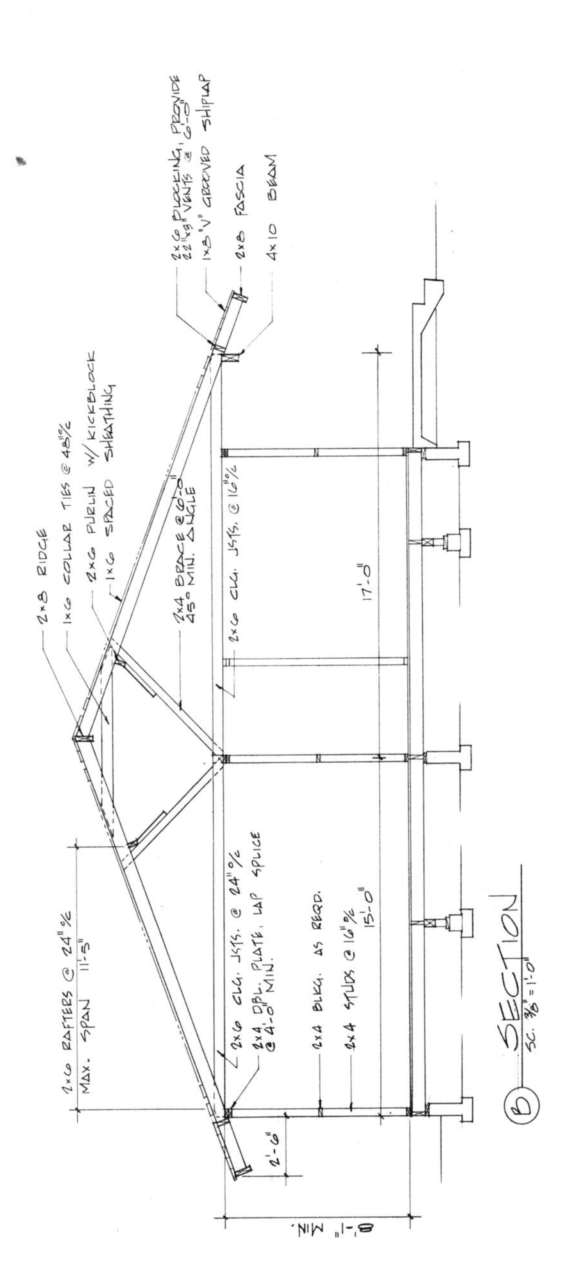
2x8 RIDGE
1x6 COLLAR TIES @ 48"o/c
2x6 PURLIN W/ KICKBLOCK
1x6 SPACED SHEATHING
2x4 BRACE @ 6'-0" 45° MIN. ANGLE
2x6 CLG. JSTS. @ 16"o/c
2x6 BLOCKING, PROVIDE 22"x8" VENTS @ 6'-0"
1x8 "V" GROOVED SHIPLAP
2x8 FASCIA
4x10 BEAM
17'-0"
2x6 RAFTERS @ 24" o/c MAX. SPAN 11'-5"
2x6 CLG. JSTS. @ 24"o/c
2x4 DBL. PLATE, LAP SPLICE @ 4'-0" MIN.
2x4 BLKG. AS REQD.
2x4 STUDS @ 16"o/c
15'-0"
2'-6"
8'-1" MIN.
B
SECTION
SC. 3/8"=1'-0"

SECTION VIEWS

A. *Door Schedule.* The number or symbol used in the illustration below will be found on the floor plan and will be repeated on the door schedule. The following symbols are used by many architects for doors and windows.

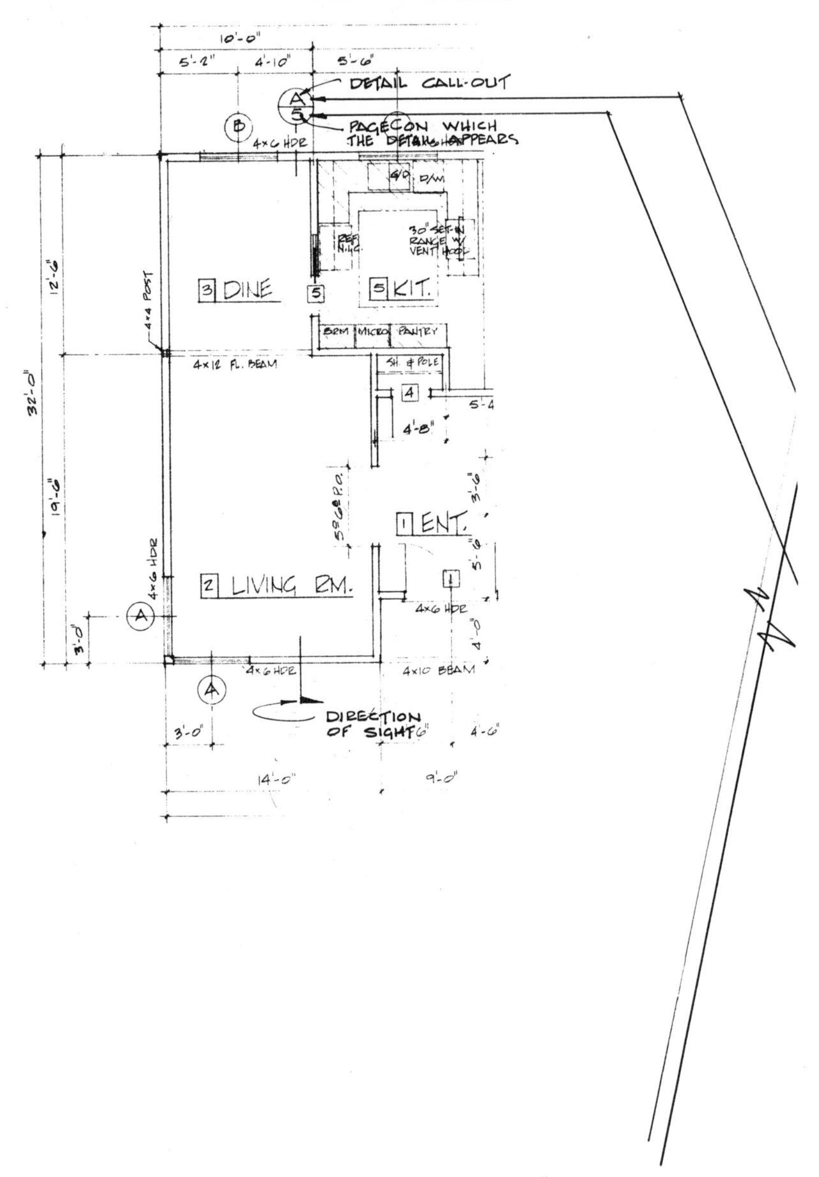

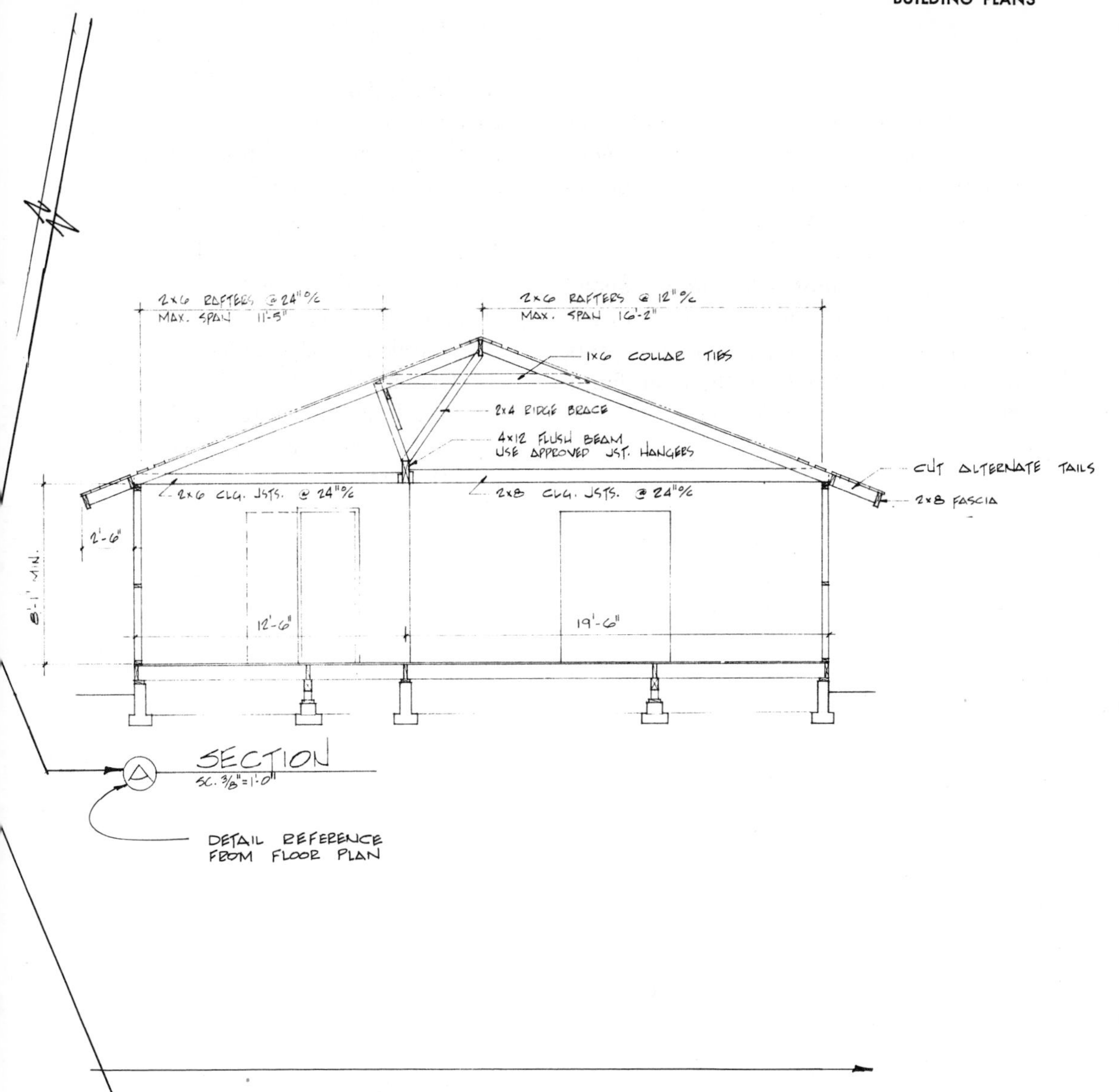
2x6 RAFTERS @ 24" O/C
MAX. SPAN 11'-5"
2x6 RAFTERS @ 12" O/C
MAX. SPAN 16'-2"
1x6 COLLAR TIES
2x4 RIDGE BRACE
4x12 FLUSH BEAM
USE APPROVED JST. HANGERS
CUT ALTERNATE TAILS
2x8 FASCIA
2x6 CLG. JSTS. @ 24" O/C
2x8 CLG. JSTS. @ 24" O/C
2'-6"
8'-1" MIN.
12'-6"
19'-6"
A
SECTION
SC. 3/8"=1'-0"
DETAIL REFERENCE
FROM FLOOR PLAN

B. *Window Schedule.* The window or opening number will be shown on the floor plan and/or on the elevation views and repeated on the window schedule.

C. *Interior Finish Schedule.* Sometimes the architect numbers or letters each room on the floor plan, and then repeats these numbers or letters on the room finish schedule or interior wall elevation.

Other drawings, depending on how extensive the project, could be included in a set of building plans:

A. Electrical plan
B. Plumbing plan
C. Heating and air conditioning plan
D. Floor framing plan
E. Roof framing plan
F. Door and window schedules
G. Room finish schedule
H. Landscaping plan
I. Miscellaneous detail plan
(Contains hard-to-classify items, such as flashing and gutters, attic louvers, planters, room dividers, access panels, letter slots, etc.).

DOOR SCHEDULE

SYM	Nº REQ'D.	SIZE	THK.	TYPE	FINISH	HARDWARE	REMARKS
1	2	3'-0" x 6'-8"	1¾"	CROSS-BUCK	PAINT GR.	KWICKSET	12 LITE
2	1	2'-8" x 6'-8"		BEL-AIRE			
3	1	2'-0" x 6'-8"	↓	H.C.			60 SQ.IN. VENTS
4	4	2'-6" x 6'-8"	1⅜"				
5	1	2'-6" x 6'-8"					POCKET
6	4	2'-0" x 6'-8"					
7	1	2'-4" x 6'-8"				↓	POCKET
8	1	8'-0" x 6'-8"					
9	1	6'-0" x 6'-8"	↓	↓	↓		

WINDOW SCHEDULE

SYM	Nº REQ'D.	SIZE	TYPE	REMARKS
A	2	5'-0" x 5'-0"	AL. SLDG.	
B	3	5'-0" x 4'-0"		
C	1	5'-0" x 3'-9"		
D	1	3'-0" x 3'-0"	↓	
E	1	8'-0" x 6'-8"	AL. SL. DR.	TEMP. GLASS
F	1	6'-0" x 6'-8"	↓	↓

INTERIOR FINISH SCHEDULE

NO.	ROOM	FLOOR	BASE	WALLS NORTH	WALLS EAST	WALLS SOUTH	WALLS WEST	CEILING	HGT	CABINETS	FINISH	REMARKS
1	ENTRY	CER. TILE	WOOD	½" D.WALL	½" D.WALL	½" D.WALL	½ D.WALL	BLOWN ACOU.	8'-0"			
2	LIVING ROOM	CARPET										
3	DINING ROOM											
4	FAMILY ROOM	↓	↓					↓	↓			
5	KITCHEN	LINO.	COVE					½" DRYWALL	7'-6"	ASH	LAC.	
6	BEDROOM Nº 1	CARPET	WOOD					BLOWN ACOU.	8'-0"			
7	BEDROOM Nº 2											
8	MASTER BEDROOM											
9	HALL								↓	PAINT GD.	ENAMEL	
10	DRESSING ROOM							↓	7'-6"			
11	BATH Nº 1	↓	↓					½" DRYWALL	8'-0"	ASH	LAC.	
12	LAUNDRY	LINO.	COVE							PAINT GD	ENAMEL	
13	BATH Nº 2	↓	↓	↓	↓	↓	↓	↓	↓	ASH	LAC.	

SHEET TITLE:

FLOOR PLAN

PROPOSED RESIDENCE FOR:

MR. & MS. I. AM ESTIMATOR
1978 SOUTH STREET
CITY OF CONSTRUCTION

DATE 9-12-77
SCALE
DRAWN T. WMS.
JOB
SHEET 3
OF 5 SHEETS

REVISIONS | BY

3

Specifications

In the building industry specifications are understood to be a written set of instructions that accompany a graphic set of building plans. They are usually printed in an 8-1/2 × 11 size booklet. They include those details of a project that normally are not handled on the working drawings.

Not all building plans have to be accompanied by a set of written specifications. Some architects or home designers, because of economic reasons, include some of the verbal description from the specifications on the graphic plans in the form of building notes. The architect must be careful when using this method, however, for too much detail will make the graphic plans become cluttered and confusing.

The precise legal definition of *specifications* is as follows: "Specifications are a written document of instruction that is supplemented *by* a graphic set of building plans." Specifications, notes, and dimensions are verbal descriptions and always take legal precedence over the graphic description, except where there has been a deliberate intent by the architect or owner to defraud. That is, when something is written into the specifications and notes or a deliberate graphic error has been made that would purposefully cause the contractor to bid the job wrong, the specifications and/or graphic plans are not legally binding.

The specifications are usually written by the architect or someone in his office familiar with the building plans. The specifications convey the ideas of the architect and the owner and provide the verbal description necessary to execute a legal contract. This legal contract in turn protects the owner. It explains what he can expect as far as quality of materials and equipment and helps insure good workmanship throughout his project.

The architect may choose to use a standardized specification format produced through many years of experience by the Construction Specification Institute (C.S.I), or he may make up his own general specification categories. Depending on the size and complexity of the project, some or all of the following categories may be used in the specifications:

1. General Conditions
2. Site work (grading and excavating)
3. Concrete work
4. Masonry
5. Rough carpentry
6. Finish carpentry
7. Millwork
8. Roofing
9. Insulation
10. Moisture protection
11. Glazing
12. Plaster or wallboard
13. Tile
14. Sheet metal
15. Electrical
16. Plumbing
17. Heating, ventilating, and air conditioning
18. Painting and finishing
19. Appliances (sometimes known as specialties)
20. Landscaping.

It must be noted that General Conditions (category no. 1), will include detailed information and requirements regarding the following:

- **A.** Owner-architect-builder contracts
- **B.** Relationship and legal rights of all parties of contracts
- **C.** Building permits
- **D.** Insurance
- **E.** Personal or surety bonds
- **F.** Contract payment provisions
- **G.** Provisions for job change
- **H.** Completion of work.

When signing contracts or bidding on work, the owner, contractor, and his subcontractors should fully understand the terms in the General Conditions in order to eliminate future job disputes.

The general contractor often uses his own employees for the rough and finish carpentry work. If this is the case, he will use the rough and finish carpentry categories from the specifications as a guide for estimating the work.

The information found in the rough carpentry category will usually include:

1. ***Scope of the Work.*** The general contractor will provide all labor, materials, equipment, transportation, and services necessary to perform all rough carpentry work to completion as indicated on drawings or specified in the specifications.

2. ***Materials.*** All lumber and plywood used for rough framing shall be grade stamped with approved standard markings. All builder's rough hardware shall be new material of standard manufacture, made in the United States of America, unit size and quantities as required by the building code, as detailed and as approved by the architect. Hardware exposed to moisture must be hot-dip galvanized steel or an approved type of nonferrous metal.

3. ***Supervision.*** The work of this section is to be under continuous supervision of a competent foreman satisfactory to the owner, who shall have authority to act for this contractor in matters relating to this work.

4. ***Inspections.*** No work of this section is to be covered until it is inspected and approved by the architect or his representative. The contractor shall cooperate with and follow any orders or requests of the architect or his representative.

5. ***Framing.*** The framing must be substantial and neatly erected, to the standard of trades and satisfactory to architect and owner. It must include all work required to be ready for work of subsequent trades and comply with all requirements of the local building code and the inspecting architect.

6. ***Cleanup.*** During the course of the work, and upon completion of all work, the contractor must keep all areas free of excessive debris, sawdust, shavings, etc. He must exercise special care to prevent accumulations of material from forming a fire or safety hazard, and he must remove excessive debris when directed.

The following is an actual excerpt from the rough carpentry category of a set of specifications.

Blocking and Bridging:

A. Joists to be blocked and bridged where shown or specified. Blocking to be of same size and materials as joists. X-bridging to consists of 1″ × 3″ bevel-cut to bear full at ends, or approved steel bridging. Nail bottom ends of X-bridging only after decking has been applied. Vent holes in blocking shall be drilled; sawcut holes not acceptable.

B. All rafters and joists shall be solid blocked over all supports, and bridged where indicated on the drawings or required by building code. Fire-blocking shall be provided where required or as directed.

C. *Metal Bridging Alternate:* In lieu of wood-cross bridging, contractor may use "Simpson," "Gibralter," or other approved metal bridging with nails supplied by manufacturer fully driven in all nailing holes. Metal bridging to be galvanized or have other approved rust-inhibitive coating.

The subcontractors will determine their work by checking the building plans and notes and by thoroughly reading their category in the specifications. If they have any questions, they should write them down and contact the general contractor, architect, or owner for the answers. When all his questions have been answered, the subcontractor is ready to compile and submit a bid for his portion of the job.

4

Building Codes

Building codes are written and used for the purpose of protecting the health, safety, and welfare of the public. Many states, counties, and cities adopt codes and ordinances to regulate the planning and construction of private and public buildings. Code regulations concern standards of construction, structural strength and restrictions of materials, light and ventilation, framing methods and requirements, installation instructions, zoning requirements, and building design limitations. It must be kept in mind, however, that codes establish only minimum requirements and that they occasionally fall short of adequate construction practices and structural standards.

Four basic codes are currently being used throughout the United States. They are published by four major professional organizations and are referred to as "Model Codes."

1. ***Basic Building Code.*** It was developed by the Building Officials and Code Administrators International Inc. (BOCA) and is used primarily in the eastern part of the United States.

2. ***National Building Code.*** It was written by the National Board of Fire Underwriters (presently known as the American Insurance Association). This code originated from the need for better fire protection standards. It is used by some midwestern and eastern states.

3. ***Uniform Building Code.*** It is published by the International Conference of Building Officials. It is used by many western states and is gaining acceptance throughout the country.

4. ***Southern Standard Building Code.*** It was developed by the Southern Building Code Conference and is used extensively in the southern states.

At present the Model Code Standardization Council (M.C.S.C.) is working toward standardization of the Model Codes. Because of the complexity and time involved in writing and updating code requirements, most states, counties, or cities will adopt and work with one of the Model Codes. Local communities may then amend certain sections of that code to meet their own needs. Many of the points not clearly defined in the code are often left to the discretion of the building inspectors, and this is where many conflicts arise between builder and city official.

A building contractor may legally challenge an amendment to the code or a judgment made by a building inspector, but it may be costly, not only in money, but also in goodwill between the builder and the building officials. On the other hand, if the builder legally wins his challenge, the cost of his project could be considerably reduced.

If the estimator is to accurately bid work, it is very important that he understand the local code and amendments. A misinterpretation or lack of knowledge by the estimator about local codes or amendments could cause a loss in profit or even eventual bankruptcy.

5

Abbreviations

Abbreviations are used by those involved in developing the building plans, notes, and specifications to save time and conserve space. Most of the abbreviations will be found on the building plans and notes, but just the most common abbreviations are used in the specifications. The ability of the estimator to interpret abbreviations is an important part of his job if a complete and accurate estimate is to be obtained.

Aluminum slider
A.S.
Anchor bolt
AB: or A.B.
Area
A
Average
Avg.
Basement
Bsmt
Bath
B
Beam
Bm
Bedroom
BR
Beveled
Bev.
Blocking
Blkg.
Board
BD
Board foot
Bd. Ft.: or fbm
Board measure
BM: or fbm
Bolt
Blt: or B
Building
Bldg.
Bundle
Bdl.
By (2 × 4)
×
Cabinet
Cab.
Caulking
Clkg.
Ceiling
Clg.
Center
Ctr.
Center to center
C to C: or C/C
Clear
Clr.
Clear gloss
Clr. gl: or CG
Closet
Clo.
Column
Col
Combination
Comb.
Concrete
Conc. PCC
Construction
Const.
Cubic
CU: or Cu
Cubic foot
CF: or cf
Cubic yard
CY: or cy
Damproofing
DP
Decking
Dkg.
Detail
Det
Diameter
Dia
Dimension
Dim
Dining room
DR
Door
Dr
Double-hung window
DH: or DHW

Douglas fir
DF: or D Fir
Drawing
DWG
Economy
Econ.
Elevation
EL: or elev
Excavate
Exc
Existing
EX: or exist
Expansion joint
EJ: or E.J.: or CJ
Exterior
Ext
Feet
Ft: or ft
Finish floor
FF: or F.F.
Flashing
FL
Floor
F: or FL
Flooring
FLG
Flush beam
FL. Bm.
Footing
FTG: or ftg
Foundation
FDN.
Furred ceiling
FC
Galvanized
Gal
Glass
GL
Grade
Gr

Hardware
HDW
Height
HGT: or H: or HT
Hemlock
H: or Hem
Hose bib
HB: or hb
Hot water heater
HWH: or WH
Hundred
C
Inch
": or in.
Insulate or insulation
INS
Interior
INT
Jamb to Jamb
J-J
Joist
JST.
Kiln dried
K.D.
Kitchen
K
Knocked down
K.d.
Laundry
LAU
Lavatory
LAV
Left hand
LH
Length
Len: or L
Level
Lev
Line
L

Lineal feet
LF: or Lin. Ft.
Linen closet
L CL
Living room
LR
Louver
LV: or LVR
Lumber
LBR
Material
Mat
Maximum
Max.
Medicine cabinet
MC
Minimum
Min
Molding
Mldg
Not to scale
N.T.S.
Number
No.: or #
On center
O.C.
Oregon pine
OP
Pair
Pr
Panel
Pnl
Penny nail size
d
Ponderosa pine
PP
Pound
Lb.: or #
Prefabricated
Prefab
Quantity
Q: or Qty
Random
Rdm: or R

Random width and length
RWL
Redwood
Rwd
Resawn
R/S: or Re/S
Right
R+: or R
Right hand
RH
Roofing
RFG
Rough
Rgh
Rough sawn
RO: or Rgh: or Ro/SAWN
Scale
Sc: or sc
Schedule
Sch.
Section
Sect.
Select
SEL
Sheathing
SH: or Shtg
Shiplap
S/L
Siding
Sdg.
Specifications
Spec.
Square
Sq.
Square foot
Sq. Ft.
Stairway
Stwy
Standard
Std
Structural
Str.
Surface 1 side, 1 edge
S1S1E

Surface 2 sides
S2S
Surface 4 sides
S4S
Thousand
M
Thousand (1,000) board feet
MBF: or M
Tongue and groove
T & G: or TG
Typical
Typ.
Utility
Util.
Vent
V
Ventilate
Vent.
Vertical
Vert.
Water closet
WC
White fir
WF
White pine
WP
Wide flange beam
W.F. Bm
Width
W: or Wth
Window
W: or Wdw
Wood
Wd

6

Estimating Forms

Personally designed or commercially printed estimating forms help a contractor-estimator organize his business and save time. The number of different forms the estimator uses depends on his company duties, the size of the company, and the volume of business. The estimator who is just doing material take-off can get by with one form. However, the estimator who has all or some of the duties listed on pp. 6–7 in Unit 1 will need various forms to keep the business organized.

Material Take-off Form (p. 35–36). This helps the estimator list the lumber, hardware, and miscellaneous materials for a project. The items should be logically listed from the start of the project to the finish. For rough framing it would be mudsills through roof sheathing. This list will help the lumberyard set up the various loads of lumber. A copy of the completed list will then be sent to the lumber supplier or suppliers for a material cost bid. Another copy of the material list may go to the foreman or superintendent on the job so that he can determine the placement of the different lengths of material.

Item Checkoff Sheet (p. 37–38). This helps keep the estimator from forgetting an item that should be included in the estimate. Keep in mind that the items listed on this form may involve the estimating of various members in that phase of construction. For example, exterior soffit material involves the estimating of the ledger material, the lookout material, and the plywood or board soffit material. Many estimators after acquiring enough experience, however, do not need to use this form.

Construction Specification Form (p. 39–40). This was produced by a room addition contractor for his personal needs. This form helps the owner understand what is and what is not going to be included in the job. Therefore, it helps to eliminate any construction disputes between the contractor and owner during or after completion of the project. When filled in, this form also helps the subcontractor because it contains much of the information needed for bidding his phase of the job. When signed by the owner and builder, this form may be a legally binding contract.

Estimate Summary Form (p. 41–42). This is a commercially printed form that lists total costs of materials, labor, subcontracted items, and costs for construction changes.

MATERIAL TAKE-OFF

DATE OF ESTIMATE	LUMBER YARD	OWNER
	SALESMAN	OWNER'S ADDRESS
CONTRACTOR		JOB LOCATION
ADDRESS		
CITY ZIP		NOTES
TELEPHONE		

CHECK	ITEM	DESCRIPTION	PCS.	SIZE	LGTH.	TOTAL FEET	PRICE	AMOUNT

CHECK	ITEM	DESCRIPTION	PCS.	SIZE	TOTAL FEET	LGTH.	PRICE	AMOUNT

ITEM CHECKOFF SHEET

CONCRETE

___ Footings
___ Foundation walls
___ Walks
___ Foundation bolts
___ Piers
___ Slabs
___ Retaining walls
___ Form material

WOOD FRAMING

___ Mudsills
___ Wood foundation posts
___ Steel columns
___ Beams (steel)
___ Girders (wood)
___ Bridging
___ Solid blocking
___ Rim joist
___ Floor joist
___ Subflooring plywood
___ Subflooring board
___ Foundation vent
___ Strong backs
___ Basement windows
___ Wall plates
___ Studs
___ Headers
___ Diagonal braces
___ Shear panels
___ Wall blocking
___ Temporary brace material
___ Ceiling joist
___ Ceiling joist blocking
___ Ceiling joist bridging
___ Ceiling backing material
___ Ceiling stay lath

(WOOD FRAMING continued)

___ Purlins

___ Ridge props

___ Collar ties

___ Drop ceiling material

___ Rafter material

___ Hip material

___ Valley material

___ Ridge board

___ Rafter freeze blocks

___ Eave vents

___ Roof sheathing

___ Starter board

___ Building paper

___ Flashing paper

___ Fascia board

___ Bargeboard

___ Shingles

___ Exterior soffit material

___ Windows

___ Siding

___ Exterior frames

INTERIOR FINISHES

___ Base

___ Base shoe

___ Doorjambs

___ Door casing

___ Door stop

___ Window trim

___ Closet shelving and trim

___ Closet doors

___ Interior doors

___ Exterior doors

___ Paneling

HARDWARE

___ Framing hardware

___ Rough hardware

___ Finish hardware

___ Cabinet hardware

___ Miscellaneous hardware

CONSTRUCTION SPECIFICATIONS

CONCRETE PREPARATION

Fill dirt ______

Demolition ______

Excavation ______

Standard footings ______

Expansive footings ______

Slab floor ______

Subfloor foundations ______

Concrete steps ______

Retaining wall ______

Concrete walk ______

Additional slab ______

New crawl hole ______

Driveway ______

Removal of surplus soil ______

Removal of trees/shrubs ______

FRAMING

Framing on slab ______

Conventional wood floor ______

Joist and plywood ______

Joist and boards ______

Conventional wall framing ______

Rake wall framing ______

Balloon framing ______

Combination of framing ______

Fill existing window ______

Fill existing door ______

New opening/concealed header ______

New opening/exposed header ______

Ceiling joist ______

Open beam ceiling ______

Open beam/false beam ______

Gable roof ______

Hip roof ______

Dutch gable ______

Other ______

Patio roof ______

WINDOWS

Type and size ______

___ x ______

___ x ______

___ x ______

___ x ______

___ x ______

___ x ______

___ x ______

___ x ______

___ x ______

___ x ______

DOORS

Hollow core stain ______

Hollow core paint grade ______

Solid core stain ______

Solid core paint grade ______

Entry door ______

Sliding glass door ______

Other ______

ROOFING

Roof tie-in as required ______

Composition shingles ______

#1 Wood shingle ______

Medium shakes ______

Heavy shake ______

Built up, w/rock ______

WOOD PANELING

4 x 8 prefinished ______

Sq. ft. of paneling ______

Prefinished mold ______

TRIM

Molding compatible with existing ______

PLUMBING

Lavatory sink ______

Toilet ______

Stall shower ______

Bath tub ______

New water heater ______

Relocate water heater ______

Gas line to fireplace ______

Bar sink ______

Appliance hook-up ______

Gas line to other ______

Kitchen sink (new) ______

Sewer line ______

Relocate water service entry ______

Laundry tub ______

Water line to ice-maker ______

Drier hook-up & vent ______

Washer hook-up ______

Drain tile ______

Connect owner's disposal ______

New hose bib ______

Plumb for dishwasher ______

Hook-up dishwasher ______

Extend roof stack ______

Lawn sprinkler: void as req'd. ______

(CONSTRUCTION SPECIFICATIONS continued)

ELECTRICAL

Connect to existing service box ______
Intr. wall plugs (dbl) ______
Waterproof wall plugs (dbl) ______
Switches ______
3-way switches ______
Ceiling outlet ______
Exhaust fan in new ______
Exhaust fan in old ______
Move elec. service box ______
Enlarge service to ______ amps

Bath ceiling heater ______
220 circuits ______
Appliance hook-up ______
Flush light fixtures ______
Connect new hood vent ______
TV outlet box ______
New door bell ______
Relocate old door bell ______
Automatic garage door closer ______
Ground fault plugs ______

HEATING

Gas heat ______
Radiant electric heat ______
Extend heat duct ______

Electric wall heater, w/fan ______
Heat pump ______

INTERIOR (Walls/Ceiling)

Plaster walls & ceiling ______
Walls drywall, ready to paint ______
(textured or smooth) ______
Ceiling smooth, ready to paint ______

Ceiling acoustic, finished ______
Patch existing where altered ______
Firewall in garage ______
Drywall nail-on only ______

EXTERIOR

Stucco (near-match exist. color) ______

Siding ______

CERAMIC TILE/FORMICA/MARBLE

Kitchen counter ______ " splash
3 sides of tub to ______ ' height

Shower to ______ ' height
Pullman top, ______ " splash

FLOOR COVERING

Underlayment ______
Vinyl asbestos (std. selection) ______
Vinyl-Montino Corlon or equal ______

Coved ______
Hardwood ______
Total allowance $ ______

MASONRY

New fireplace: brick/stone face; to ceiling/mantle height ______
Fireplace chimney to be brick/stone ______

Raised hearth ______
Floor-level hearth ______
Planter ______
Raise exis. chimney as req'd. ______

CABINETS (see drawings)

Kitchen cabinets: hardwood face ______
paint grade ______
Bookcase ______

Pullman cabinet: hardwood face ______
paint grade ______
Other ______

SPECIAL OR OTHER ITEMS

Plans by builder ______
Permits by builder ______
Fencing ______
Shower door ______
Shower or tub enclosure ______
Medicine cabinet: allows $ ______
Relocate sliding glass door ______
Relocate window ______
Relocate door ______
Insulation: new outside walls ______
new ceiling area ______
New sky light ______

Gutters/downspouts ______
New stairway ______
Turnedposts ______
Ornamental iron ______
Kitchen soffit ______ ft.
Hall soffit ______ ft.
Bathroom soffit ______ ft.
Cleanup of debris & scraps ______
New plate glass mirror ______
New aluminum threshold ______
Insulation board over roof boards ______

SPECIAL HARDWARE

Cabinet knobs allowance ______

Other ______

PAINTING

Paint exterior ______
Paint interior ______

Stain ______
Other ______

Above specifications, together with final working plans, form basis for work to be accomplished. Any change in plans and specifications to be agreed upon between owner and builder as to addition or deductions.

APPROVAL: ______ Owner
Date: ______

APPROVAL: ______ Contractor
Date: ______

ESTIMATE SUMMARY

PROJECT		TOTAL AREA	SHEET NO.
LOCATION		TOTAL VOLUME	ESTIMATE NO.
ARCHITECT		COST PER S.F.	DATE
OWNER		COST PER C.F.	NO. OF STORIES
QUANTITIES BY	PRICES BY	EXTENSIONS BY	CHECKED BY

NO.	DESCRIPTION	MATERIAL	LABOR	SUBCONTRACT	TOTAL	ADJUSTMENT
	SITE WORK					
	Demolition					
	Excavation & Fill					
	Caissons & Piling					
	Drainage & Utilities					
	Sewage Treatment					
	Roads, Walks, & Walls					
	Lawns & Plantings					
	Termite Control					
	CONCRETE					
	Formwork					
	Reinforcing Steel & Mesh					
	Foundations					
	Superstructure					
	Floors & Finish					
	Precast Concrete					
	Cementitious Decks					
	MASONRY					
	Brick					
	Block					
	Stonework					
	Mortar & Reinforcing					
	METALS					
	Joists					
	Structural Steel					
	Miscellaneous Metal					
	Ornamental Metal					
	Fasteners, Rough Hardware					
	Metal Decks					
	CARPENTRY					
	Rough Carpentry					
	Finish Carpentry					
	Cabinets & Counters					
	Laminated Construction					
	MOISTURE PROTECTION					
	Water & Dampproofing					
	Insulation					
	Roofing					
	Siding					
	Sheet Metal Work					
	Roof Accessories					
	DOORS, WINDOWS, GLASS					
	Doors & Entrances					
	Windows					
	Glass & Glazing					
	Weather Stripping					
	Finish Hardware					
	PAGE TOTAL					

(ESTIMATE SUMMARY continued)

NO.	DESCRIPTION	MATERIAL	LABOR	SUBCONTRACT	TOTAL	ADJUSTMENT
	TOTAL FROM FRONT PAGE					
	FINISHES					
	Lath, Plaster & Stucco					
	Drywall					
	Tile, Marble & Stone					
	Terrazzo					
	Acoustical Ceilings					
	Floor Covering					
	Painting					
	Wall Covering					
	SPECIALTIES					
	Chalkboard & Tackboard					
	Partitions					
	Chutes					
	Lockers					
	Toilet Accessories					
	Flagpoles					
	EQUIPMENT					
	Kitchen					
	Refrigeration					
	Incinerator					
	FURNISHINGS					
	Blinds					
	Carpet					
	Mats					
	Seating					

NO.	DESCRIPTION	MATERIAL	LABOR	SUBCONTRACT	TOTAL	ADJUSTMENT
	SPECIAL CONSTRUCTION					
	Pedestal Floors					
	Integrated Ceilings					
	Prefab, Rooms & Bldgs.					
	CONVEYING SYSTEMS					
	Elevators					
	Dumb Waiters					
	Escalators, Moving Ramps					
	Pneumatic Tube System					
	MECHANICAL					
	Sprinklers					
	Plumbing					
	Heating & Ventilating					
	Air Conditioning					
	ELECTRICAL					
	Electric Heat					
	Intercom & T.V.					
	Alarm Systems					
	Motor Generator					
	Exterior Lighting					
	TOTAL DIRECT COSTS					
	CONTRACTOR'S OVERHEAD					
	Insurance, Taxes, Fringes					
	Equipment & Tools					
	Performance Bond					
	Profit & Contingencies					

SCHEDULING OF OPERATIONS SHEET

Many estimators get involved with the scheduling of the jobs. A form as found below is especially important when the contractor is running many jobs at one time, for it allows the contractor or estimator to know exactly what phase of construction each job is in, and he can schedule the inspections and subcontractors accordingly.

The various forms used in this unit may not fit the personal needs of every contractor-estimator, but if revised they may provide a logical starting point for the organizing of a well-run construction business.

Job ______________________ No. __________

SCHEDULE OF OPERATIONS

Operations/Remarks	Scheduled Start	Actual Start	Completion
Layout			
Trench & grade			
Set forms & rebar			
Plumbing/soil and copper water			
Plumbing inspection			
Forms & footings inspection			
Pour footings & slab strip forms			
Frame/walls, ceiling joists Plumbing/top out, run gas and water			
Cut & stack rafters Starter & fascia Lay & nail sheathing Set tub or shower Lay-up fireplace			
Sheathing inspection Rough heating, electrical & prewire phone			
Set windows & patio door Roofing & sheet metal Set wood jambs			
Rough plumbing, electrical & heating inspection			
Framing inspection			
Exterior lath Insulation			
Hang drywall			

(continued)

Operations/Remarks	Scheduled Start	Actual Start	Completion
Lath & drywall nailing inspection			
Drywall tape			
Drywall fill			
Drywall sand & texture			
Plaster scratch cost Paint overhangs & siding			
Set cabinets (formica & pullmans) Hang doors/shelf & pole Interior trim & base			
Plaster brown coat Painting interior Cabinets stain & lacquer			
Hard surface floors			
Finish plumbing, electrical & hardware			
Color coat stucco			
Clean windows & replace screens Scrape floors Final cleanup Final grading			
Final inspection			
Lay carpet			
FINAL CLOSING DATE: ______			

FEDER & BOYD CONTRACTORS
Fullerton, California

7

Lumber Grading

The estimator has to be familiar with the size and grades of rough framing and finish lumber. In 1970 new lumber grading rules became effective which incorporated the National Dimension Rule unifying lumber sizes, grade names, and moisture content requirements. The new standard eliminated a multiplicity of grade names and descriptions, and assured that (green) unseasoned lumber would shrink to the same size as seasoned lumber of the same relative dimension.

Product Standard 20–70 defines dry lumber as being 19 percent or less in moisture content and unseasoned lumber as 19 percent or more. Rough lumber that has been run through a surface planer for smoothing and dimension sizing is known as *dressed* or *surfaced lumber.* The tradesman commonly calls this material *framing lumber* and/or *yard lumber.* It is designated *S4S* or surfaced four sides material.

Framing and clear grade lumber 2 × 4 to 2 × 12 inches wide may be ordered in even 2-foot lengths usually starting at 8′-0″. Most lumberyards will commonly stock this

TABLE 7-1 FRAMING LUMBER SIZES (SOFTWOOD)

Product Classification	*Minimum Dressed Sizes*	
(Nominal Size)	*Unseasoned (Actual Size)*	*Dry (Actual Size)*
Inches	Inches	Inches
Dimension Lumber		
2 × 4	1-9/16 × 3-9/16	1-1/2 × 3-1/2
2 × 6	1-9/16 × 5-5/8	1-1/2 × 5-1/2
2 × 8	1-9/16 × 7-1/2	1-1/2 × 7-1/4
2 × 10	1-9/16 × 9-1/2	1-1/2 × 9-1/4
2 × 12	1-9/16 × 11-1/2	1-1/2 × 11-1/4
Board Lumber		
1 × 4	25/32 × 3-9/16	3/4 × 3-1/2
1 × 6	25/32 × 5-5/8	3/4 × 5-1/2
1 × 8	25/32 × 7-1/2	3/4 × 7-1/4
1 × 10	25/32 × 9-1/2	3/4 × 9-1/4
1 × 12	25/32 × 11-1/2	3/4 × 11-1/4

material in up to 20- or 22-foot lengths. Framing lumber in lengths from 22 to 28 feet will often have to be special ordered and may take a few days for delivery. Pine is usually available in up to 16-foot maximum lengths, redwood and cedar in up to 20-foot lengths. Timber material may be special ordered up to 36′-0″ long. It may sometimes take weeks to months for delivery of these specially ordered items, and the contractor-estimator should keep this in mind when scheduling materials for his jobs.

Table 7-2 gives the category, grade designation, and nominal sizes for dimension lumber.

If any questions should arise on lumber grades or sizes, the estimator must call the local supplier for the answer.

Not all the lumber the estimator orders for a job will be usable in its present length. Depending on the number and size of knots, method of drying, and length of time lying around on the job, lumber will have a tendency to warp, twist, or crook. Pieces of material not usable in their present length may be cut into shorter lengths and used for framing members such as drop ceilings, header cripples, fire-blocks, etc. For this reason, the estimator must add a certain amount of waste lumber to his estimate.

TABLE 7-2 GRADE DESIGNATION FOR DIMENSION LUMBER

Category	*Grades*	*Sizes*
Light Framing	Construction, Standard, Utility, Economy	2″ to 4″ thick 2″ to 4″ wide
Studs	Stud, Economy	2″ to 4″ thick 2″ to 4″ wide
Structural Light Framing	Select Structural No. 1, No. 2, No. 3, Economy	2″ to 4″ thick 2″ to 4″ wide
Appearance Framing	Appearance	2″ to 4″ thick 2″ and wider
Structural Joists and Planks	Select Structural No. 1, No. 2, No. 3, Economy	2″ to 4″ thick 6″ and wider
Decking	Selected Decking, Commercial Decking	2″ to 4″ thick 4″ and wider
Beams and Stringers, Posts and Timbers	Select Structural No. 1, No. 2, No. 3	5″ and thicker 5″ and wider

TABLE 7-3 Grade Designation for Douglas Fir and Pine Appearance Grade

Category	*Douglas Fir (Finish Grades)*	*Pine (Select Grades)*
Clears	*C and Better* *(Superior) Finish* *D Prime Finish*	*C and Better* *Selects* *D Selects*
Factory		
Douglas Fir and Pine Shop Grade (Material to be remanufactured into moldings, casings, base jambs, etc.)	Molding No. 3 Clear No. 1 Shop No. 2 Shop No. 3 Shop	Molding No. 3 Clear No. 1 Shop No. 2 Shop No. 3 Shop
Boards	Select Construction Standard Utility Economy	No. 1 Common No. 2 Common No. 3 Common No. 4 Common No. 5 Common

It has become very popular in recent years to use rough lumber for exterior and interior finishes. The estimator has two alternatives for ordering rough-textured lumber.

1. ***Rough-sawn Lumber.*** Lumber that has been cut from the log at a lumber mill is known as *rough-sawn* lumber. *Dimension* lumber is manufactured by planing smooth rough-sawn lumber. A piece of 2 × 6 rough-sawn lumber is supposed to be a full 2 inches by 6 inches, but because of moisture and cutting conditions, the size will often vary between pieces by as much as 3/8 inch. For this reason, rough-sawn lumber often does not match up with adjoining members.

2. ***Resawn Lumber.*** This may be cut from rough-sawn lumber or S4S-dimensioned material. The texture of resawn lumber is not quite as rough as rough-sawn lumber. If resawn lumber is to be cut from rough-sawn material, all four sides would be resawn. A 2 × 6 resawn member for example, would finish dimension at 1-1/2 inches by 5-1/2 inches.

When resawing S4S-dimension lumber, usually only one side and one edge (1S1E) are resawn, and the finish dimension size would be approximately 1-3/8 inches by 5-3/8 inches.

Resawn lumber should be used where members adjoining each other are to match in width and thickness, such as fascia and bargeboard. Resawing cleans up lumber, making it suitable for visual exposure, with or without the use of stain or finish.

8

Estimating Lumber Costs

Lumberyards buy framing lumber and plywood from the mill by the board foot or square foot and sell it by the board foot, square foot, lineal foot, or occasionally, individual unit or *stick* price. In order to estimate lumber costs, the estimator will first have to convert the various sizes of lumber from his material list to board feet. This procedure is illustrated below:

BOARD FEET (bd. ft.)

One board foot equals a board that measures 1″ × 12″ × 12″.

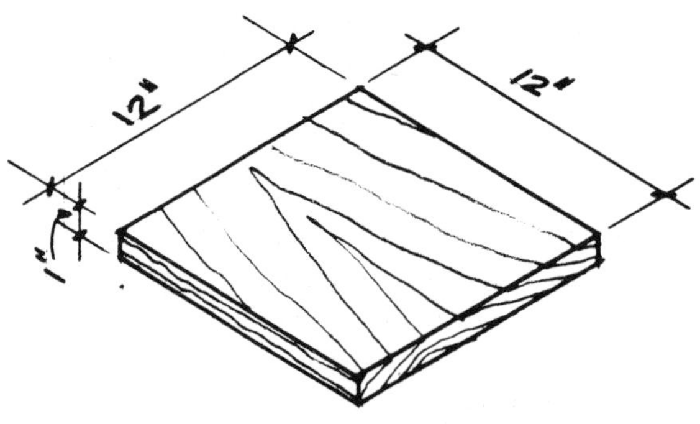

Note: Boards less than one inch thick are considered one inch.

Framing lumber is changed to board feet by the following formula:

$$\text{Board Feet} = \frac{\text{Number of Pieces} \times \text{Thickness in Inches} \times \text{Width in Inches} \times \text{Length in Feet}}{12}$$

EXAMPLE: How many board feet in one 2″ × 4″ × 18′?

$$\frac{1 \times 2 \times 4 \times 18}{12} = \frac{144}{12} = 12 \text{ bd. ft.}$$

Cancellation may be used to simplify the problem.

$$\frac{1 \times \overset{1}{\cancel{2}} \times 4 \times \overset{3}{\cancel{18}}}{\underset{1}{\underset{\cancel{6}}{\cancel{12}}}} = 12 \text{ bd. ft.}$$

Lumberyards usually round off to the next higher board foot for any fraction of a foot. This will be the case in the following problem.

EXAMPLE: How many board feet of lumber in 7 pieces 2″ × 4″ × 16′?

$$\frac{7 \times 2 \times \overset{1}{\cancel{4}} \times 16}{\underset{3}{\cancel{12}}} = \frac{224}{3} = 74\text{-}2/3 \text{ or } 75 \text{ bd. ft.}$$

BOARD MEASURE

Estimate the board feet in the following problems:

	Quantity	*Description*	*Answer*
1.	16	2″ × 6″ × 16′	______
2.	8	2″ × 4″ × 12′	______
3.	20	2″ × 6″ × 14′	______
4.	4	2″ × 8″ × 18′	______
5.	32	4″ × 4″ × 14′	______
		TOTAL BOARD FEET	______

	Quantity	*Description*	*Answer*
1.	13	4″ × 6″ × 16′	______
2.	140	1″ × 6″ × 18′	______
3.	38	1″ × 8″ × 16′	______
4.	7	1″ × 12″ × 16′	______
5.	14	4″ × 8″ × 20′	______
		TOTAL BOARD FEET	______

BOARD MEASURE

Estimate the board feet in the following problems:

	Quantity	*Description*	*Answer*
1.	5	1″ × 6″ × 18′	______
2.	24	1″ × 8″ × 14′	______
3.	6	2″ × 3″ × 12′	______
4.	14	1″ × 4″ × 18′	______
5.	34	2″ × 4″ × 16′	______
		TOTAL BOARD FEET	______

	Quantity	*Description*	*Answer*
1.	54	1″ × 3″ × 20′	______
2.	28	2″ × 6″ × 10′	______
3.	32	1″ × 10″ × 16′	______
4.	22	2″ × 12″ × 22′	______
5.	62	1″ × 10″ × 18′	______
		TOTAL BOARD FEET	______

Once the estimator has converted his list of materials to board feet, he can price it out. The lumberyard sells lumber to the contractor or tradesman in terms of a thousand board foot price. The abbreviation for one thousand is M.

EXAMPLE: \$220.00 per thousand bd. ft. or \$220.00/M

To find the cost of a single board foot, the cost per thousand is divided by 1,000 board feet; this just moves the decimal three places to the left.

EXAMPLE: $\frac{\$170.00}{1{,}000}$ = \$.17 or 17 cents a board foot

Example Problem:

How much would 350 pieces of 2″ × 6″ × 16′ board cost if 2 × 6 material is selling for \$192.50/M?

PROCEDURE

1. Estimate the board feet.

$$\frac{350 \times \overset{1}{\cancel{2}} \times \overset{1}{\cancel{6}} \times 16}{\underset{1}{\cancel{\underset{\cancel{6}}{12}}}} = \frac{5{,}600}{1} = 5{,}600 \text{ bd. ft.}$$

2. Divide \$192.50 by 1,000 to find the cost of one board foot. (Move the decimal three places to the left.)

$$\frac{\$192.50}{1{,}000} = \$.1925 \text{ or } 19\text{-}1/4 \text{ cents a bd. ft.}$$

3. To find the total cost, multiply 5,600 bd. ft. times the cost per board foot (\$.1925).

```
      5,600 bd. ft.
    × $.1925
      28000
     11200
    50400
    5600
  ----------
Answer .... $1,078.00
```

BOARD FOOT AND LUMBER COSTS

Estimate the board feet and lumber costs in the following problems:

	Quantity	*Description*	*Cost/M*	*Board Feet*	*Cost*
1.	240	2 × 4 × 10	$300.00	______	______
2.	125	2 × 4 × 14	$300.00	______	______
3.	84	2 × 4 × 18	$300.00	______	______
4.	47	2 × 6 × 16	$310.00	______	______
5.	108	2 × 6 × 20	$310.00	______	______
				TOTAL COST	______

	Quantity	*Description*	*Cost/M*	*Board Feet*	*Cost*
1.	54	2 × 8 × 14	$300.00	______	______
2.	97	2 × 8 × 16	$300.00	______	______
3.	44	2 × 8 × 18	$300.00	______	______
4.	7	4 × 6 × 20	$360.00	______	______
5.	12	4 × 8 × 18	$360.00	______	______
				TOTAL COST	______

BOARD FOOT AND LUMBER COSTS

Estimate the board feet and lumber costs in the following problems:

	Quantity	Description	Cost/M	Board Feet	Cost
1.	2	2 × 6 × 10 Pressure Treated	$320.00	______	______
2.	2	2 × 6 × 14 Pressure Treated	$320.00	______	______
3.	8	2 × 6 × 16 Pressure Treated	$320.00	______	______
4.	1	2 × 6 × 18 Pressure Treated	$320.00	______	______
5.	54	2 × 8 × 16 Const. Std. D.F.	$300.00	______	______
				TOTAL COST	______

	Quantity	*Description*	*Cost/M*	*Board Feet*	*Cost*
1.	52	2 × 8 × 18 Const. Std. D.F.	$300.00	________	______
2.	16	2 × 8 × 20 Const. Std. D.F.	$300.00	________	______
3.	6	2 × 6 × 16 Const. Std. D.F.	$310.00	________	______
4.	2	2 × 6 × 18 Const. Std. D.F.	$310.00	________	______
5.	7	2 × 6 × 20 Const. Std. D.F.	$310.00	________	______
6.	100	2 × 4 × 8 Const. Std. D.F.	$285.00	________	______
				TOTAL COST	______

Lumberyards also buy and sell various materials by the square foot or surface measure.

A square foot is the surface area covering 12″ by 12″.

Note: Surface area measure is bound by straight or curved lines and has no thickness.

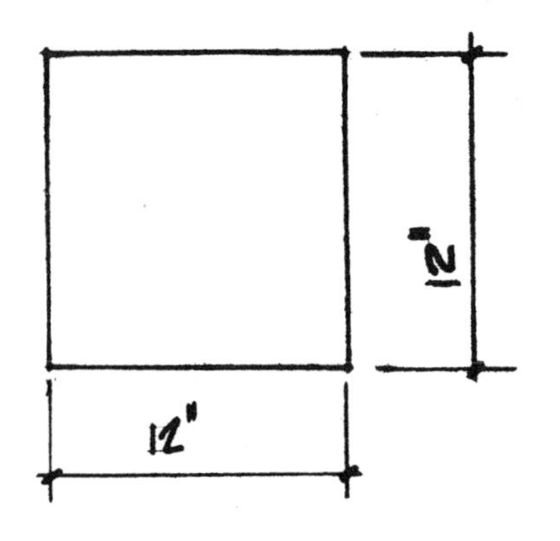

Plywood is one of the most common items that is bought by the square foot. Lumberyards will buy and sell plywood with a unit price of a thousand square feet (M).

EXAMPLE: $360.00 per thousand sq. ft. or $360.00/M

Most lumberyards deal with two types of customers: one being the contractor or tradesman who understands the unit pricing of thousand bd. ft. or sq. ft. and will usually be quoted prices in this manner.

The other customer is the layman who doesn't understand these units of pricing. For the layman, the plywood dealer or lumberyard will break down the 1,000 sq. ft. price and give a price per unit of material as in cost per sheet.

Example Problem:

How much will one sheet of 1/2″ × 4′ × 8′ A-B interior plywood cost at $410.00/M?

PROCEDURE:

1. Calculate the cost per sq. ft. This is done by dividing the cost per thousand sq. ft. by 1,000 or just moving the decimal three places to the left.

$$\frac{410.00}{1000} = \$.41 \text{ per sq. ft.}$$

2. Determine the surface of the sheets used. In this case:

4 ft. × 8 ft. or 32 sq. ft.

3. Multiply the surface area of one sheet (32 sq. ft.) by the cost per sq. ft. ($.41).

$$32 \times .41 = \$13.12 \text{ per sheet}$$

SQUARE FOOT LUMBER COSTS

	Number of Sheets	Size & Grade	Cost Per/M Sq. Ft.	Answer
1.	30	5/8″ × 4′ × 8′ A-D Int.	$387.00/M	______
2.	15	3/4″ × 4′ × 8′ A-B Int.	$546.00/M	______
3.	56	5/8″ × 4′ × 8′ A-C Int.	$413.00/M	______
4.	7	1/2″ × 4′ × 8′ A-B Int.	$480.00/M	______
5.	38	3/8″ × 4′ × 8′ CDX	$223.00/M	______
			TOTAL COST	______

	Number of Sheets	Size & Grade	Cost Per/M Sq. Ft.	Answer
1.	34	3/8 × 4 × 8 Rough-sawn Redwood	$562.00/M	______
2.	25	2/3 × 4 × 8 CDX Doug Fir	$298.00/M	______
3.	16	3/8 × 4 × 8 Plain Rough-sawn Doug Fir	$348.00/M	______
4.	44	3/4 × 4 × 8 A-D Int. Doug Fir	$471.00/M	______
5.	6	5/8 × 4 × 10 T-1-11 Doug Fir	$559.00/M	______
			TOTAL COST	______

Estimate the square foot lumber cost in the following problems.

	Number of Sheets	*Size & Grade*	*Cost Per/M Sq. Ft.* (*Check with lumberyard for present prices*)	*Answer*
1.	54	5/8 × 4 × 8 A-D Int.	________	______
2.	20	3/4 × 4 × 8 A-B Int.	________	______
3.	18	1/2 × 4 × 8 A-B Int.	________	______
4.	12	3/8 × 4 × 8 CDX	________	______
5.	72	5/8 × 4 × 8 Int.	________	______
6.	13	1/2 × 4 × 8 CDX	________	______
			TOTAL COST	______

Lumberyards will often convert board foot prices of lumber to lineal foot prices. This usually helps the contractor-estimator and especially the layman to understand the cost of material better. To convert from board foot to lineal foot prices, use Table 8-1 to find how many board feet there are in one lineal foot of a certain size material.

TABLE 8-1 Board Feet per Lineal Foot of Standard Size Framing Lumber

Lumber Size	*Board Feet per Lineal Foot*	*Decimal Equivalent*
1 × 2	1/6	.1666
1 × 3	1/4	.2500
1 × 4	1/3	.3333
1 × 6	1/2	.5000
1 × 8	2/3	.6666
1 × 10	5/6	.8333
1 × 12	1	1.0000
2 × 2	1/3	.3333
2 × 3	1/2	.5000
2 × 4	2/3	.6666
2 × 6	1	1.0000
2 × 8	1-1/3	1.3333
2 × 10	1-2/3	1.6666
2 × 12	2	2.0000
2 × 14	2-1/3	2.3333
3 × 3	3/4	.7500
3 × 4	1	1.0000
3 × 6	1-1/2	1.5000
3 × 8	2	2.0000
3 × 10	2-1/2	2.5000
3 × 12	3	3.0000
4 × 4	1-1/3	1.3333
4 × 6	2	2.0000
4 × 8	2-2/3	2.6666
4 × 10	3-1/3	3.3333
4 × 12	4	4.0000
4 × 14	4-2/3	4.6666
6 × 6	3	3.0000
6 × 8	4	4.0000
6 × 10	5	5.0000
6 × 12	6	6.0000

Example Problem:

Estimate the lineal foot price and the total cost of the following material:

Quantity	*Description*	*Price*
70 pc.	2 × 4 × 12	$190/M

PROCEDURE:

1. Find the number of board feet per lineal foot from Table 8-1. (One lin. ft. of 2 × 4 material has 2/3 of a bd. ft.)
2. Find the cost for one board foot:

$$\frac{\$190.00}{1000} = 19 \text{ cents a bd. ft.}$$

3. Multiply the cost per board foot (19¢) times the number of board feet in one lineal foot of 2 × 4 material (2/3 bd. ft.).

$$\frac{.19}{1} \times \frac{2}{3} = \frac{.38}{3} = 12\text{-}2/3 \text{ cents per lin. ft.}$$

4. Determine the number of lineal feet of 2 × 4 material.

(70 pc. each 12 ft. long)

$$\begin{array}{r} 70 \\ \times\ 12 \\ \hline 840 \text{ lin. ft.} \end{array}$$

5. Multiply the 840 lineal feet times the cost (12-2/3 cents) per lin. ft.

$$12\text{-}2/3 \times 840 = \frac{.38}{3} \times \frac{840}{1} = \frac{319.20}{3} = \$106.40$$

Note: If decimals are used, they must be carried out four places for accuracy.

EXAMPLE: The decimal equivalent of 2/3 is 0.6666.

12-2/3 cents from problem above = $0.126666

$$\begin{array}{r} \$0.126666 \\ \times\ 840 \text{ lin. ft.} \\ \hline 000000 \\ 506664 \\ 1013328 \\ \hline 106.399440 \end{array} \qquad \$106.40$$

OR $106.40 (**Note:** Any fractional part of a cent is raised to the next even cent.)

LINEAL FOOT PRICING

Estimate the lineal foot costs in the follow ing problems.

	Number of Pieces	*Description*	*Lineal Feet*		*Cost per Lin. Ft.*	*Total Cost per Item*
1.	36	2 × 4 × 12	______	×	20¢	______
2.	42	2 × 4 × 14	______	×	20¢	______
3.	18	2 × 4 × 16	______	×	20¢	______
4.	26	2 × 4 × 20	______	×	22¢	______
5.	100	2 × 6 × 18	______	×	31¢	______
6.	152	2 × 6 × 20	______	×	33¢	______
7.	52	2 × 8 × 16	______	×	41¢	______
8.	74	2 × 10 × 16	______	×	56¢	______
9.	8	2 × 12 × 14	______	×	78¢	______
10.	12	4 × 4 × 20	______	×	53¢	______
11.	10	4 × 6 × 18	______	×	80¢	______
12.	38	4 × 8 × 16	______	×	$1.07	______
					TOTAL COST	______

LINEAL FOOT COSTS

Convert the following materials to lineal foot prices.

Note: Follow the procedure on previous page.

Item	*Material Size*	*Price per Thousand*	*Cost per Board Foot*		*Number of Bd. Ft. per Lin. Ft.*		*Lineal Foot Price*
1.	1 × 4	$260.00/M	______	×	______	=	______
2.	1 × 6	$260.00/M	______	×	______	=	______
3.	1 × 8	$280.00/M	______	×	______	=	______
4.	2 × 4	$300.00/M	______	×	______	=	______
5.	2 × 6	$310.00/M	______	×	______	=	______
6.	2 × 8	$310.00/M	______	×	______	=	______
7.	2 × 10	$335.00/M	______	×	______	=	______
8.	2 × 12	$340.00/M	______	×	______	=	______
9.	4 × 4	$400.00/M	______	×	______	=	______
10.	4 × 6	$410.00/M	______	×	______	=	______
11.	4 × 8	$420.00/M	______	×	______	=	______
12.	6 × 6	$450.00/M	______	×	______	=	______

II

FOUNDATIONS

9

Concrete for Foundations

When estimating the amount of concrete for footings, foundation walls, piers and slabs, one must deal in lineal lengths, square foot areas, and cubic volumes. The estimator must convert volume to cubic feet and then to cubic yards, for this is the procedure for ordering concrete. This unit will deal with various methods of acquiring volume measures and time-saving tables used to estimate concrete volumes.

The unit of measure for concrete, or some of the ingredients of concrete (coarse and fine aggregate), is the cubic yard, which contains 27 cubic feet.

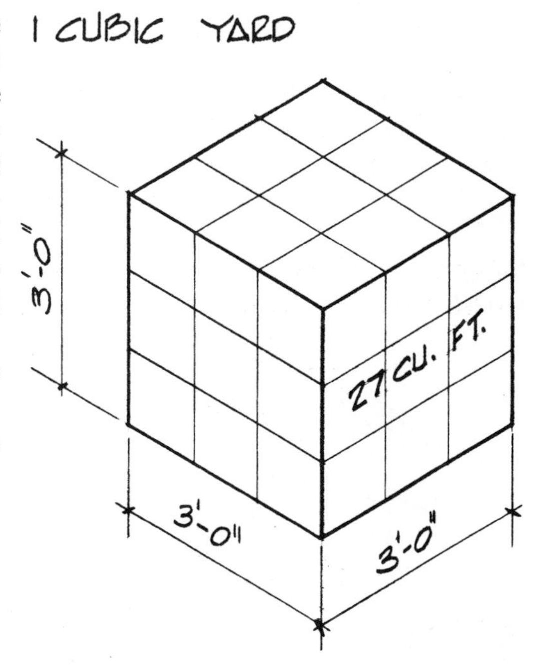

FOUNDATIONS

Foundations for residential structures are made up of four basic parts:

1. Footings
2. Foundation walls
3. Piers and column footings
4. Concrete slabs.

1. ***Footings.*** Usually concrete, they transmit the superimposed load of a structure to the soil.

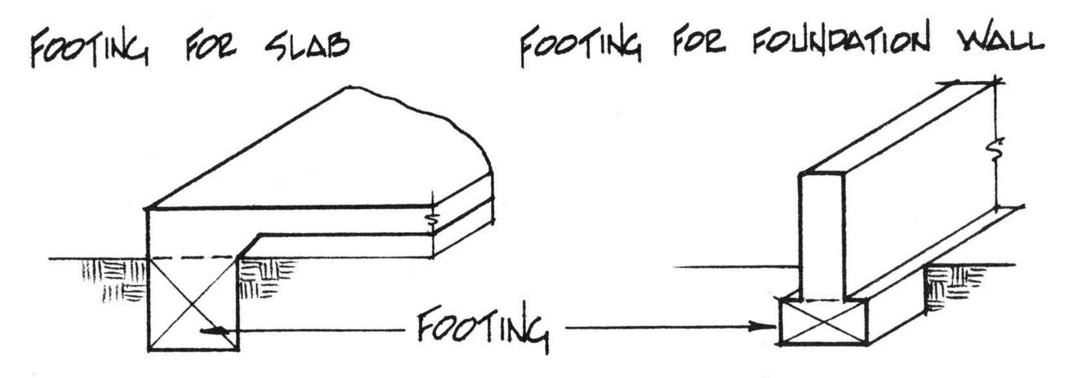

2. ***Foundation Walls.*** Of concrete or masonry block, they form the perimeter of a basement or crawl space and transmit the load from the structure above to the footings. Proper footings are usually as deep as the thickness of the foundation wall they are supporting and twice as wide.

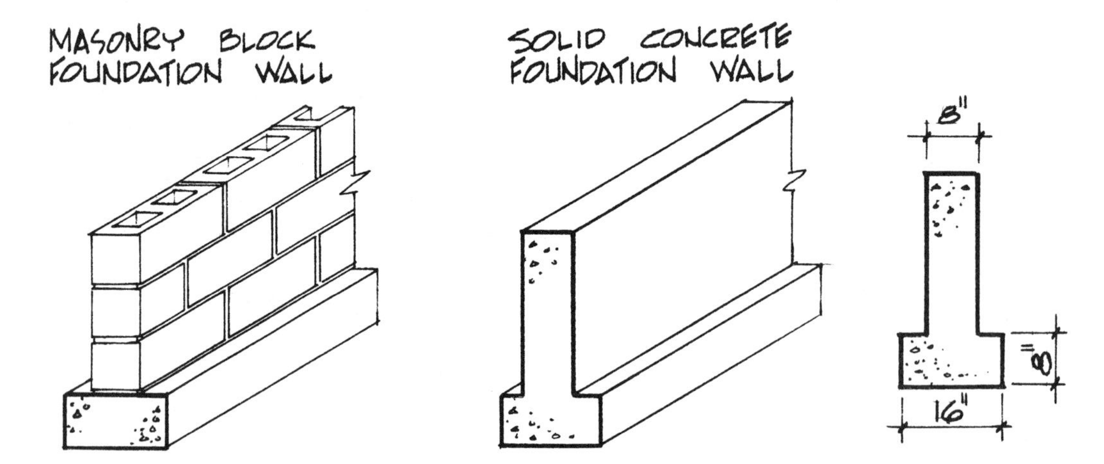

3. ***Pier and Column Footings.*** These are concrete and usually square in shape but may also be round. They also help support the building load through wood or steel posts placed on top of them.

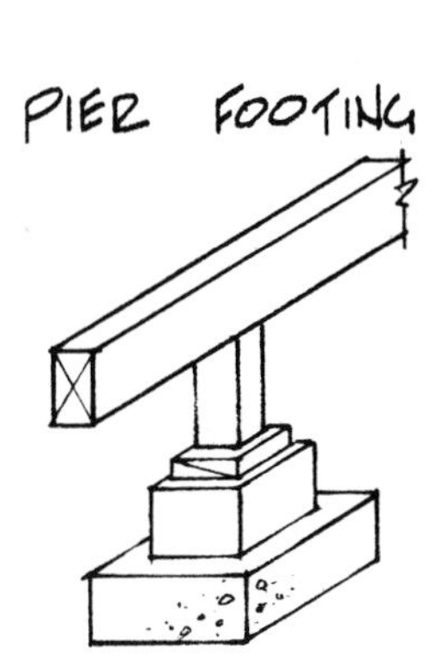

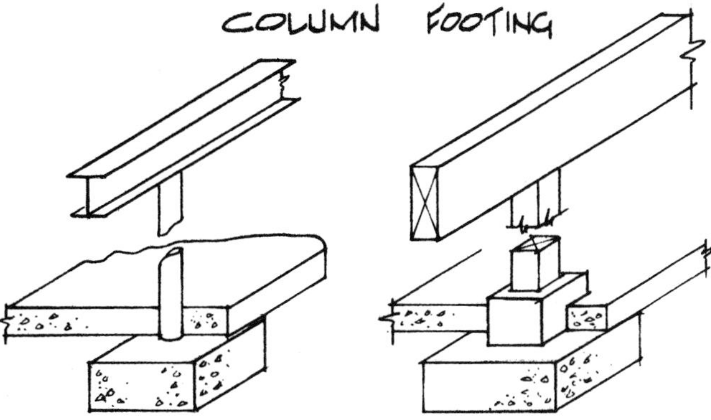

4. ***Concrete Slabs.*** These are used for residential construction primarily in temperate climates, such as found in western and southern states. They carry the live and dead load weights of people, furniture, and equipment placed in the home.

10

Estimating Concrete Volume

The following formula can be used to determine the amount of concrete needed for any square or rectangular area.

$$\text{Cubic Yard Volume} = \frac{\text{Thickness in feet} \times \text{width in feet} \times \text{length in feet}}{27}$$

For example, for a volume of concrete that is 4″ thick by 40′ wide and 60′ long, how many cubic yards of concrete would be required?

Note: The thickness of 4″ is 1/3 of a foot and may be expressed in the problem as follows:

$$\text{Cu. yd.} = \frac{1 \times 40' \times \overset{20}{\cancel{60'}}}{\underset{1}{\cancel{3}} \times 27} = \frac{800 \text{ cu. ft.}}{27 \text{ cu. ft./per yd.}}$$

$$= 29.629 \text{ or } 29.6 \text{ cu. yds. of concrete*}$$

---

* **Note:** If the estimator prefers to use decimals, the problem illustrated above would be expressed as follows:

$$\text{Cubic yard volume} = \frac{.3333 \times 40 \times 60}{27}$$

$$= \frac{799.92}{27} = 29.626 \text{ or } 29.6 \text{ cu. yds.}$$

Table 10-1 will help change several common slab thicknesses to fractions and decimal parts of a foot.

TABLE 10-1 Converting Slab Thickness Measures

Thickness in Inches	*Fractional Part of a Foot*	*Decimal Part of a Foot*
2″	2/12 or 1/6	.1667
3″	3/12 or 1/4	.2500
4″	4/12 or 1/3	.3333
5″	5/12	.4167
6″	6/12 or 1/2	.5000
7″	7/12	.5833
8″	8/12 or 2/3	.6667
9″	9/12 or 3/4	.7500
10″	10/12 or 5/6	.8333
11″	11/12	.9167
12″	1	1.0000

Estimate the amount of cubic yardage in the following problems:

1. A concrete slab 4″ thick by 28′-6″ wide by 42′-0″ long.

2. A concrete retaining wall 8″ thick by 8′-0″ high by 120′-0″ long.
3. A concrete shuffleboard slab 6″ thick by 6′-0″ wide by 40′-0″ long.
4. A concrete sidewalk 4″ thick by 4′-0″ wide by 450′-0″ long.
5. A concrete slab 3″ thick by 20′-9″ wide by 25′-3″ long.

Although the above formula is adequate for estimating concrete volume, in certain cases the estimator also has various tables available to help cut his estimating time. These tables will estimate volumes by using square foot area and lineal foot measurement. Methods of acquiring these will be discussed below.

SQUARE FOOT AREA

A square foot is the surface area covered by a 12-inch by 12-inch square without regard to its thickness.

12″ 12″ ONE SQUARE FOOT

There are two methods that can be used to find the number of square feet in a building. In the first method, the positive areas are added together as in the example below:

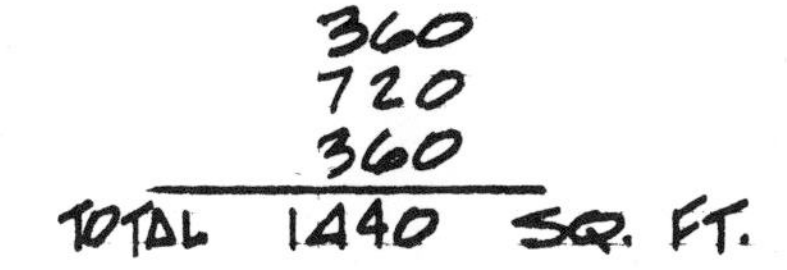

EXAMPLE 1—Positive Method:

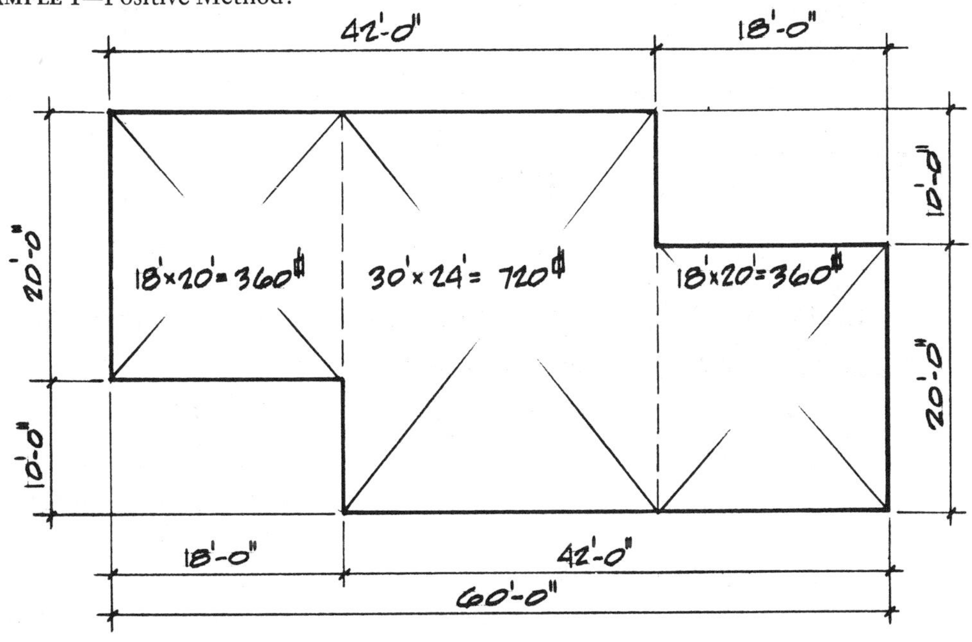

In the second method, the largest width dimension must be multiplied by the largest dimension, and the negative areas must be subtracted.

EXAMPLE 1—Negative Method:

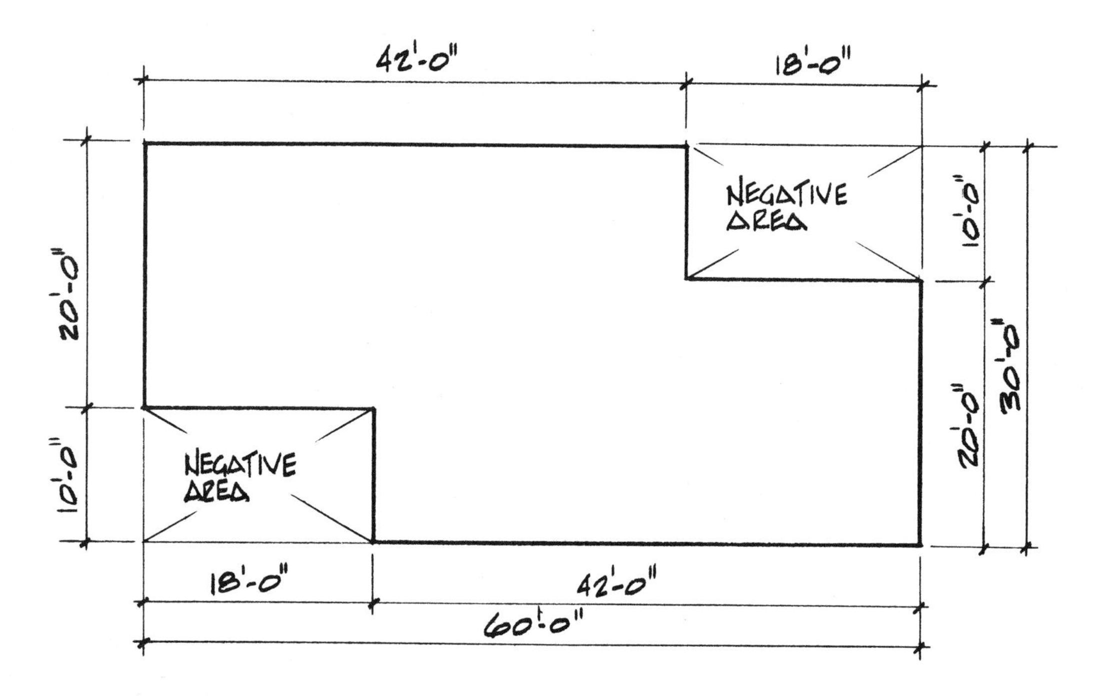

LARGEST WIDTH & LENGTH DIMENSION 30×60 = 1800 SQ. FT.

NEGATIVE AREAS 10×18 = 180

10×18 = 180

−360

1800

− 360

TOTAL 1440 SQ. FT.

The estimator can double-check the square footage by using both methods.

EXAMPLE 2—Positive Method:

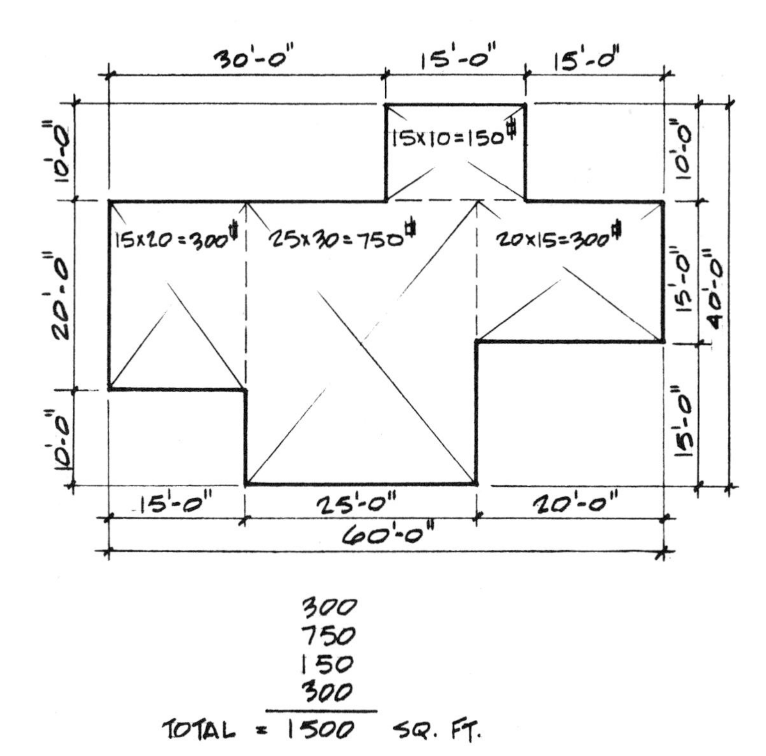

EXAMPLE 2—Negative Method:

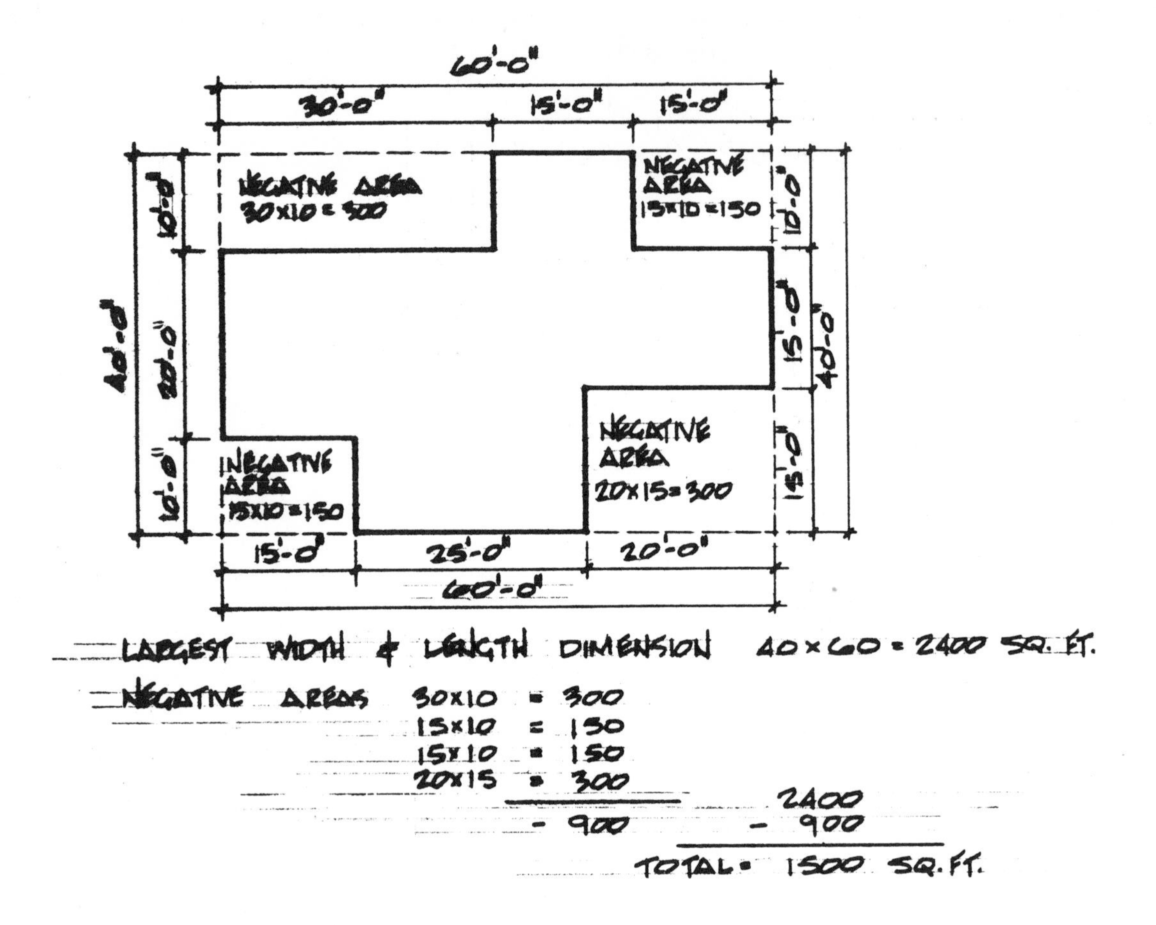

The building shapes in the previous examples have been given in even foot measurement. The following table will help an estimator find the square footage of buildings that are not dimensioned in even feet. The table changes the inches in a foot to decimal parts of a foot.

TABLE 10-2 CONVERTING INCHES TO DECIMAL EQUIVALENTS

Inches	Decimal Parts	Inches	Decimal Parts
1	.0833	7	.5833
2	.1667	8	.6667
3	.2500	9	.7500
4	.3333	10	.8333
5	.4167	11	.9167
6	.5000	12	1.0000

EXAMPLE: Find the square footage in the following building:

1. Check Table 10-2 to find the decimal equivalent of 6″ (.5000) and 3″ (.2500).
2. Multiply width times length:

```
       48.50
     × 28.25
     -------
       24250
       9700
      38800
      9700
     -------
Answer .... 1,370.1250 sq. ft.
```

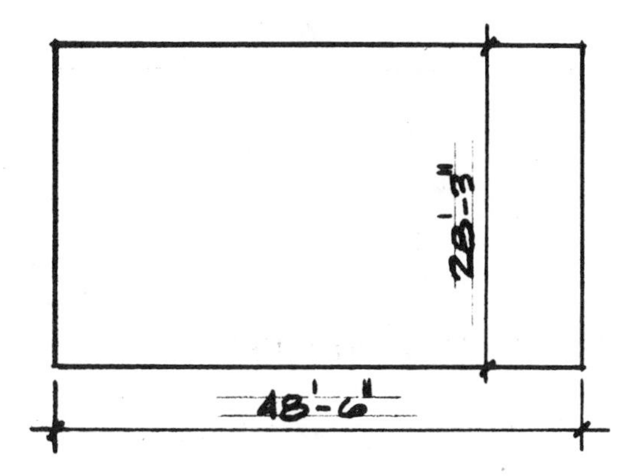

Find the square footage in the following practice problems.

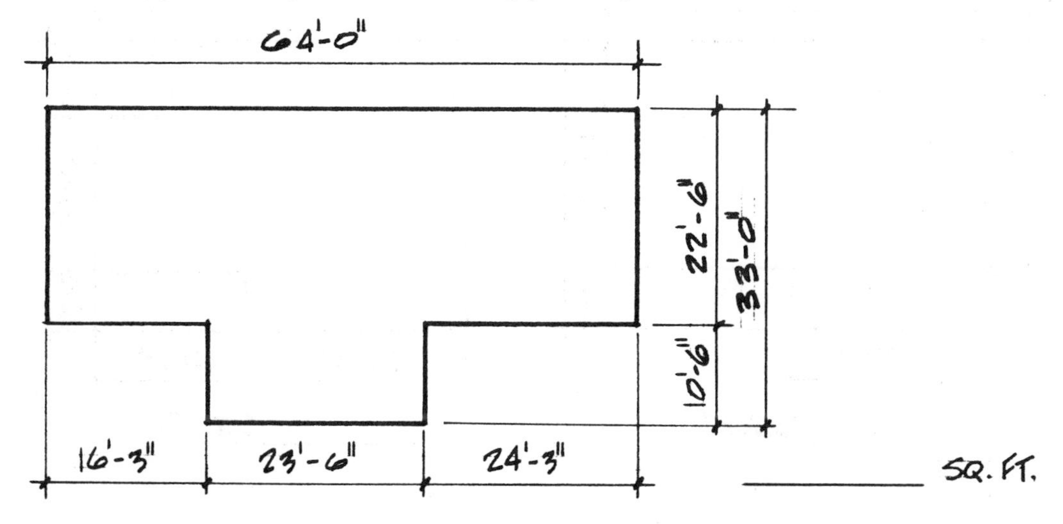

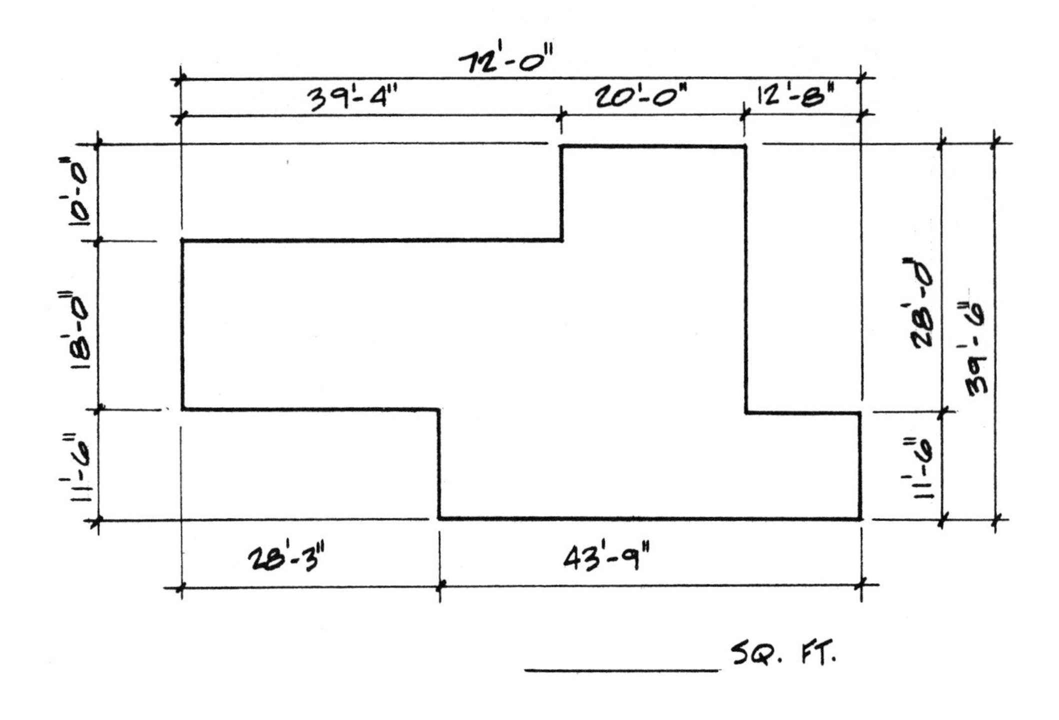

SQUARE FOOT MEASURE

Find the square footage of the following buildings.

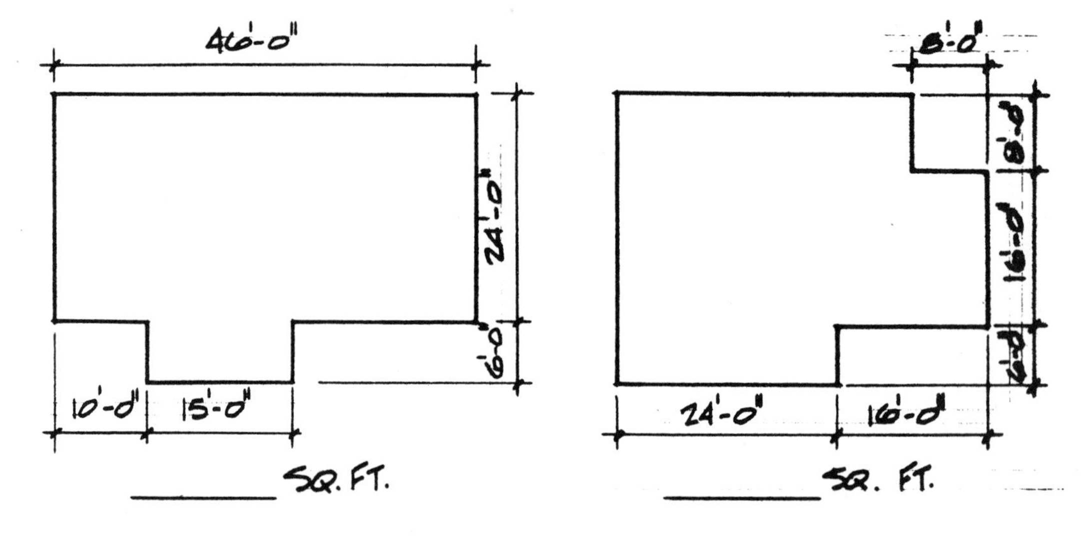

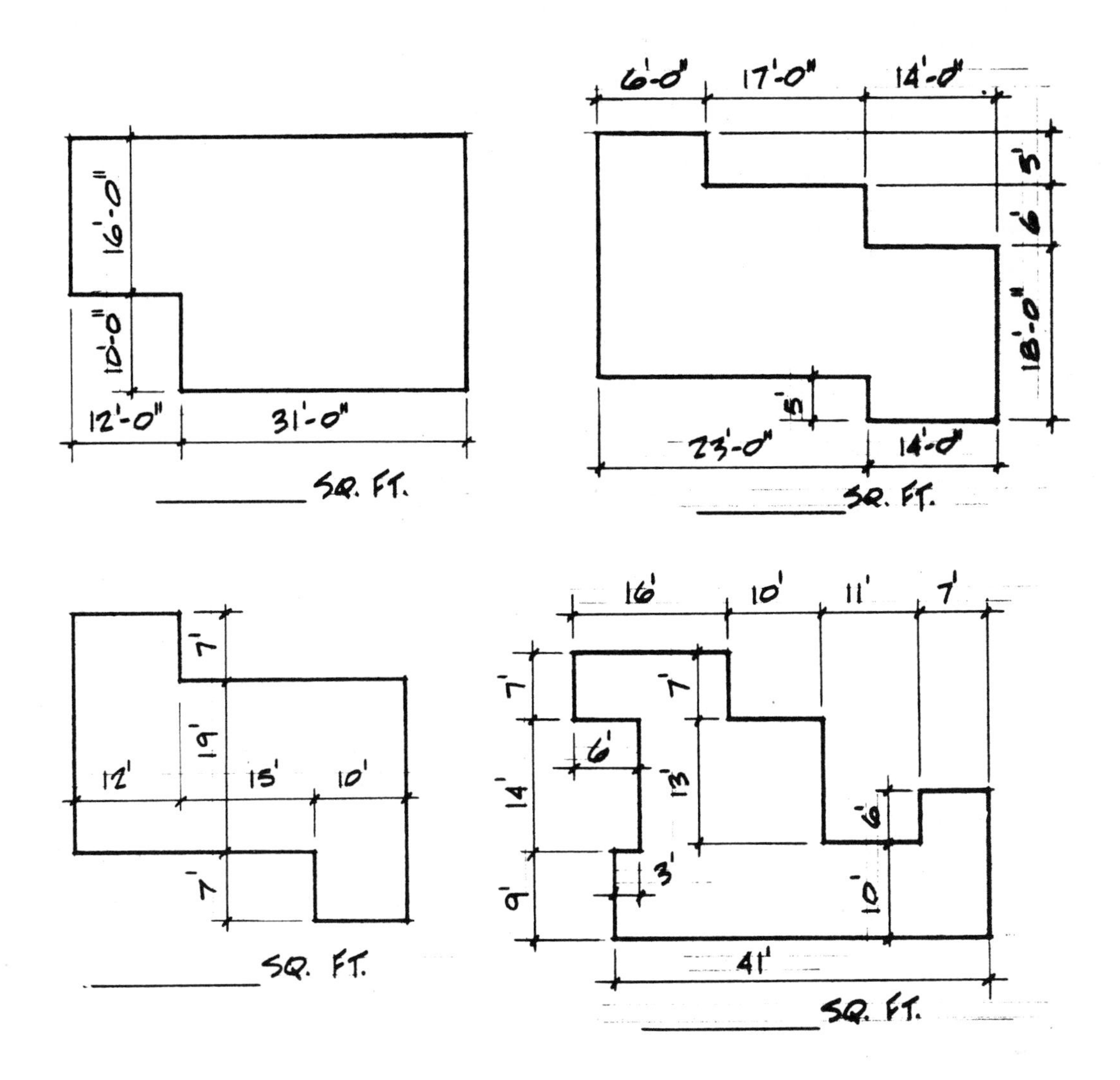

LINEAL FOOT MEASUREMENT

A lineal foot is a line designated by length in feet.

10′-0″

10 lineal feet or 10 lin. ft.

Lineal feet is the measurement used to designate the perimeter of a building. (The perimeter is the distance around the outside of the building.)

The perimeters of the following buildings are calculated by adding together the lengths of the sides.

EXAMPLE 1:

$$\begin{array}{r} 48 \\ 48 \\ 28 \\ +\ 28 \\ \hline 152 \text{ lin. ft.} \end{array}$$

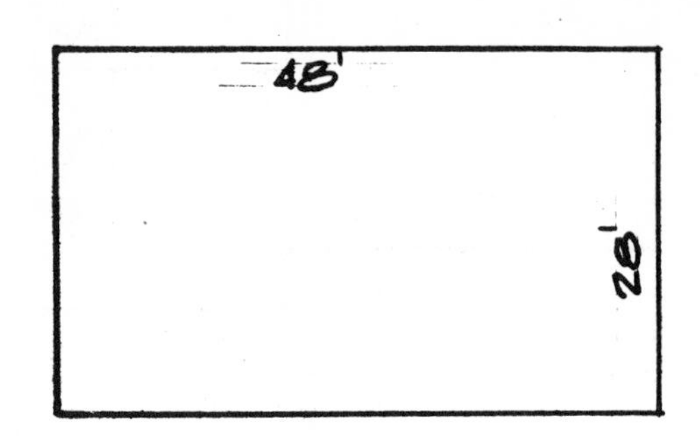

EXAMPLE 2:

$$\begin{array}{r} 50 \\ 15 \\ 22 \\ 15 \\ 72 \\ +\ 30 \\ \hline 204 \text{ lin. ft.} \end{array}$$

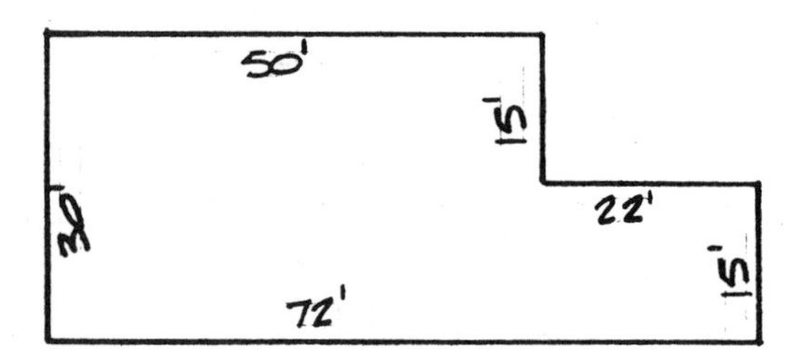

Find the perimeter of the following practice problems.

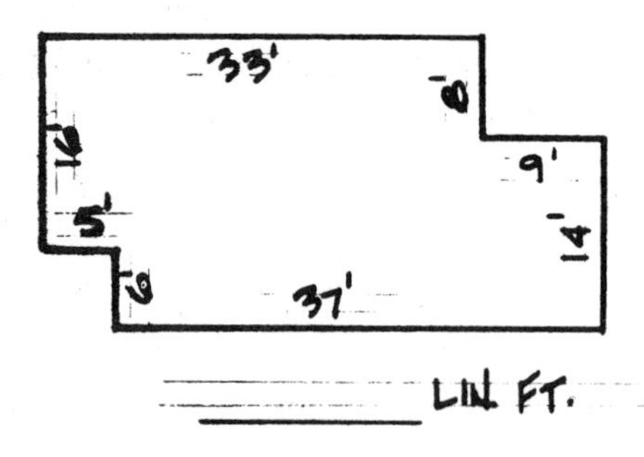

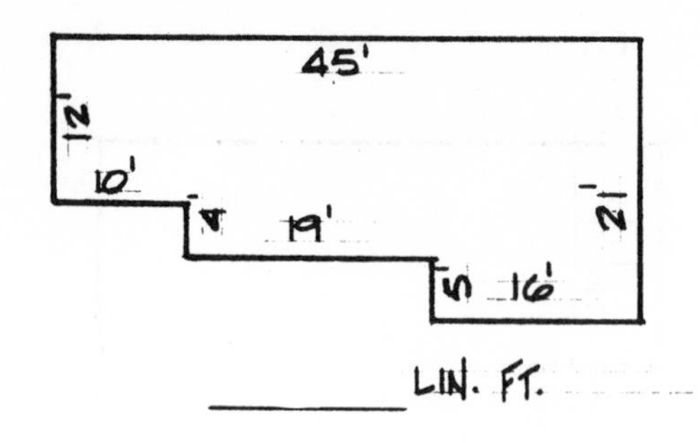

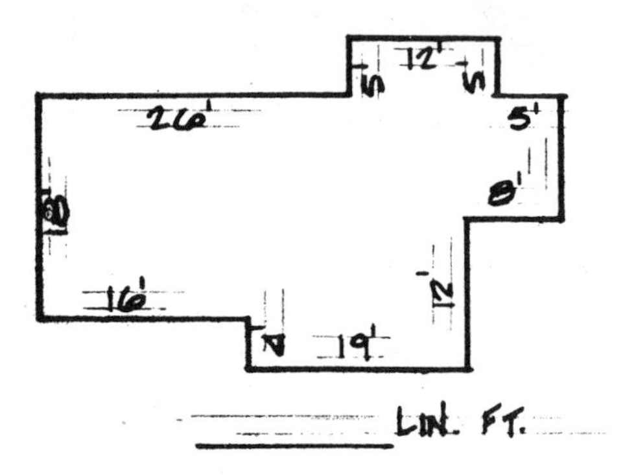

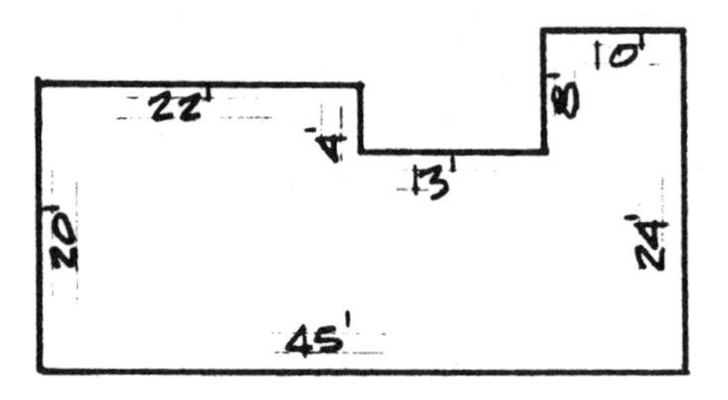

Find the perimeter of the following buildings.

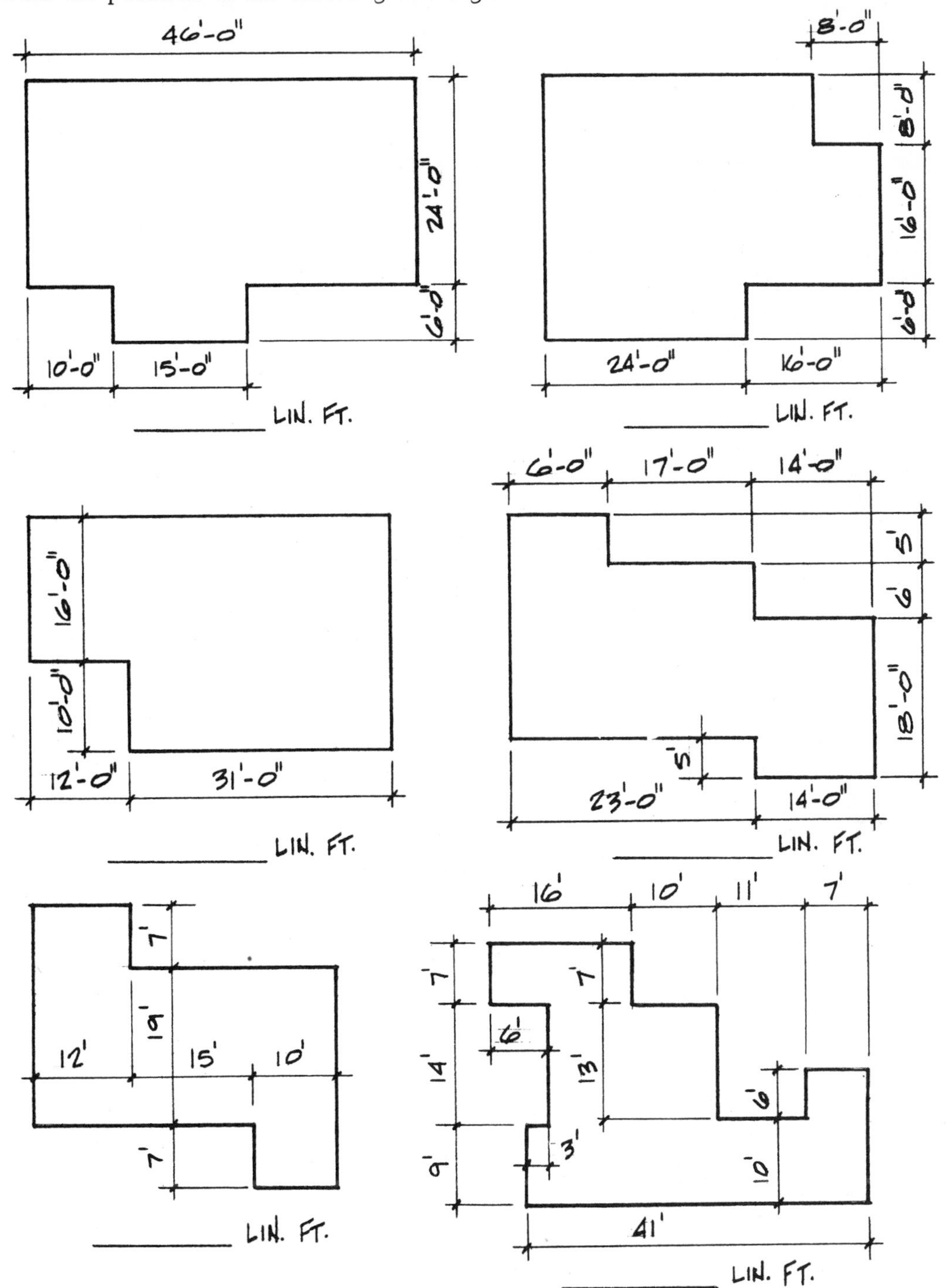

Now that the method for finding square foot and lineal foot measure is established, the estimator can use the following rules and tables for finding concrete volume.

PROCEDURE FOR ESTIMATING CONCRETE VOLUME OF FOOTINGS AND FOUNDATION WALLS

1. Check the foundation plan for the size of footings and foundation walls.
2. From Table 10-3, select the cubic foot volume that corresponds to the footing and foundation wall sizes.
3. Calculate the lineal feet of footing and foundation wall.
4. Multiply the lineal feet of footing and foundation wall by the cubic foot volumes per lineal foot.
5. Divide the cubic feet of form content by 27 for cubic yards.

Example Problem:

Find the cubic feet and cubic yards of concrete for the following foundation problem.

PROCEDURE:

1. Determine the size of the footing and foundation wall when the footing size is 6″ × 12″ and the foundation wall size is 6″ × 22″.
2. The cubic foot volume from Table 10-3 is:

   6″ × 12″ footing (.5)
   6″ × 22″ foundation wall (.9167)

3. The lineal footage of footing and foundation wall is 120 lineal feet.

   Note: The extra concrete from doubling the corners is offset by the losses that occur in placing the concrete.
4. Multiply the number of lineal feet of footing and foundation wall (120 lin. ft.) by the cubic foot volume per lineal foot (.5) and (.9167).

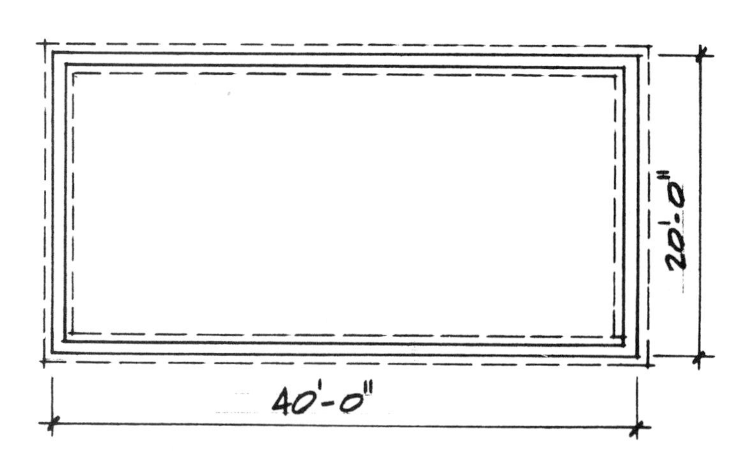

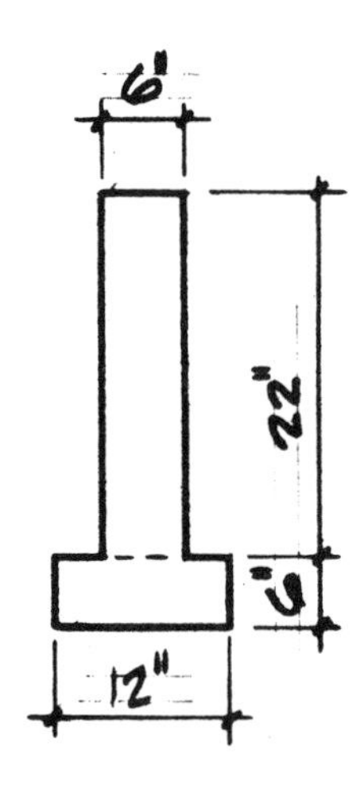

```
  120            120
×  .5          × .9167
60 cu. ft.     110.0040 cu. ft.
   .               .
   .               .
   .               .
60 cu. ft. plus  110.0040 cu. ft.
 = 170.0040 or 170 cu. ft.
```

5. Divide the cubic feet (170) by 27 for cubic yards: 170 ÷ 27 = 6.29 cubic yards. Concrete from a ready-mix yard is usually ordered to the nearest one-quarter yard, so 6-1/2 yards would be ordered.

If there is no need to separate the volume of concrete in the footings from that in the foundation wall, the .5 and .9167 cu. ft. per lineal foot may be added together, multiplied by the length (120 ft.), then divided by 27 to acquire the cu. yd. volume.

EXAMPLE: .5 + .9167 = 1.4167 × 120 = 170.0040 or 170 ÷ 27 = 6.29 cu. yds.

TABLE 10-3 CUBIC FOOT VOLUME FOR CONCRETE FOOTINGS AND WALLS

Width of Concrete	*Depth of Concrete*									
	6″	*8″*	*10″*	*12″*	*14″*	*16″*	*18″*	*20″*	*22″*	*24″*
6″	.25	.3333	.4167	.50	.5833	.6667	.75	.8333	.9167	1.00
8″	.3333	.4444	.5555	.6667	.7778	.8889	1.00	1.1111	1.2222	1.3333
10″	.4167	.5555	.6944	.8333	.9722	1.1111	1.25	1.3889	1.5278	1.6667
12″	.50	.6667	.8333	1.00	1.1667	1.3333	1.50	1.6667	1.8333	2.00
14″	.5833	.7778	.9722	1.1667	1.3611	1.5555	1.75	1.9444	2.1389	2.3333
16″	.6667	.8889	1.1111	1.3333	1.5555	1.7778	2.00	2.2222	2.4444	2.6667
18″	.75	1.00	1.25	1.50	1.75	2.00	2.25	2.50	2.75	3.00
20″	.8333	1.1111	1.3889	1.6667	1.9444	2.2222	2.50	2.7778	3.0555	3.3333
22″	.9167	1.2222	1.5278	1.8333	2.1389	2.4444	2.75	3.0555	3.3611	3.6667
24″	1.00	1.3333	1.6667	2.00	2.3333	2.6667	3.00	3.3333	3.6667	4.00
30″	1.25	1.6667	2.0833	2.50	2.9167	3.3333	3.75	4.1667	4.5833	5.00
32″	1.3333	1.7778	2.2222	2.6667	3.1111	3.5555	4.00	4.4444	4.8889	5.3333
36″	1.50	2.00	2.50	3.00	3.50	4.00	4.50	5.00	5.50	6.00

Note: If a member is larger than 36″, add the two volumes together.

EXAMPLE: What is the cubic volume per lineal ft. of a wall 8″ wide by 72″ high?

The volume is found by taking 8″ × 36″ (one-half the height) which is 2.0 cu. ft. per lineal foot, then doubling it to arrive at the volume of 4.0 cu. ft. per lineal foot for a wall 8″ × 72″.

RULES FOR ESTIMATING CONCRETE WALLS AND FLOOR SLABS

Another method used to estimate the volume of a concrete wall or floor slab is to consult Table 10-4, which designates how many square feet a cubic yard of concrete will cover at a certain thickness.

TABLE 10-4 Estimating Slab Area per Cubic Yard of Concrete

Thickness (Inches)	*No. Sq. Ft.*	*Thickness (Inches)*	*No. Sq. Ft.*
1	324	7	46
1-1/4	259	7-1/4	45
1-1/2	216	7-1/2	43
1-3/4	185	7-3/4	42
2	162	8	40
2-1/4	144	8-1/4	39
2-1/2	130	8-1/2	38
2-3/4	118	8-3/4	37
3	108	9	36
3-1/4	100	9-1/4	35
3-1/2	93	9-1/2	34
3-3/4	86	9-3/4	33
* 4	81	10	32
4-1/4	76	10-1/4	32
4-1/2	72	10-1/2	31
4-3/4	68	10-3/4	30
5	65	11	29-1/2
5-1/4	62	11-1/4	29
5-1/2	59	11-1/2	28
5-3/4	56	11-3/4	27-1/2
6	54	12	27
6-1/4	52	12-1/4	26-1/2
6-1/2	50	12-1/2	26
6-3/4	48	12-3/4	25-1/2

* A 4″ slab is a very common thickness for house pads, patio slabs, sidewalks, and driveways. At 4″ thick, a yard of concrete will actually cover 81 sq. ft.; but for small jobs, 80 sq. ft. is used for convenience and speed of dividing.

Example Problem:

Estimate the cubic yards of concrete in the slab illustrated below.

PROCEDURE:

1. Calculate the square foot area of the slab: 20 ft. × 40 ft. = 800 sq. ft.
2. Divide the total square foot area of the slab by the sq. ft. area a cubic yard of concrete will cover at the designated thickness: 800 sq. ft. ÷ 81 sq. ft. = 9.87 or 10 cu. yds. This size slab is considered a small slab, and the total sq. footage (800) could have been divided by 80 for convenience.

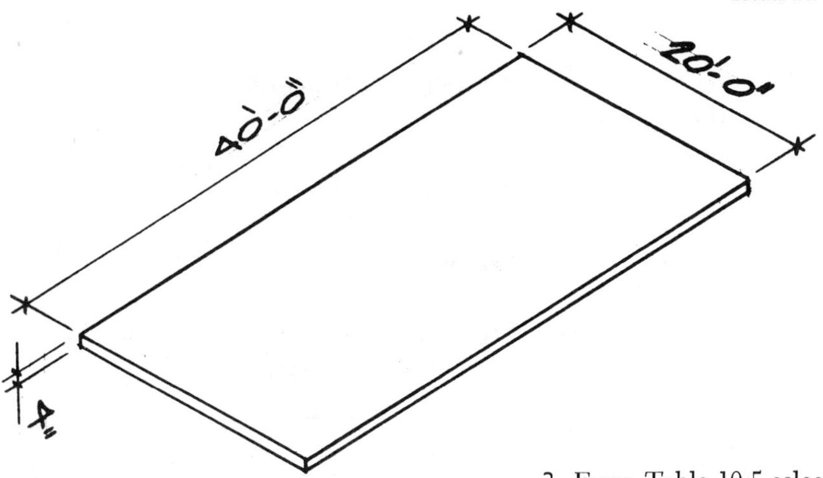

RULES FOR ESTIMATING CONCRETE PIERS

PROCEDURE:

1. From the foundation plan take off the size and depth of concrete piers.
2. From Table 10-5 select size and read the cu. ft. volume per pier.
3. Multiply the number of piers by the cu. ft. volume per pier. The result equals pier content in cubic feet.
4. To convert to cubic yards, divide cu. ft. by 27.

TABLE 10-5 CUBIC VOLUME FOR SQUARE CONCRETE PIERS

Pier Depth	Size 10″×10″	12″×12″	14″×14″	16″×16″	18″×18″
6″	.3	.5	.7	.9	1.1
7″	.4	.6	.8	1.0	1.3
8″	.5	.7	.9	1.2	1.5
9″	.5	.8	1.0	1.3	1.7
10″	.6	.8	1.1	1.5	1.9
11″	.6	.9	1.2	1.6	2.1
12″	.7	1.0	1.4	1.8	2.2
14″	.8	1.2	1.6	2.1	2.6
16″	.9	1.3	1.8	2.4	3.0
18″	1.0	1.5	2.0	2.7	3.4
20″	1.2	1.7	2.3	3.0	3.8
22″	1.3	1.8	2.5	3.3	4.1
24″	1.4	2.0	2.7	3.6	4.5

Note: The volumes in the table above have been rounded off to the nearest tenth of a cubic foot.

Example Problem:

How many cubic yards of concrete are there in the piers illustrated below?

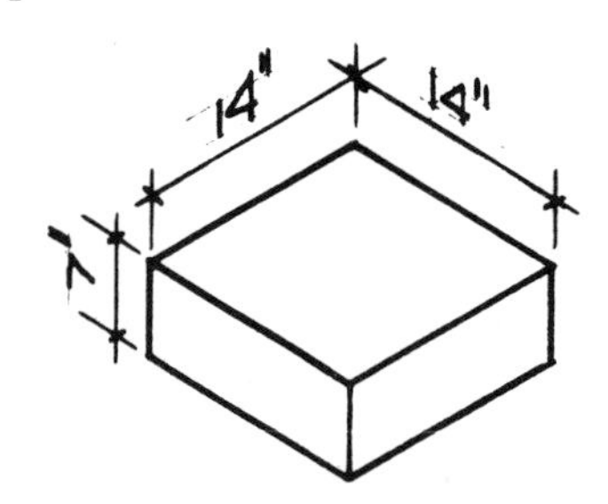

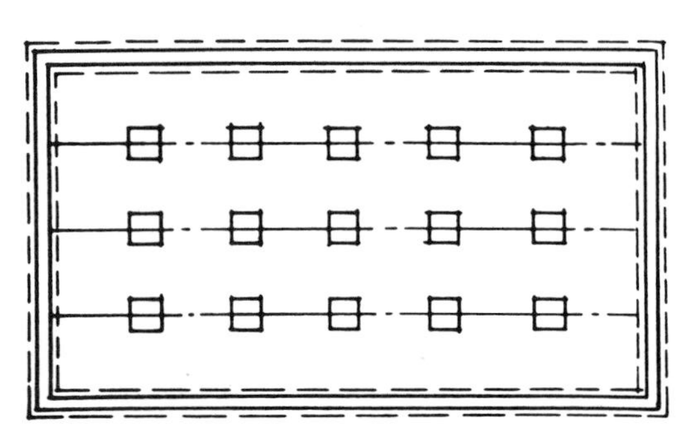

PROCEDURE:

1. The piers are 14" × 14" square and 7" deep.
2. The volume from the concrete pier table is 0.8 cu. ft. per pier.
3. Multiply the number of piers (15) by the cu. ft. per pier: 15 × 0.8 = 12 cu. ft.
4. Divide the cu. ft. of pier content (12) by 27 for cubic yards: 12 ÷ 27 = 0.44 cu. yds.

Note: The piers would take approximately 1/2 cubic yard of concrete.

TRIANGULAR AREAS

The area of a triangle equals 1/2 the area of a rectangle of the same length and width.

Example Problem:

Estimate the amount of concrete needed in the following triangular slab:

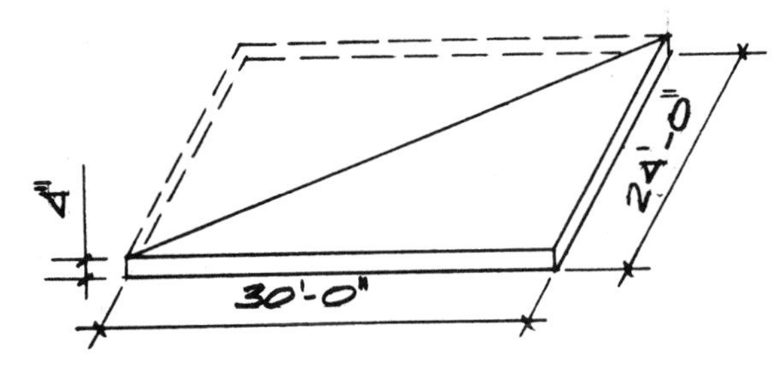

PROCEDURE:

1. Multiply the height by the width: 30′ × 24′ = 720 sq. ft.
2. Divide the square footage above by 2 to find the sq. ft. area of the triangle: 720 ÷ 2 = 360 sq. ft.
3. The slab thickness is 4" (1/3 of a foot). Divide the sq. ft. area of the triangle by 1/3: 360 ÷ 3 = 120 cu. ft.
4. Divide 120 cu. ft. by 27 for cu. yd. volume: 120 ÷ 27 = 4.44 cu. yd.

11

Estimating Foundation Bolts

In some localities the building code may require that the mudsill or the bottom wall plate on a slab be held down with bolts embedded in the concrete. The usual recommendation is for one bolt within one foot of each corner, exterior door openings or wall plate splice, and a maximum span of 6′-0″ for bolts falling between corners.

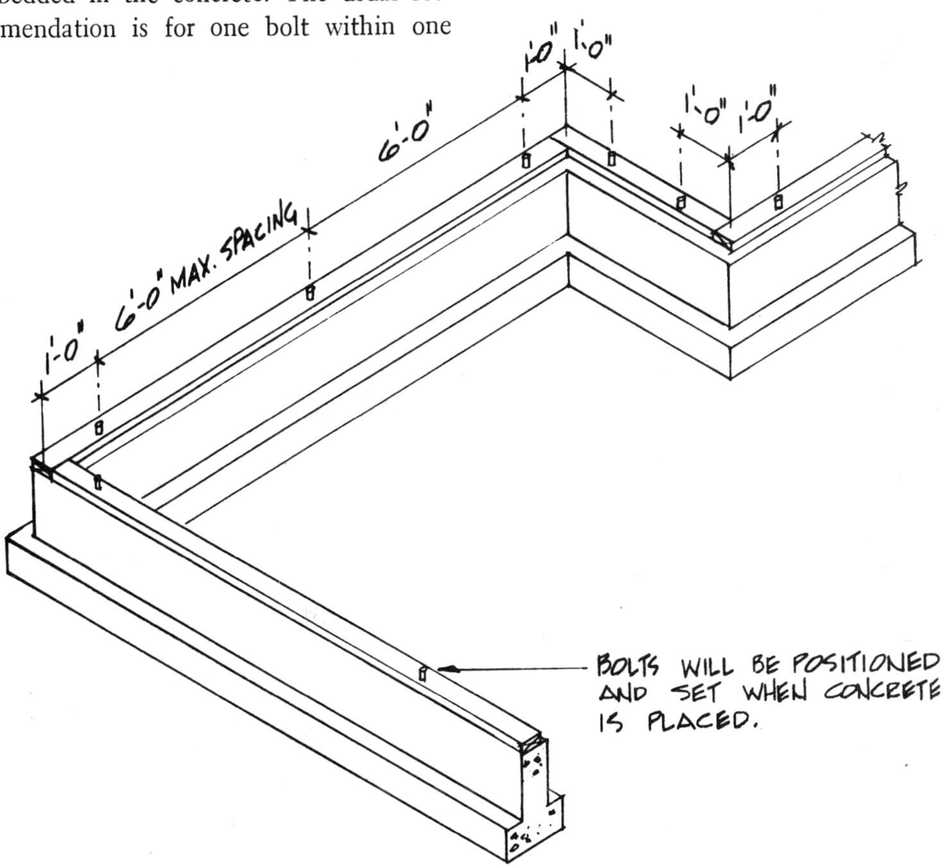

The building inspector may require that the mudsill or exterior bottom wall plate be fastened with a power-activated tool within one foot of any sill splice.

SLAB CONSTRUCTION

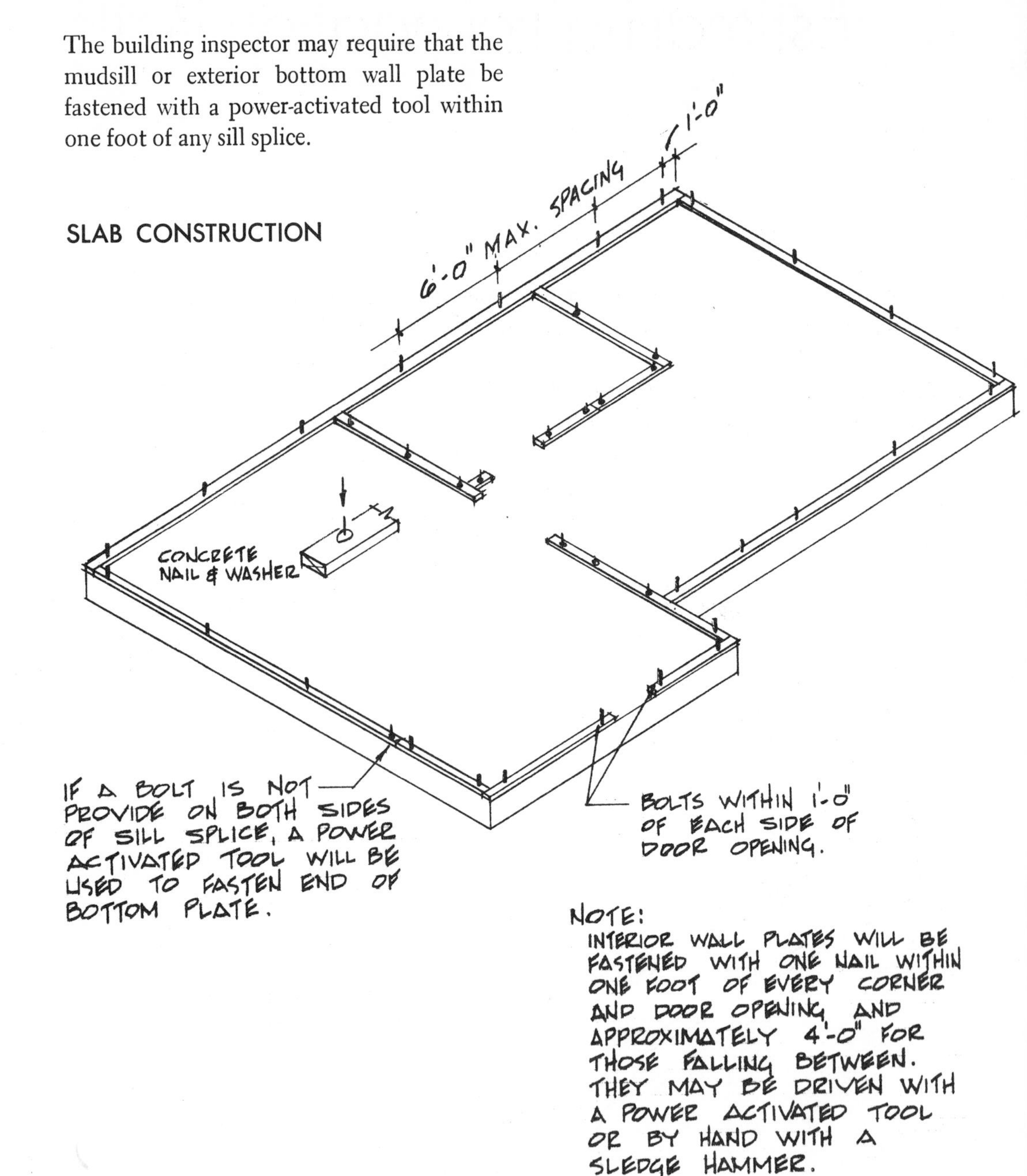

PROCEDURE FOR ESTIMATING FOUNDATION BOLTS

1. Calculate the number of bolts for the corners of the building. Two bolts are used for each interior and exterior corner.
2. For the bolts between the corners, divide the wall perimeter by six. (Six-foot maximum spacing allowed.)
3. Add the number of bolts for the corners and between the corners. The result will be the number of bolts needed. If the building has long walls with many plate splices and if there are more than the usual number of exterior door openings, add a few extra bolts.

12

Problems—Concrete Volume

PROBLEM 1: SLAB AND FOOTING

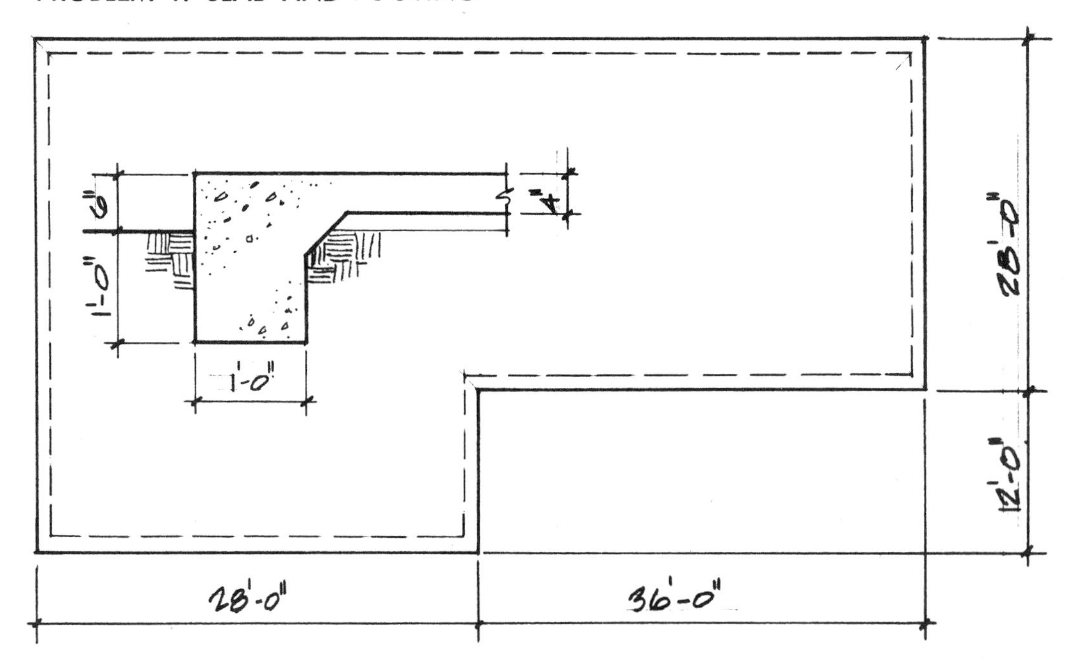

ESTIMATE THE FOLLOWING:

1. Perimeter of building __________
2. Cubic feet in footings __________
3. Square foot area of building __________

4. Cubic feet in slab ______________
5. Total cubic feet ______________
6. Cubic yards ______________
7. Cost per cubic yard of concrete in local area ______________
8. Total cost of concrete ______________

PROBLEM 2: SLAB AND FOOTING

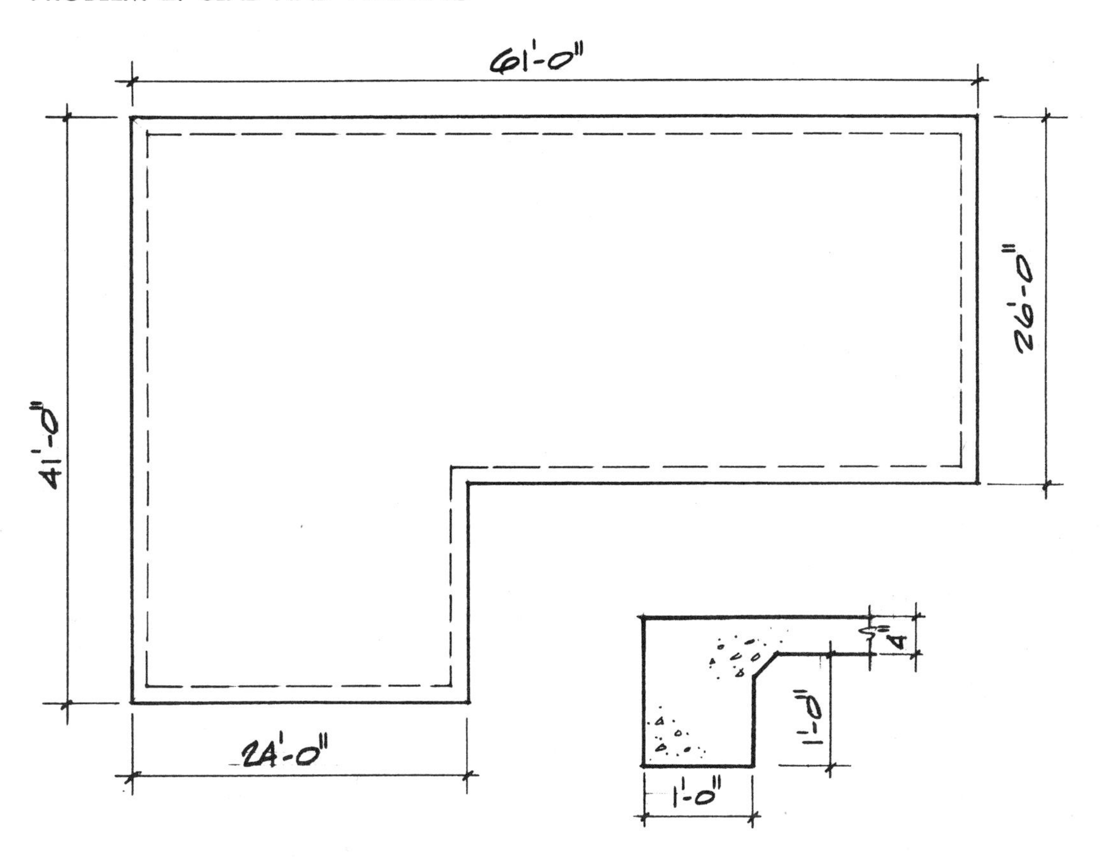

ESTIMATE THE FOLLOWING:

1. Perimeter of building ______________
2. Cubic feet in footings ______________
3. Square foot area of building ______________
4. Cubic feet in slab ______________
5. Total cubic feet ______________
6. Cubic yards ______________
7. Cost per cubic yard of concrete in local area ______________
8. Total cost of concrete ______________

PROBLEM 3: GARAGE FOOTING AND SLAB

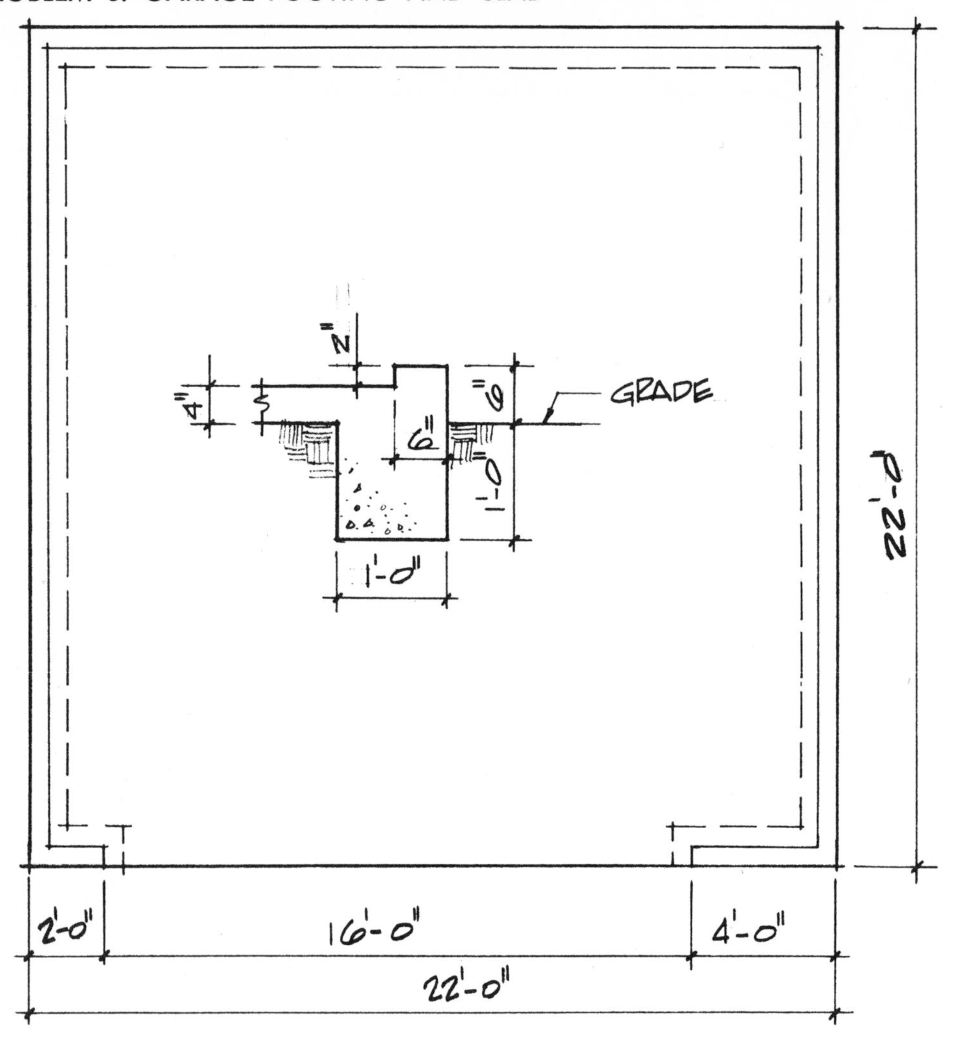

ESTIMATE THE FOLLOWING:

1. Lineal feet of footing ________
2. Cubic feet in footings ________
3. Square foot area of slab ________
4. Cubic feet in slab ________
5. Lineal feet of foundation curb ________
6. Cubic feet in foundation curb ________
7. Total cubic feet ________
8. Cubic yards ________
9. Cost per cubic yard of concrete in local area ________
10. Total cost of concrete ________

PROBLEM 4: CONVENTIONAL "T" FOOTING AND PIERS

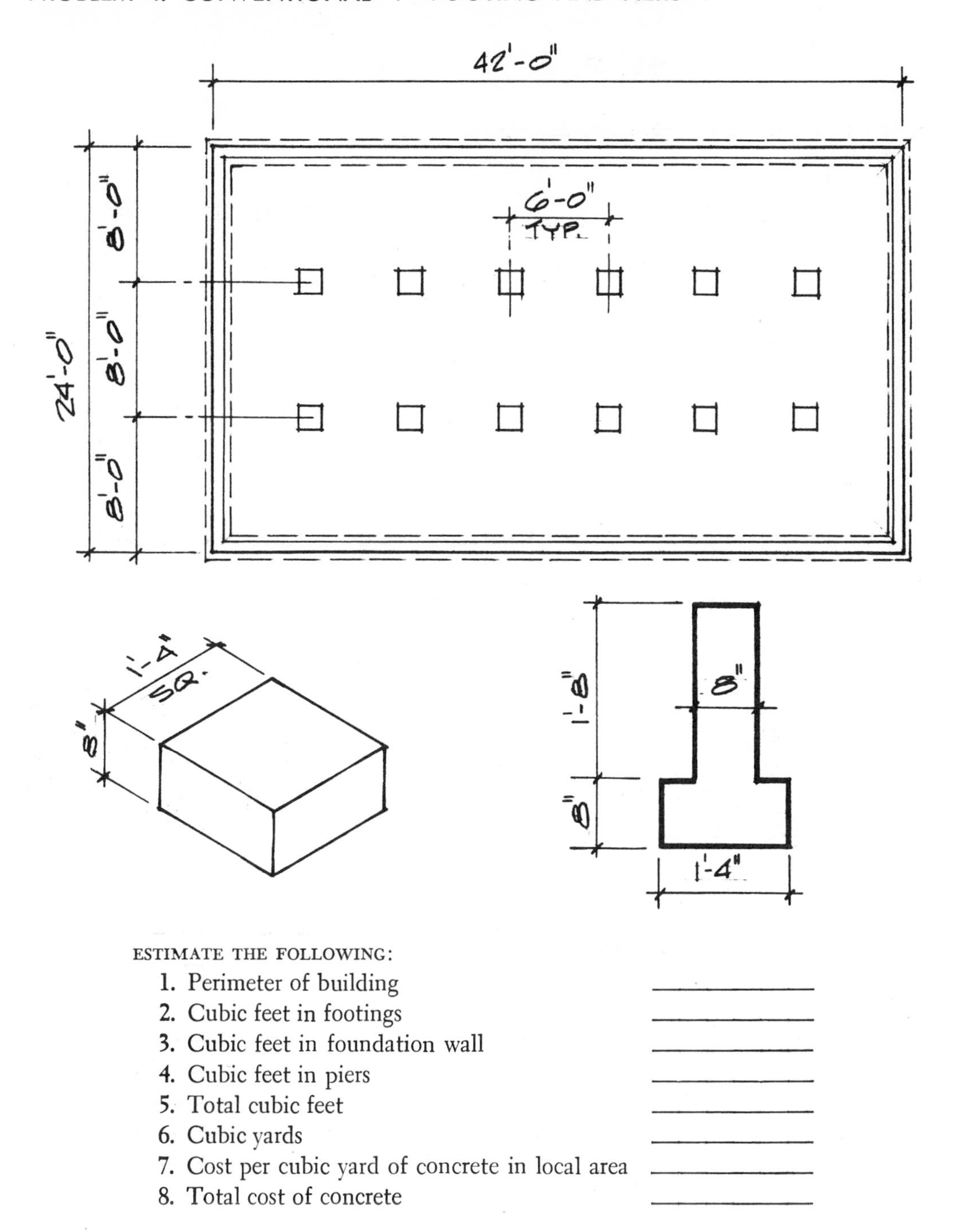

ESTIMATE THE FOLLOWING:

1. Perimeter of building ____________
2. Cubic feet in footings ____________
3. Cubic feet in foundation wall ____________
4. Cubic feet in piers ____________
5. Total cubic feet ____________
6. Cubic yards ____________
7. Cost per cubic yard of concrete in local area ____________
8. Total cost of concrete ____________

PROBLEM 5: CONVENTIONAL "T" FOOTING AND PIERS

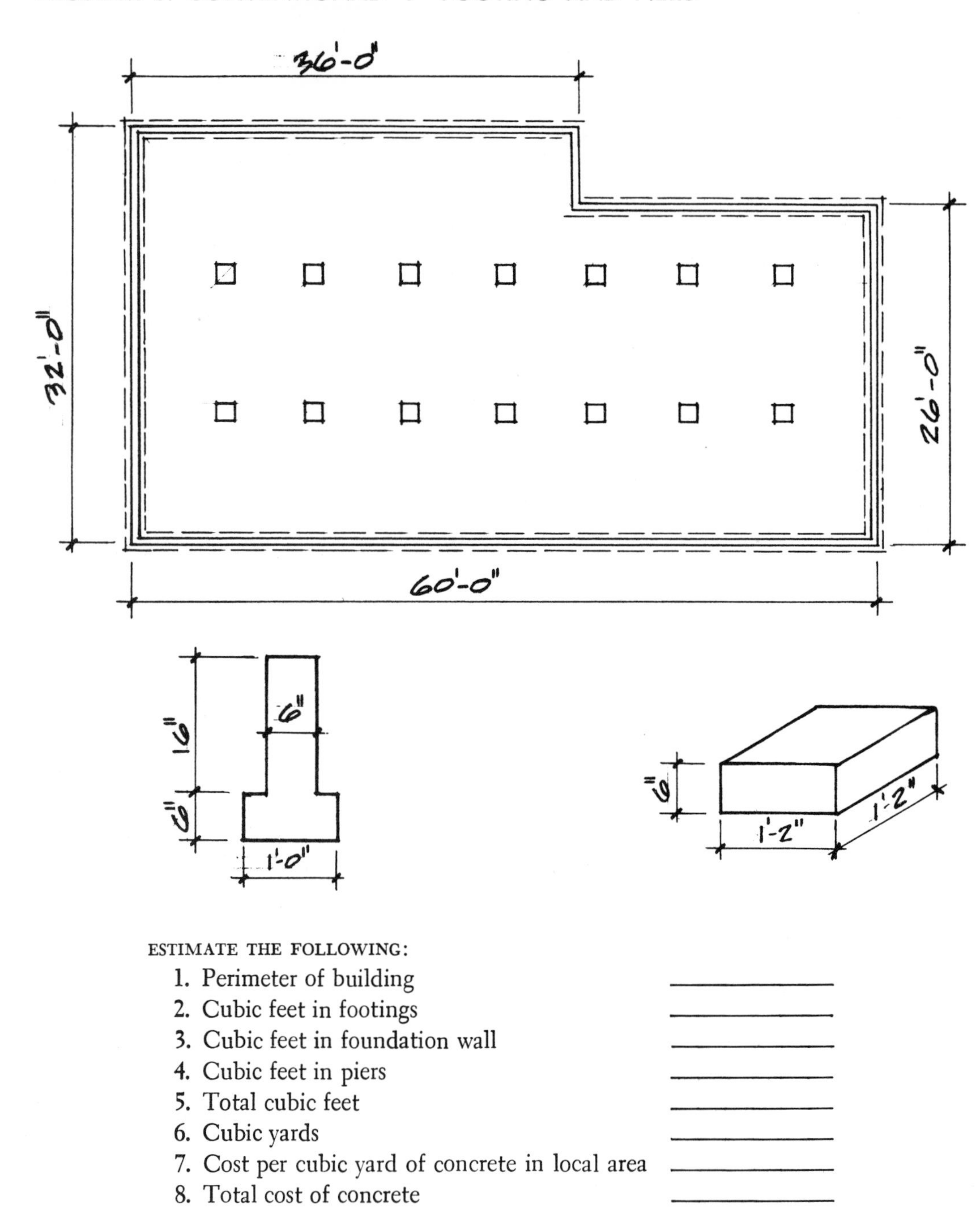

ESTIMATE THE FOLLOWING:

1. Perimeter of building ________
2. Cubic feet in footings ________
3. Cubic feet in foundation wall ________
4. Cubic feet in piers ________
5. Total cubic feet ________
6. Cubic yards ________
7. Cost per cubic yard of concrete in local area ________
8. Total cost of concrete ________

PROBLEM 6: CONVENTIONAL "T" FOOTING AND DWARF WALL

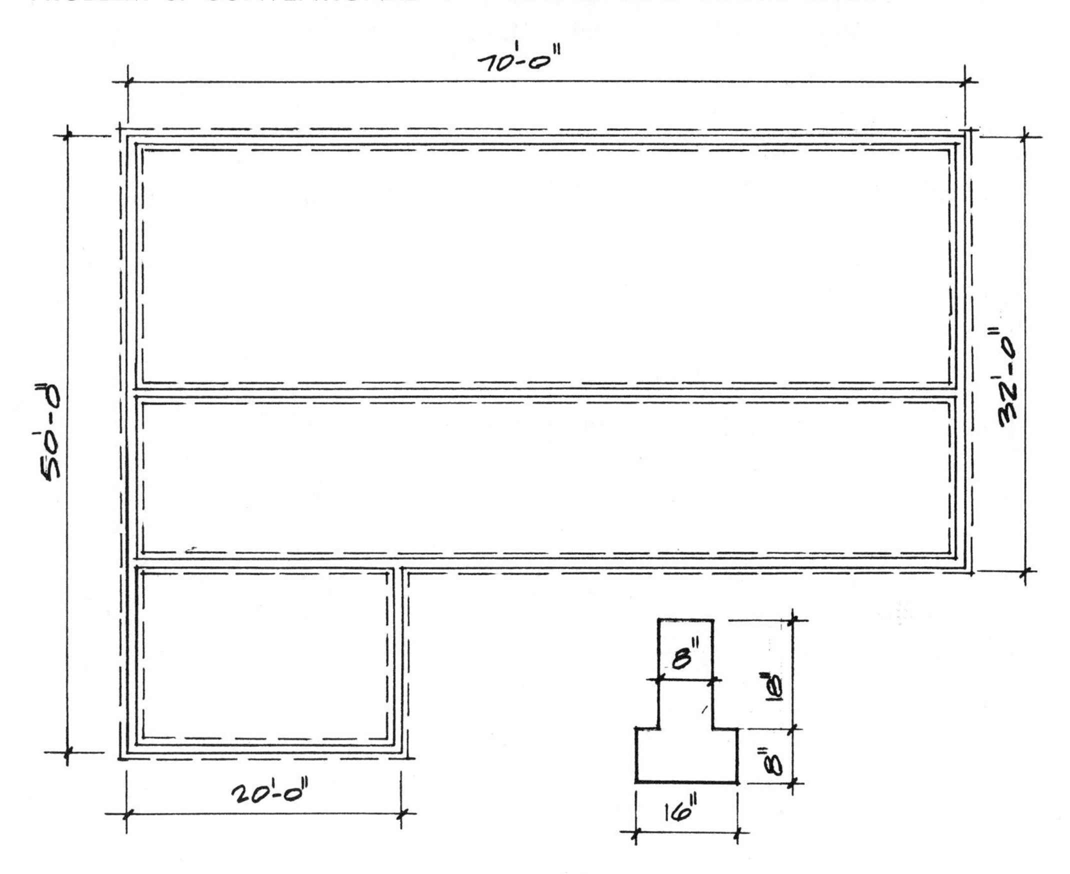

ESTIMATE THE FOLLOWING:

1. Perimeter of building	__________
2. Lineal feet of dwarf wall	__________
3. Total lineal feet of footings including dwarf wall	__________
4. Cubic feet in footings	__________
5. Total lineal feet of foundation wall including dwarf wall	__________
6. Cubic feet in foundation wall	__________
7. Total cubic feet	__________
8. Cubic yards	__________
9. Cost per cubic yard of concrete in local area	__________
10. Total cost of concrete	__________

PROBLEM 7: PERIMETER FOOTINGS, FOUNDATION WALLS, AND COLUMN FOOTINGS

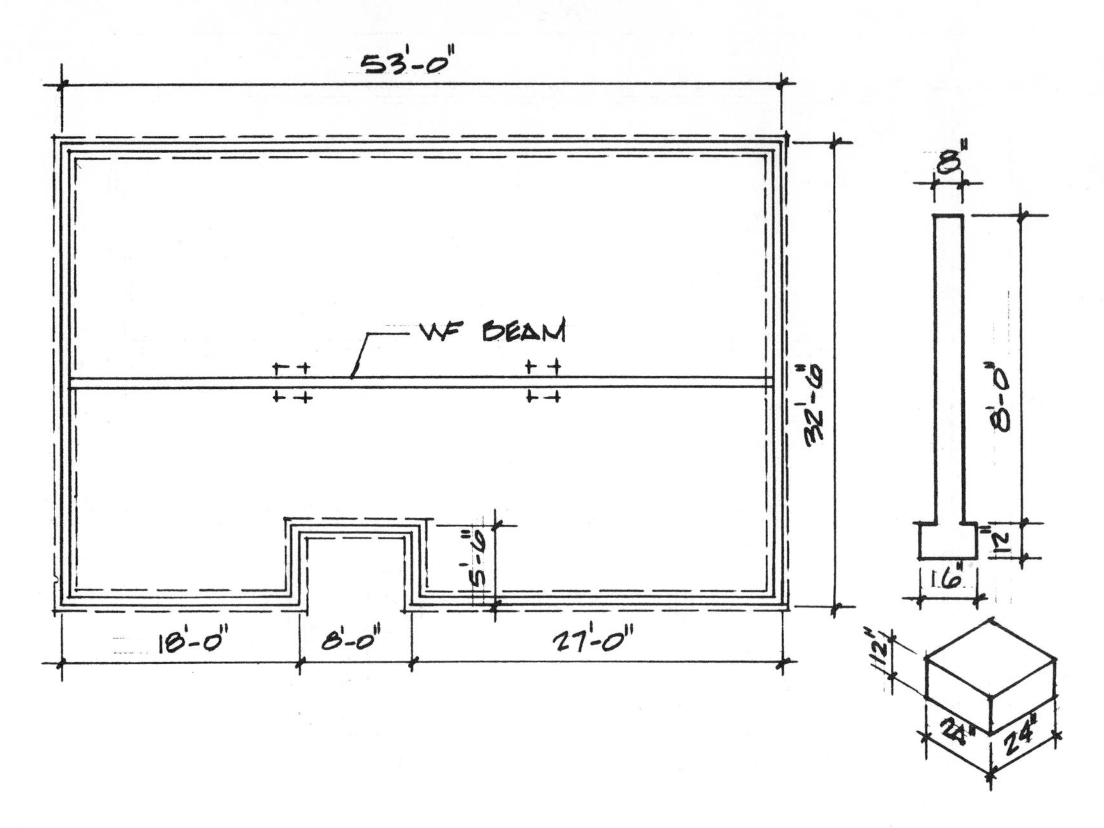

ESTIMATE THE FOLLOWING:

1. Perimeter of building ____________
2. Cubic feet in footings ____________
3. Cubic feet in foundation walls ____________
4. Cubic feet in column footings ____________
5. Total cubic feet ____________
6. Cubic yards ____________
7. Cost of a cubic yard of concrete in local area ____________
8. Total cost of concrete ____________

PROBLEM 8: PERIMETER FOOTINGS, FOUNDATION WALLS, AND COLUMN FOOTINGS

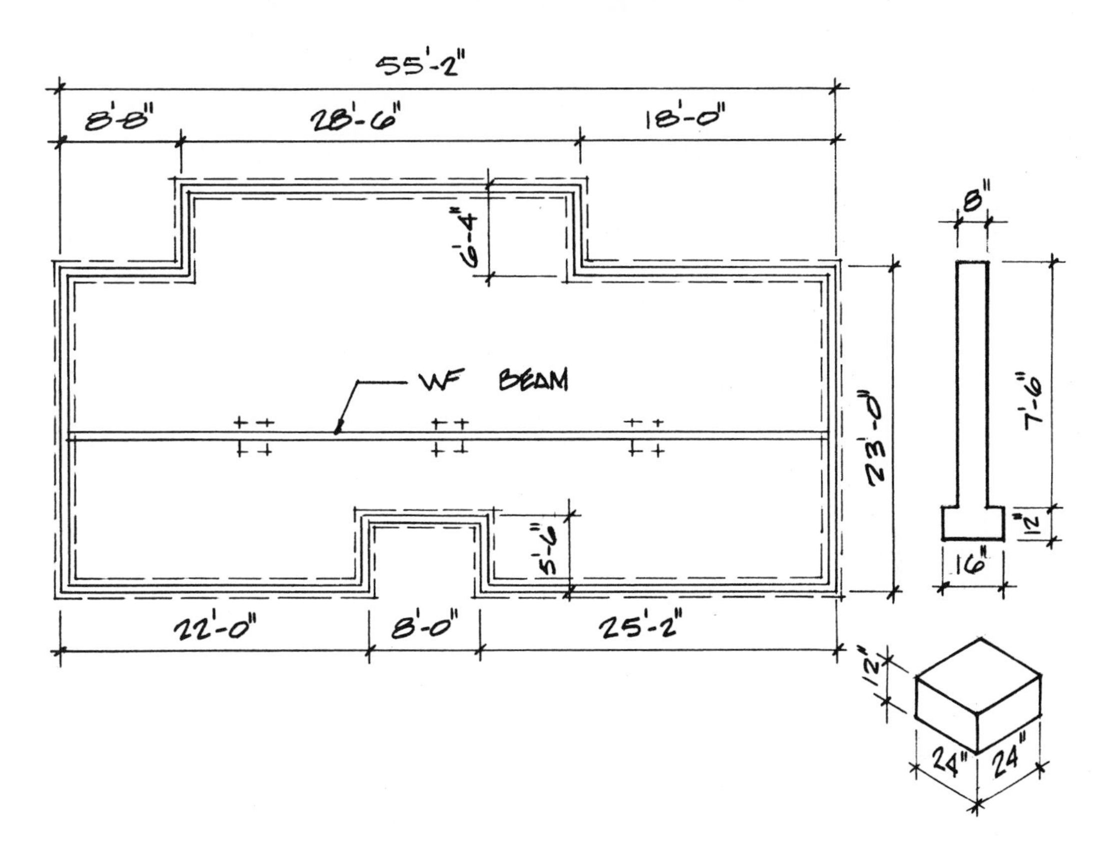

ESTIMATE THE FOLLOWING:

1. Perimeter of building __________
2. Cubic feet in footings __________
3. Cubic feet in foundation walls __________
4. Cubic feet in column footings __________
5. Total cubic feet __________
6. Cubic yards __________
7. Cost of a cubic yard of concrete in local area __________
8. Total cost of concrete __________

III

CONVENTIONAL WOOD FLOOR FRAMING

13

Estimating Mudsills

The mudsill rests on top of the foundation wall and is 2″ or thicker, the most common sills being 2 × 4 or 2 × 6 in size. In better construction the sill is bolted to the foundation wall and may have a sill sealer material between the sill and the foundation wall. Sill sealer is used on basement construction to take care of any height irregularities in the concrete or cement block and to prevent air leakage into the basement. Sill sealer may be of a felt, insulation type, or caulking material. In certain parts of the United States, building codes require sills to be of a certain type of wood, such as redwood, cedar, cypress, etc., or to be treated with a material that will also resist deterioration from moisture and wood-boring insects.

PROCEDURE FOR ESTIMATING MUDSILL FOR CONVENTIONAL WOOD FLOOR FRAMING

1. From the foundation plan estimate the lineal feet of perimeter foundation wall.
2. Add to this the lineal feet of sill needed for dwarf walls.
3. Add 10% for waste, and the result will be the number of lineal feet of sill needed.

Note: To cut costs, certain lengths of sill material may be picked out for each specific wall. These lengths should be marked on the plan so the carpenter will know their placement without loss of time. Pressure-treated material may be ordered in even lengths, usually up to 20′-0″. It is wise to order long lengths of material, placing the long lengths first and then cutting the shorter lengths from the remaining material.

Some codes or inspectors may require that a bolt be within one foot of sill plate splices. If bolts are not placed right, it may be necessary to fasten the plates down with a power-activated tool. (The rental of this equipment should be kept in mind and should be added to the estimate.)

On concrete slab floors, the mudsill becomes the bottom wall plate. Because it is in direct contact with the concrete, this sill

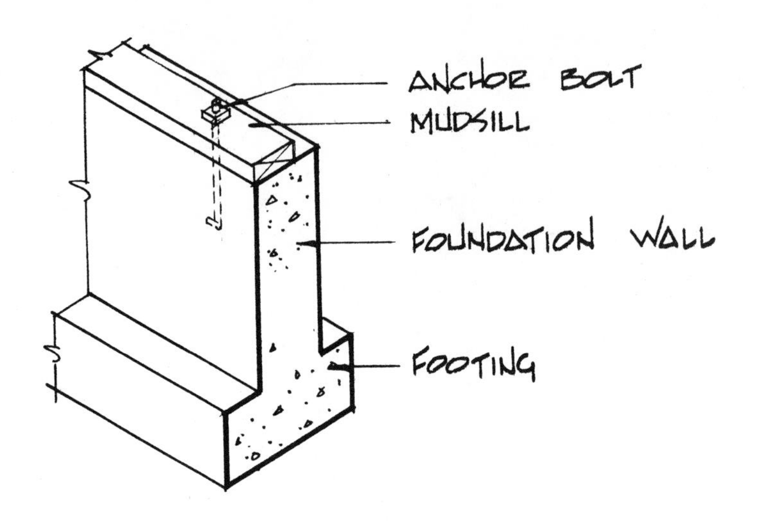
ANCHOR BOLT
MUDSILL
FOUNDATION WALL
FOOTING

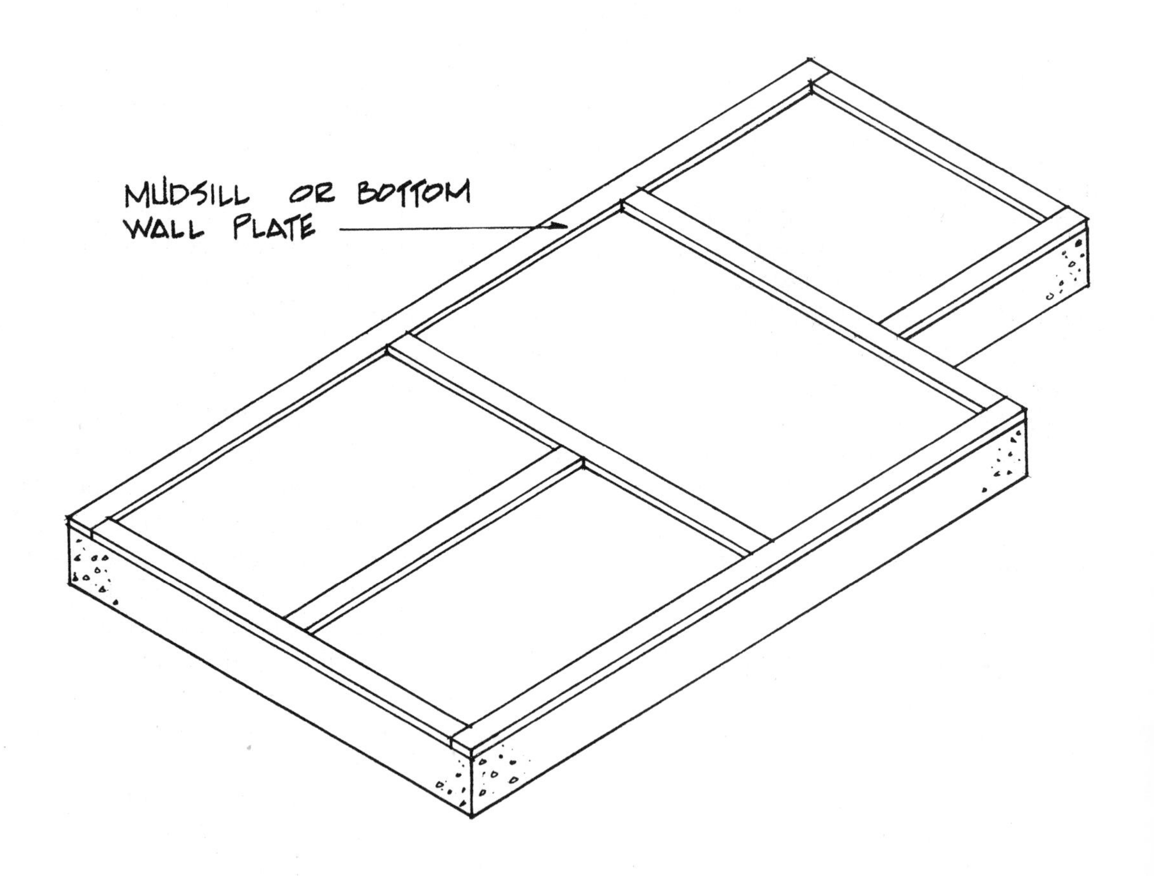
MUDSILL OR BOTTOM
WALL PLATE

plate must be made of a special type of wood or treated to resist deterioration from moisture or wood-boring insects.

PROCEDURE FOR ESTIMATING WALL SILL PLATE

1. From the floor plan estimate the lineal feet of building perimeter.
2. To this add the length of all walls running horizontally on the floor plan and then the length of all the walls running vertically.
3. Add 10% for waste. The result will be the lineal feet of sill plate needed.
4. Check the floor plan for the size and length of special sill plates—those that may be wider or narrower than the common four-inch wall, such as plumbing walls, closet partitions, sound barriers, etc.

14

Estimating Foundation Posts

Foundation posts are used on top of the pier blocks to support the girder. Concrete piers are spaced according to the size of the girder. Usual code requirements call for 5-foot maximum spacing between pier posts for 4″ × 4″ girders and 7-foot maximum spacing between the pier posts for 4″ × 6″ girders.

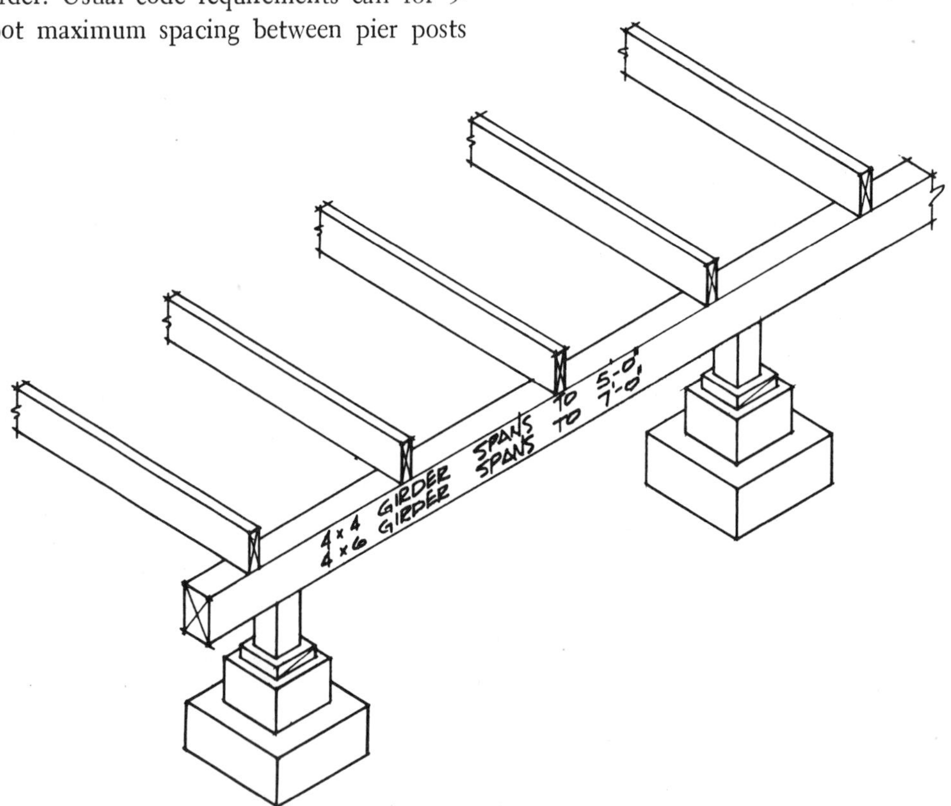

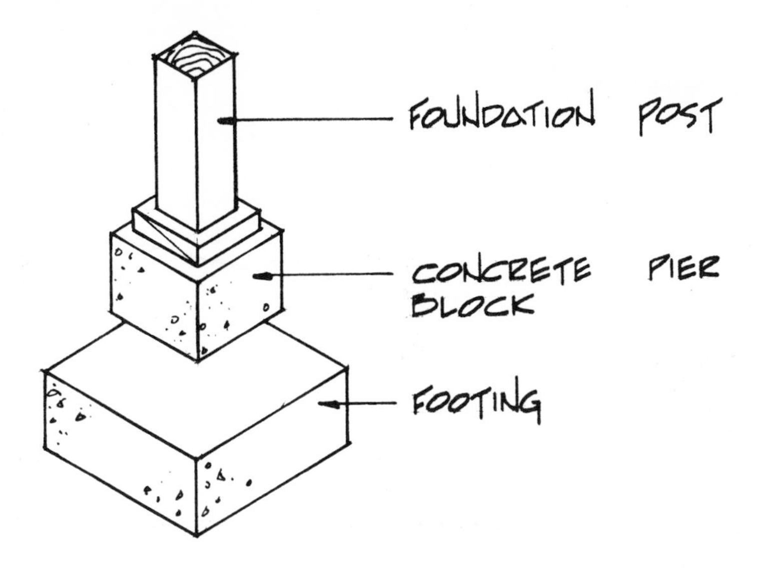

The following diagram illustrates the usual minimum code requirements for conventional floor framing.

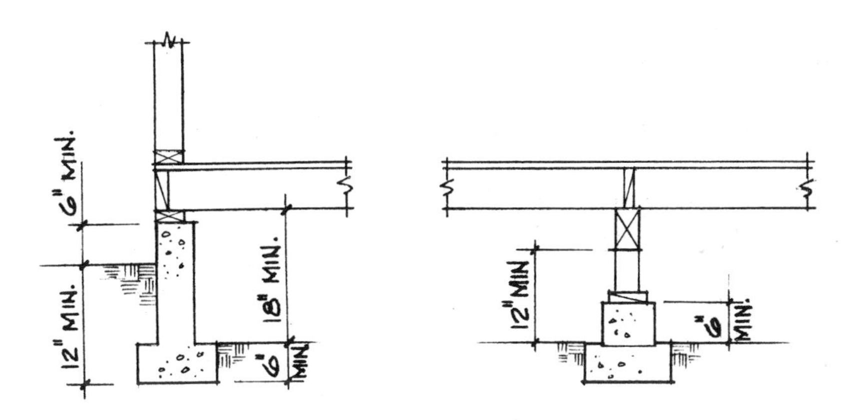

The length of pier posts may be estimated arithmetically or scaled from the section view of the plans. Most pier posts will be the same length, but in special cases, where the structure is built on a slope, the posts may be of different lengths as illustrated below. In this case a rough or average estimate should be made of the posts' length and the material needed. Long lengths of material should be ordered and carefully cut for the least amount of waste.

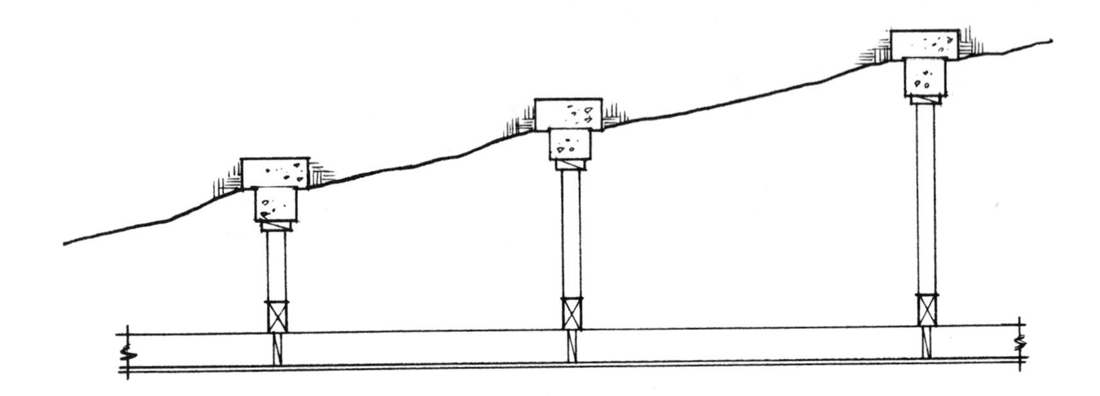

PROCEDURE FOR ESTIMATING PIER POSTS

1. Check the building plan details for the length or average length of the pier posts.
2. Multiply the length times the number of posts needed.
3. Combine the pieces in lengths that will give the least amount of waste.

15

Estimating Concrete Pier Blocks

Concrete pier blocks are precast and usually include a redwood or chemically treated wood block fastened to the top. They can be special ordered with various metal hold-down anchors as illustrated below.

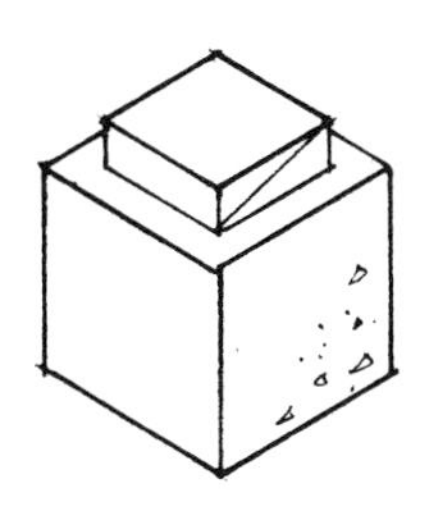

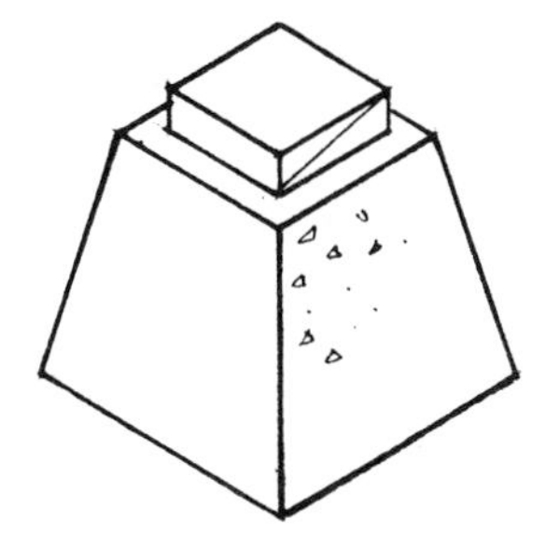

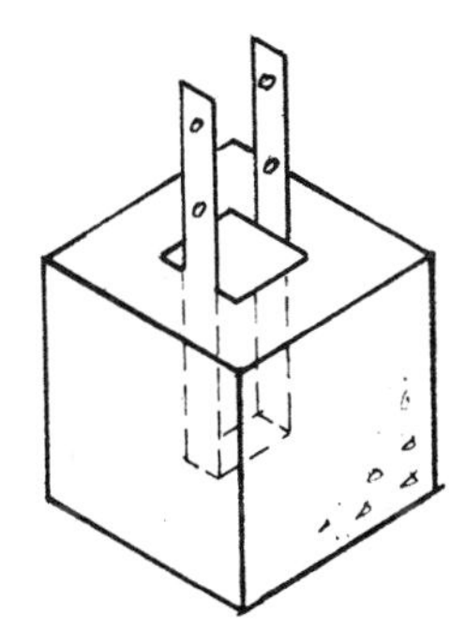

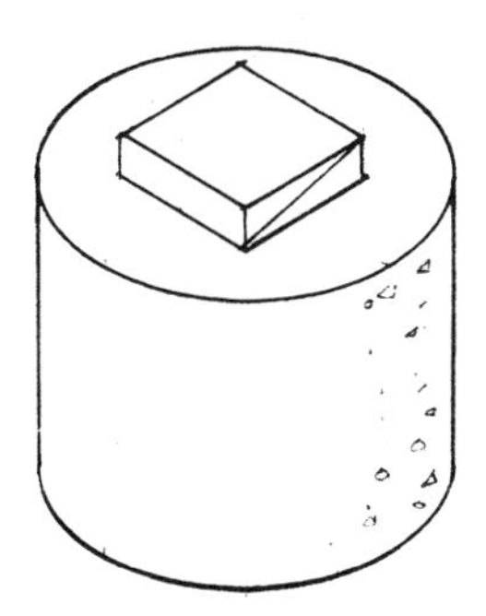

These blocks are set in the concrete in the footing hole or form while it is still plastic. They are leveled and aligned to receive pier posts and girders. To save money, some contractors cast their own concrete pier blocks, using 6″ wide flashing paper.

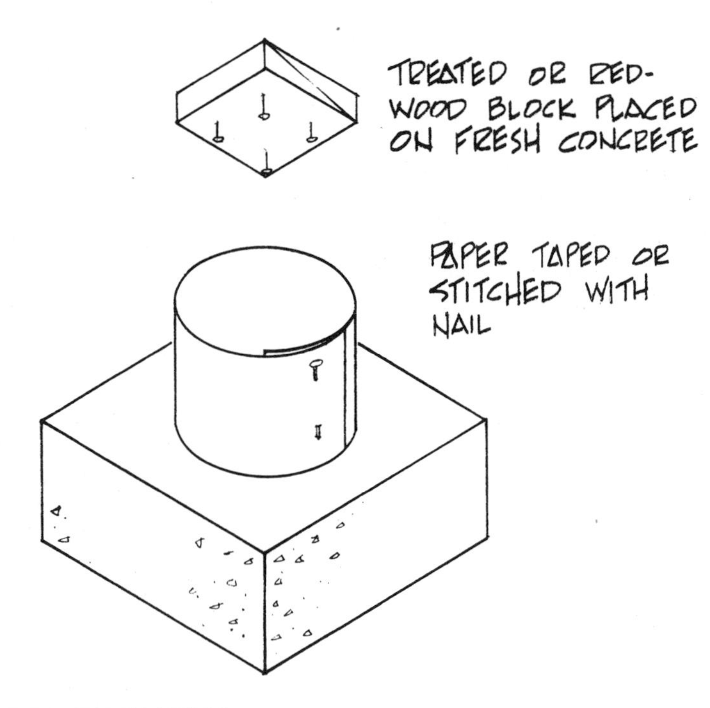

PROCEDURE FOR ESTIMATING CONCRETE PIER BLOCKS

1. Check the foundation plan for the number of pier footings.
2. One concrete pier block is needed for every pier footing.

16

Estimating Girders

Girders are members that are placed horizontally on the pier posts and support the first floor joists. Girders must be spliced over the center of the pier posts as illustrated below.

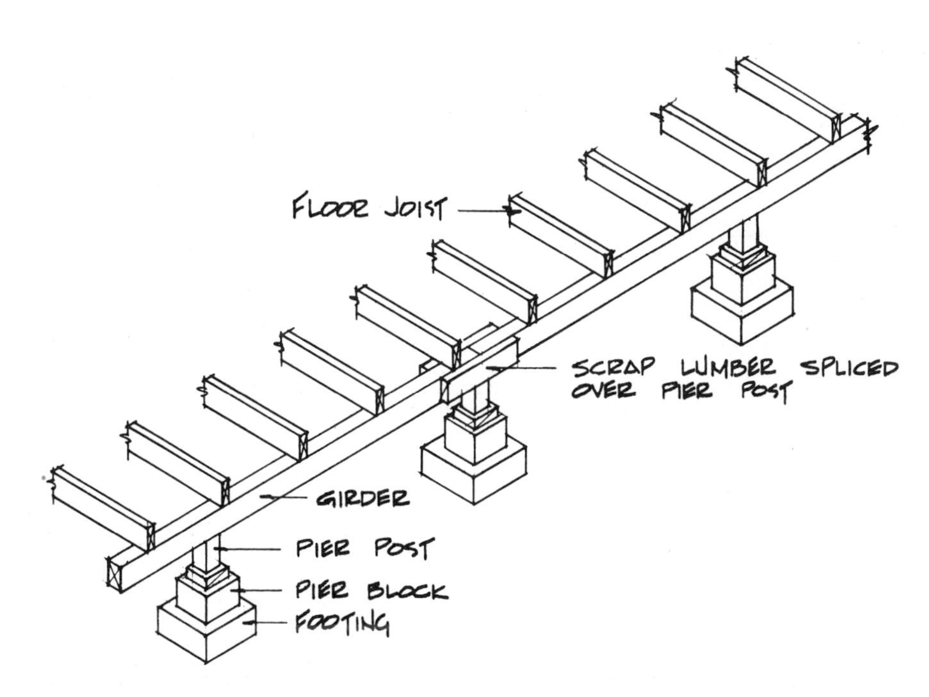

PROCEDURE FOR ESTIMATING GIRDERS

1. The size of the girder may be found on either the foundation plan or section view.
2. The length of each girder is determined by adding up the pier or column spacing sizes. (Some estimators prefer to scale from the foundation plan the various lengths of girder material.)

Example Problem:

Estimate the girder material for the following foundation plan.

PROCEDURE:

1. Check foundation plan for size of girder: 4 × 6
2. Add the various pier spacing dimensions and determine a practical length of girder material: Pier spacing sizes 6′-6″ + 6′-0″ + 6′-0″ = 18′-6″, or a 20-foot length of material is needed.
3. Determine other lengths of girder material: 2 pieces 4 × 6 × 20′
   Using two 20′-0″ lengths of material involves one splice and 3 lineal ft. of waste material. The waste can be reduced if two 12′-0″ lengths and a 14′-0″ length are used. By using these lengths and three splices, there is only one foot of waste material. The estimator may find that the cost in changing the lengths of material may be offset by the labor and materials used to make three splices. In other words, the estimator will have to decide which is the most economical method.

FOUNDATION PLAN

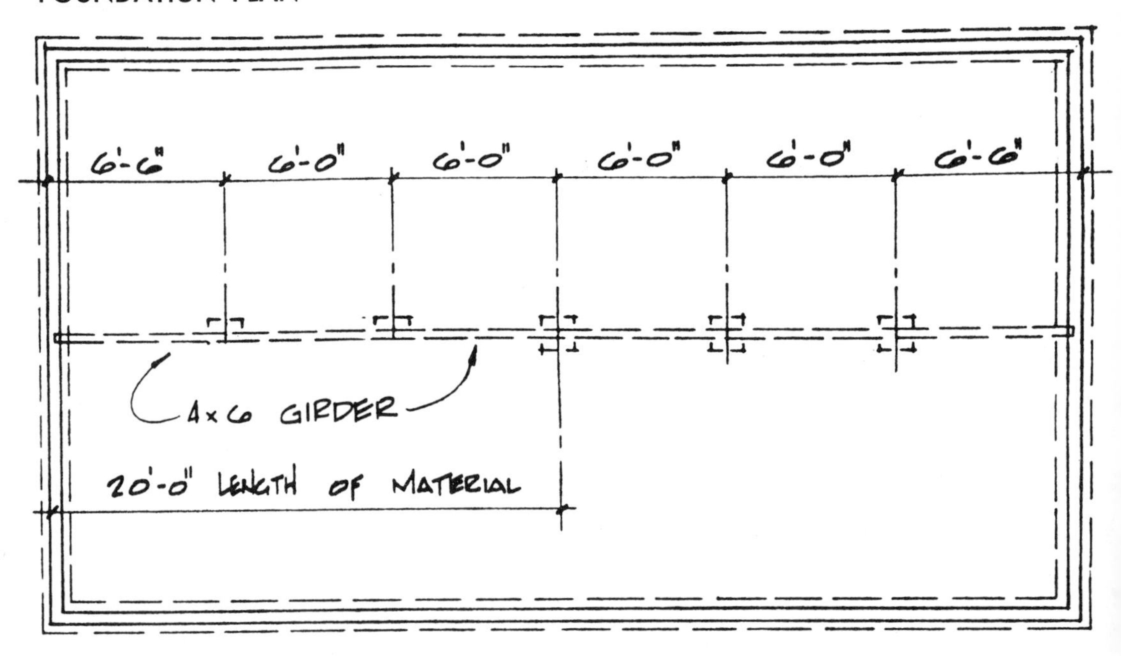

17

Estimating Floor Joists

Floor joists are horizontal members in the floor system that span from mudsill to mudsill or are lapped over a central girder or beam as illustrated below.

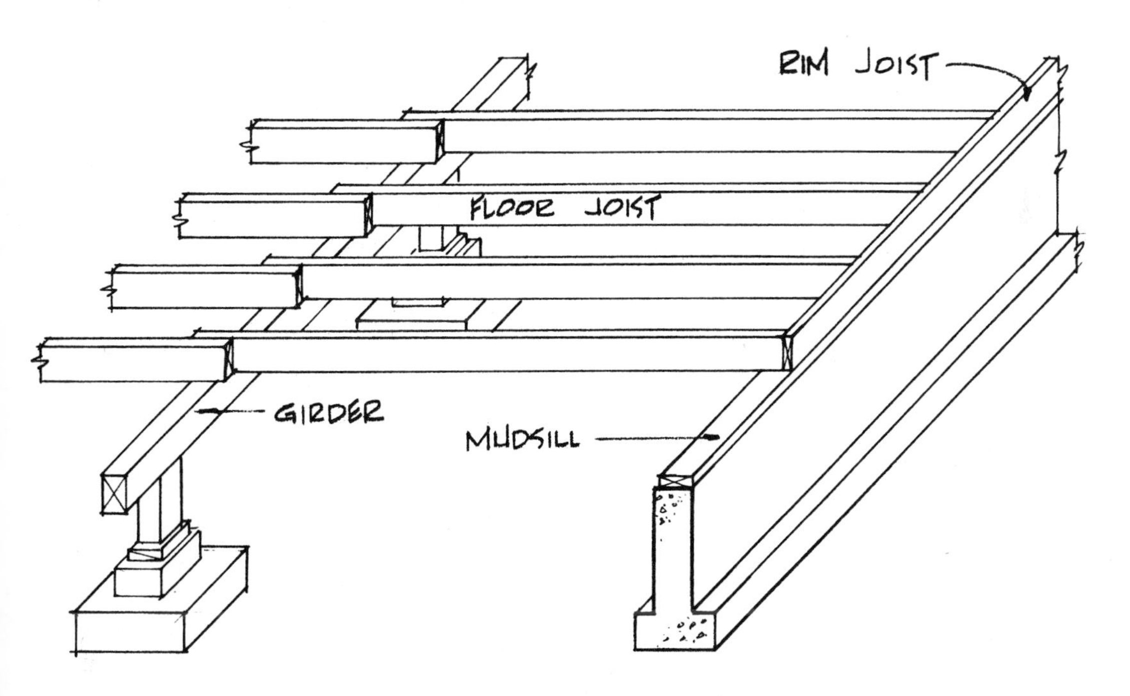

The size and spacing of floor joists is usually indicated on the foundation plan or the floor plan. They are usually designated as follows.

2 × 8 16″ O.C.
←――――――――→

The size of joist here is 2 × 8, the spacing of the joist is 16″ on center, and the arrow gives the direction of the joist run. If the direction of the joist is not given, it may be determined from the directional run of the beams or girders.

Table 17-1 helps the estimator determine the number of floor joists or various other members in a structure that are at different center-to-center spacings. The constant from the table is multiplied by the lineal feet over which the particular center-to-center spacings span. One is added because there is always one more member than there are spaces.

TABLE 17-1 CENTER-TO-CENTER FLOOR JOIST SPACING

Spacing	*Constant*	*Decimal Equivalent*
12″ O.C.	1.0	1.00
* 16″ O.C.	3/4	.75
18″ O.C.	2/3	.6667
20″ O.C.	3/5	.60
24″ O.C.	1/2	.50
32″ O.C.	3/8	.375
36″ O.C.	1/3	.3333

* Usual floor joist spacing

Explanation of Table:

When multiplying by the constant (3/4 for 16″ center-to-center spacing), one is merely changing feet to inches and dividing by the spacing (12/16 = 3/4).

Note: The center-to-center spacing constants from this table will be referred to and used in various units throughout this book.

PROCEDURE FOR ESTIMATING FLOOR JOISTS

1. Determine from the foundation plan the length of the wall (in feet) that the floor joist rests on.
2. Multiply this length by the spacing constant from the center-to-center spacing table.
3. Add one.
4. The resulting figure indicates the number of floor joists required.

Example Problem 1:

Estimate the number of floor joists in the following problem.

PROCEDURE:

1. Determine from the foundation plan below the length of wall (in lin. ft.) on which the floor joist rests: 36 lineal feet.
2. The number of lineal feet (36) is multiplied by the center-to-center spacing constant 3/4 (for members spaced 16″ on center): 3/4 × 36 = 108/4 = 27.
3. Add one more floor joist because there is always one more floor joist than there are spacings: 27 + 1 = 28 pieces 2 × 10 × 18′.

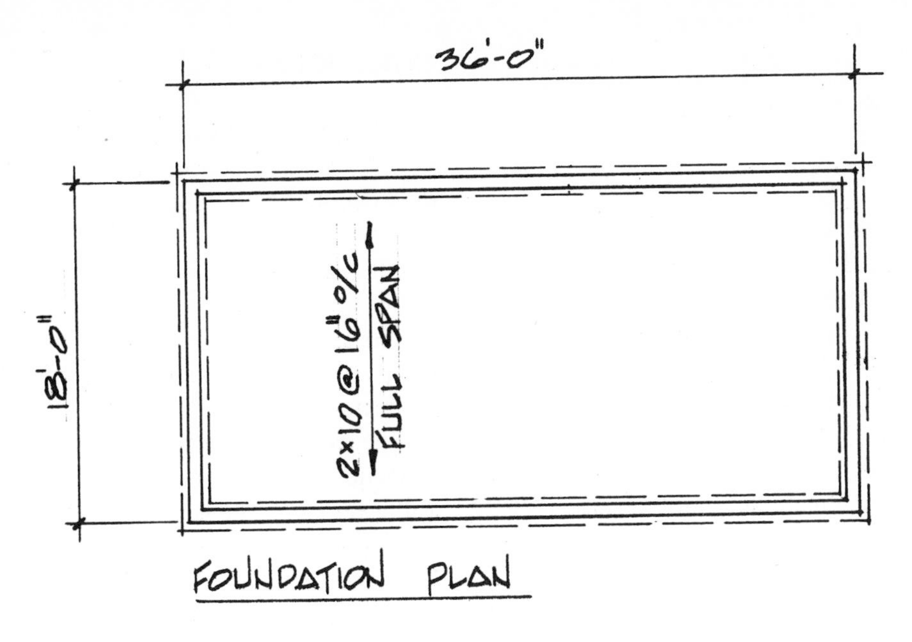

Example Problem 2:

Estimate the number of floor joists in the following problem.

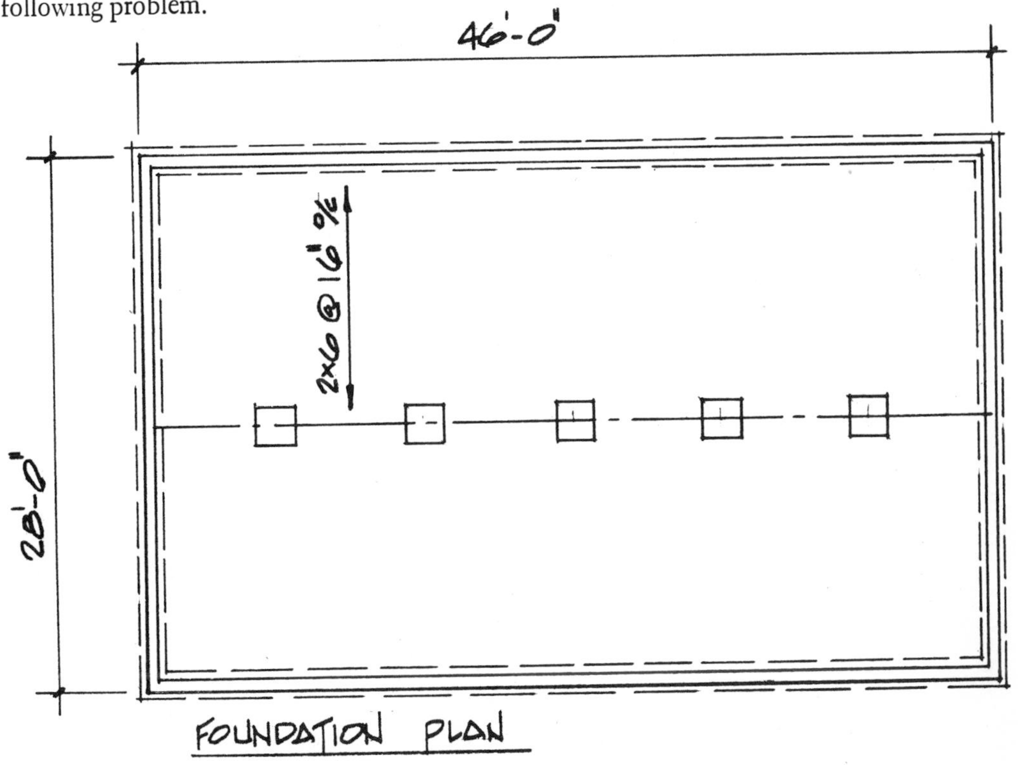

PROCEDURE:

1. From the foundation plan above, determine the length of wall (in lin. ft.) on which the floor joist rests: 46 ft.
2. The number of lineal feet (46) is multiplied by the center-to-center spacing constant (3/4): $3/4 \times 46 = 138/4 = 34\text{-}1/2$ or 35 floor joists.
3. Add one more floor joist because there is always one more floor joist than there are spacings: $35 + 1 = 36$ floor joists.
4. Double the number of floor joists to cover both sides of the girder: $36 \times 2 = 72$ pieces $2 \times 6 \times 14'$.

EXTRA JOISTS

Extra joists have to be added to the materials list for any bearing wall partition that runs parallel to the floor joist. The code in some areas may also require doubling the number of joists for nonbearing wall partitions.

The architect may provide the number of extra double joists on the foundation plan as illustrated below.

The architect may provide the number of extra double joists on the foundation plan as illustrated on the facing page. If the foundation plan does not provide for the extra joists, the estimator will have to calculate these joists from the floor plan as illustrated.

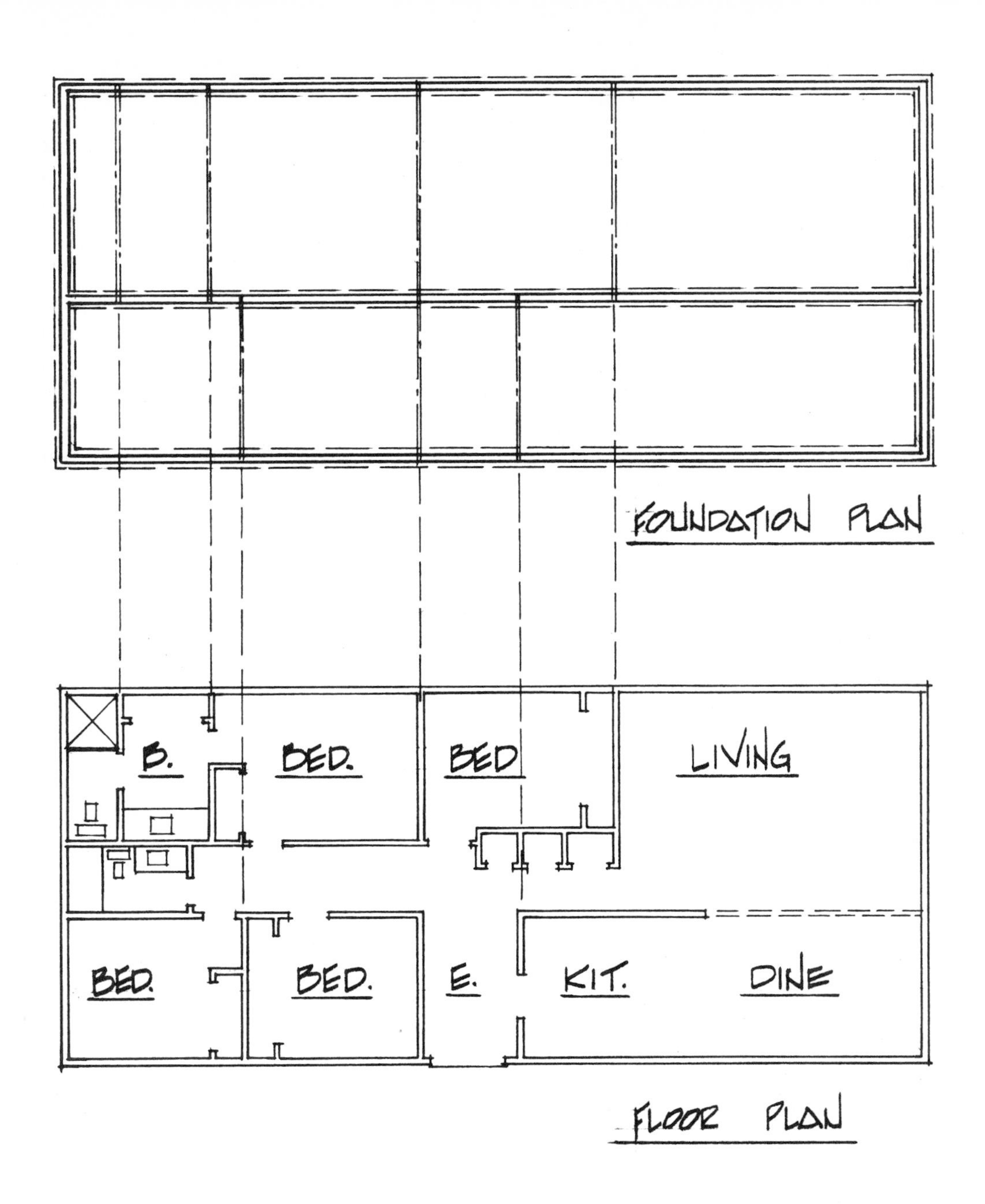
FOUNDATION PLAN
B.
BED.
BED
LIVING
BED.
BED.
E.
KIT.
DINE
FLOOR PLAN

18

Estimating Header Joist

The header joist is also known as the *skirt* or *rim-joist*. The floor joist is supported in an upright position by nailing through the header joist.

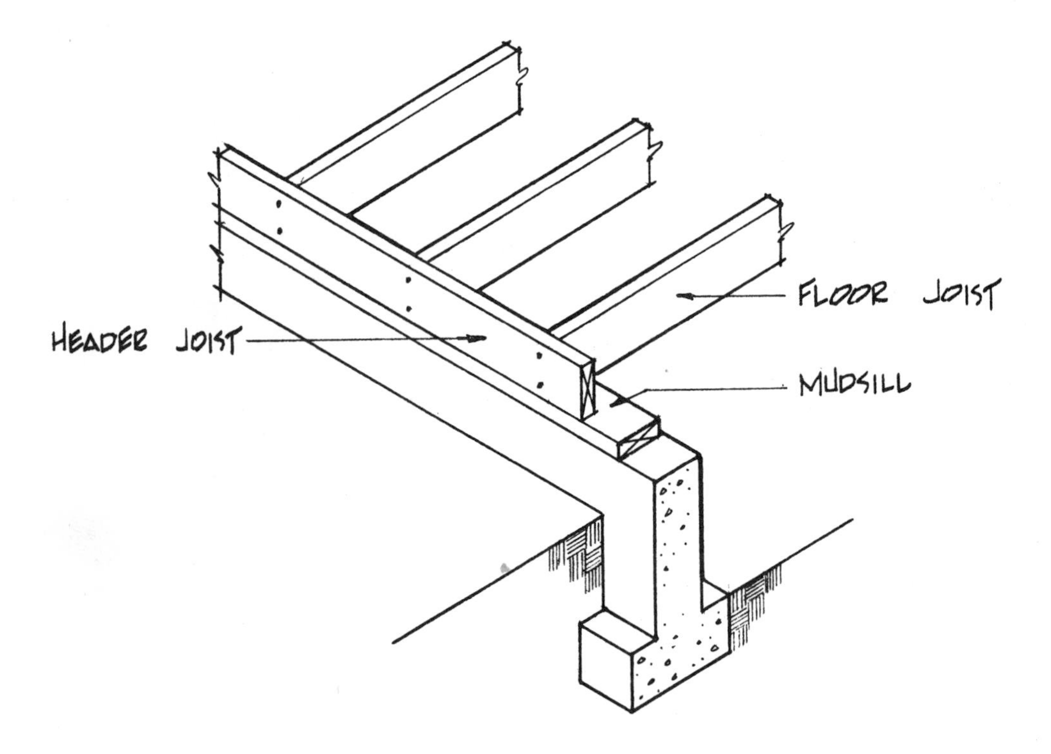

Continuous blocking is another method used to hold the floor joist in position.

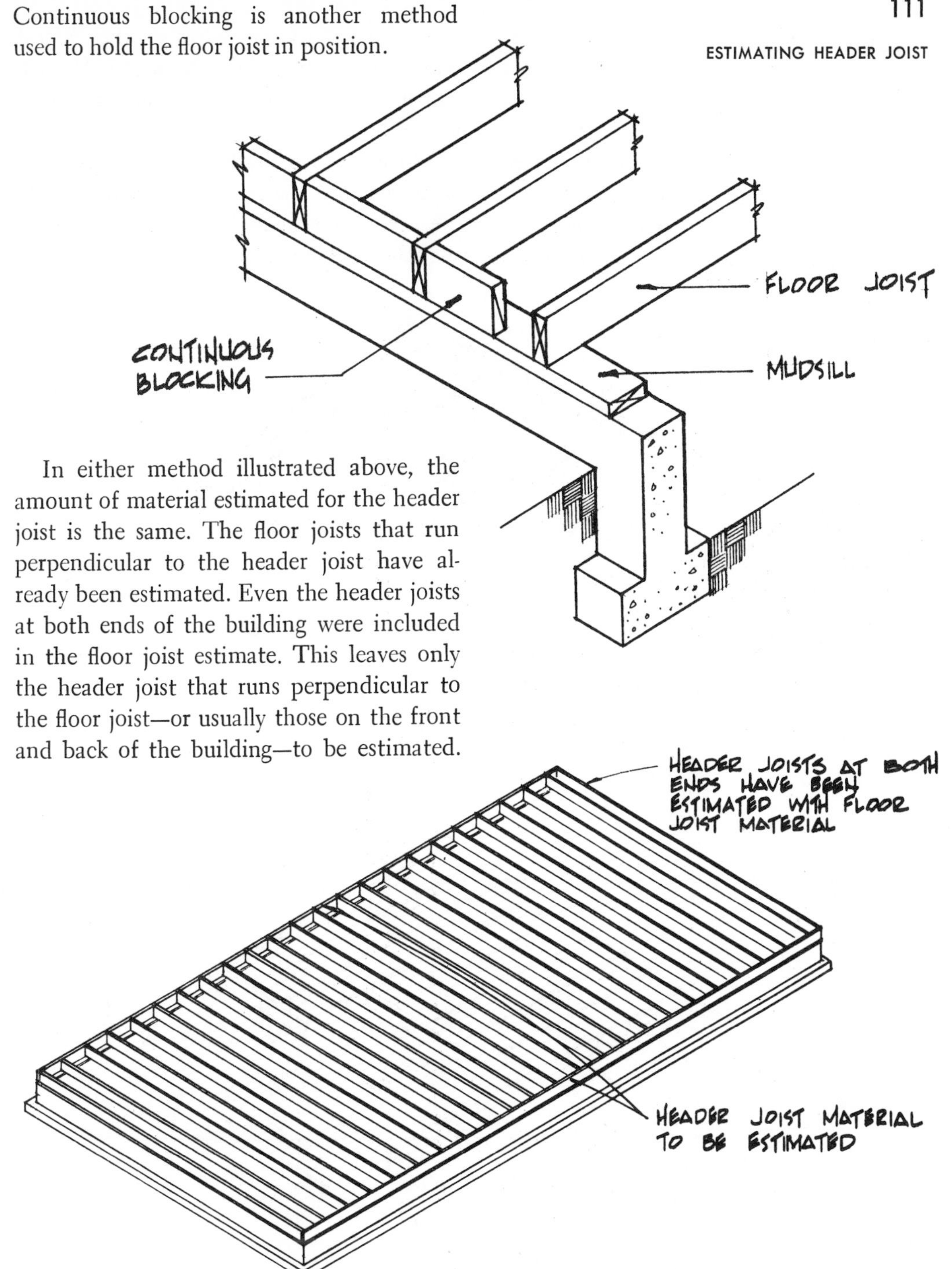

In either method illustrated above, the amount of material estimated for the header joist is the same. The floor joists that run perpendicular to the header joist have already been estimated. Even the header joists at both ends of the building were included in the floor joist estimate. This leaves only the header joist that runs perpendicular to the floor joist—or usually those on the front and back of the building—to be estimated.

ESTIMATING HEADER JOIST

The lineal footage of header joist material needed is taken from the foundation or floor plan. Long lengths of material are usually ordered, but the header joist could also be ordered in the same length as the longest floor joist, allowing the carpenter to pick and use the straightest material for header joist.

The next two illustrations are examples of header joist material for estimating. List the header joist material needed in each of the following buildings.

EXAMPLE 1:

EXAMPLE 2:

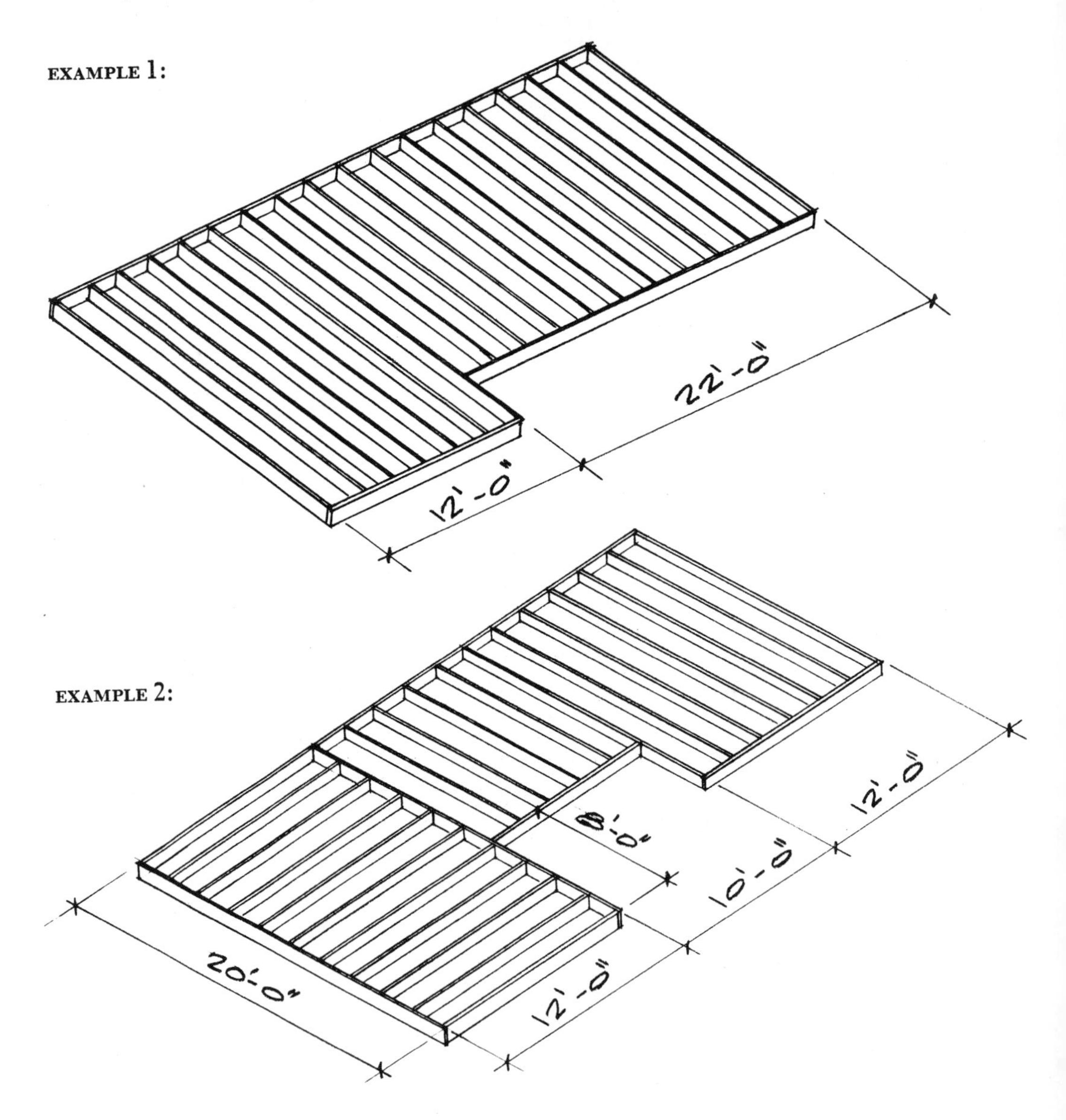

19

Estimating Floor Bridging and Blocking

Bridging and blocking are members inserted between floor joists to stiffen, align, and strengthen the floor system and help distribute the load weight.

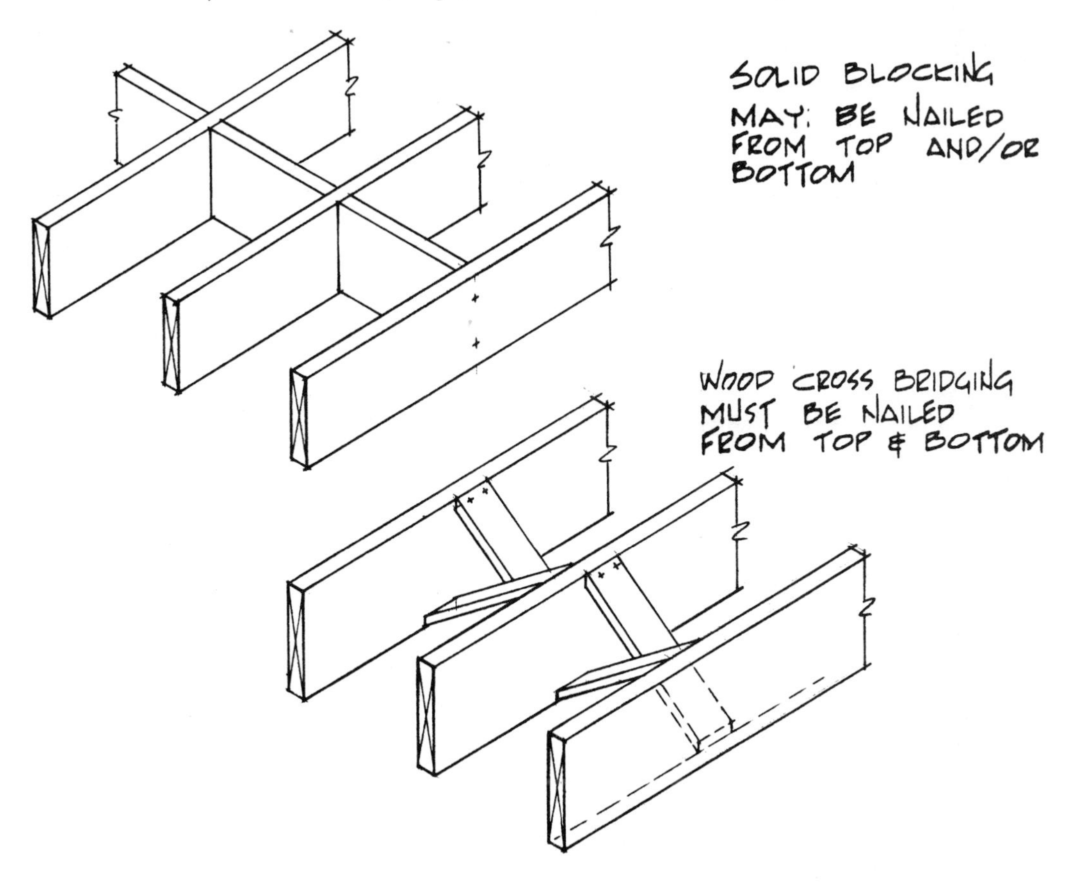

Bridging or solid blocking is required for any floor joist span over 8′-0″. If the span exceeds 16′-0″, two rows are needed. In other words, the distance between wall bearing and bridging or between bridging and bridging should never exceed 8′-0″. The example below illustrates the number of rows of bridging necessary for various spans (joist size 2 × 6 @ 16″ O.C.).

PROCEDURE FOR ESTIMATING WOOD DIAGONAL BRIDGING

1. Calculate the lineal feet of each row of bridging and add them.
2. From Table 19-1, determine the lineal feet of bridging material needed per lineal foot of bridging row.
3. Multiply the constant by the total lineal feet of bridging row. The result will be the amount of bridging material needed.

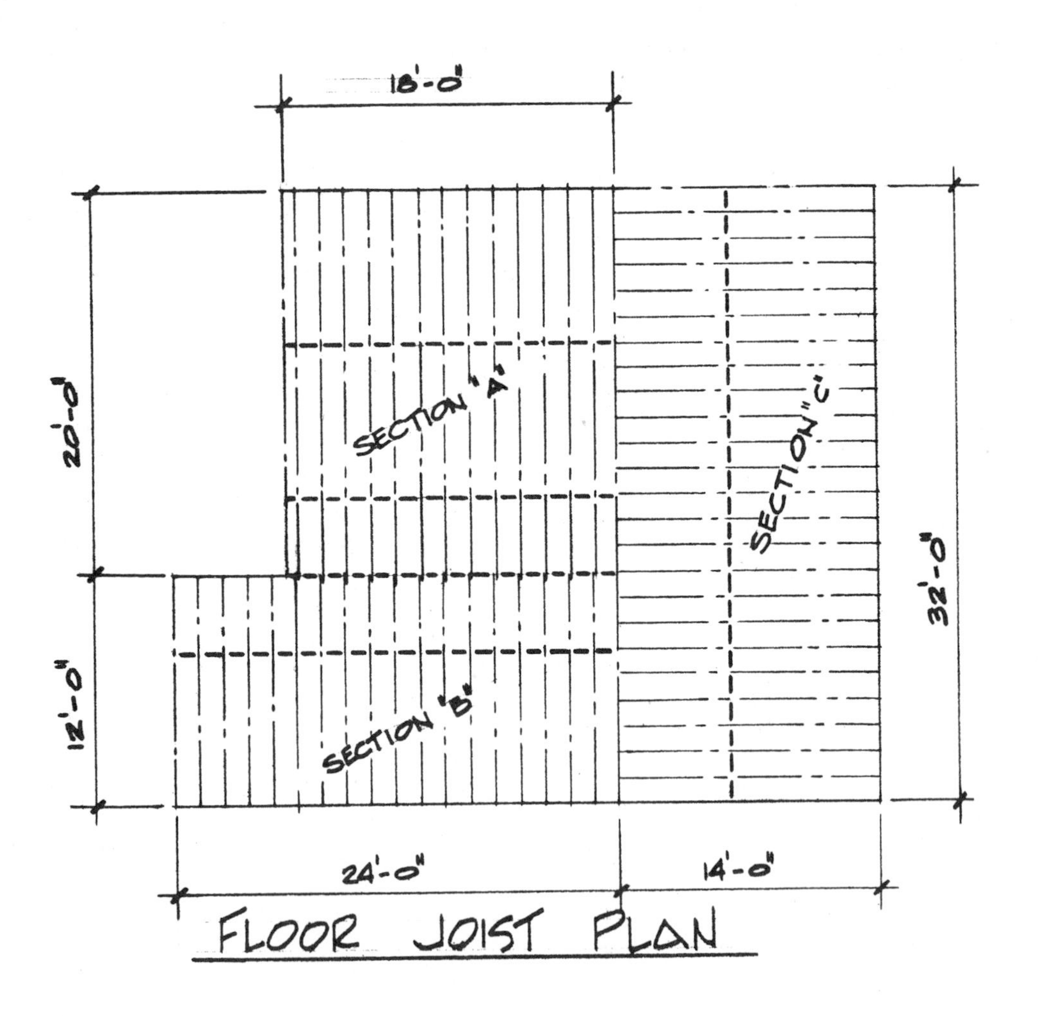

TABLE 19-1 Estimating Wood-Bridging Material in Lineal Feet

Joist Size	*Joist Spacing*	*Constant*
2 × 6	16″ O.C.	2.0
2 × 8	16″ O.C.	2.0
2 × 10	16″ O.C.	2.0
2 × 12	16″ O.C.	2.25
2 × 14	16″ O.C.	2.50

EXAMPLE: Estimate the amount of bridging needed for the previous joist layout plan.

1. Determine the lineal feet of each row of bridging and add them.

   Section A 2 rows 18′ each = 36′
   Section B 1 row 24′ = 24′
   Section C 1 row 32′ = 32′
   Total 92 lin. ft.

2. The constant from the bridging table for a 2 × 6 joist is 2.
3. Multiply the constant (2) by the total lineal feet of the bridging rows (92): 2 × 92 = 184 lin. ft. of bridging material.

METAL BRIDGING

Metal bridging may be ordered for 2 × 8, 2 × 10, 2 × 12, 2 × 14 and 2 × 16 material, spaced at 12″, 16″, or 24″ on center. Two pieces of bridging are required for each space between joists.

Metal bridging may be installed either from the top or bottom by locating the bend line approximately 1″ from the joist corner, before or after the sheathing is installed.

PROCEDURE FOR ESTIMATING METAL BRIDGING

1. Multiply the lineal feet of wall on which the joists rest by the center-to-center spacing constant. (For joists placed 16″ O.C., the spacing constant

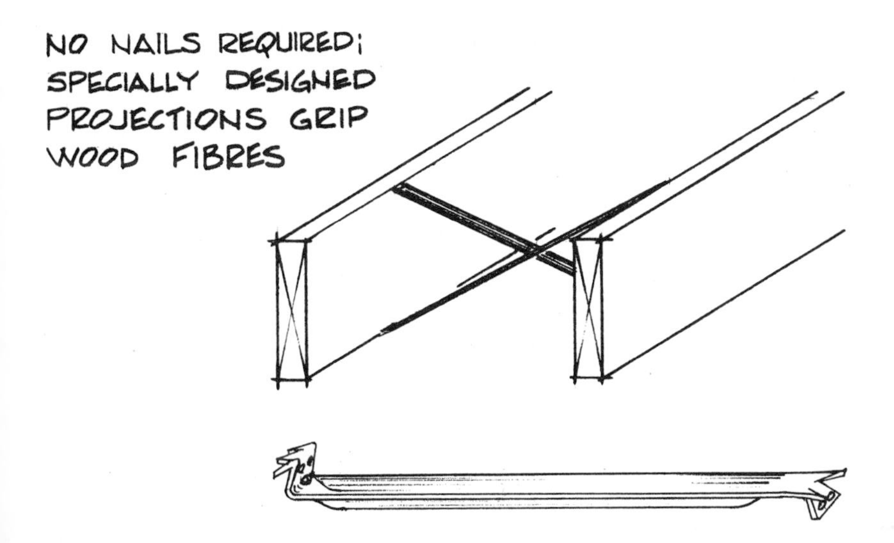

is 3/4 or .75.) This will give the number of spaces that need bridging.
2. Multiply the number of spaces by 2 for the number of pieces of metal bridging.

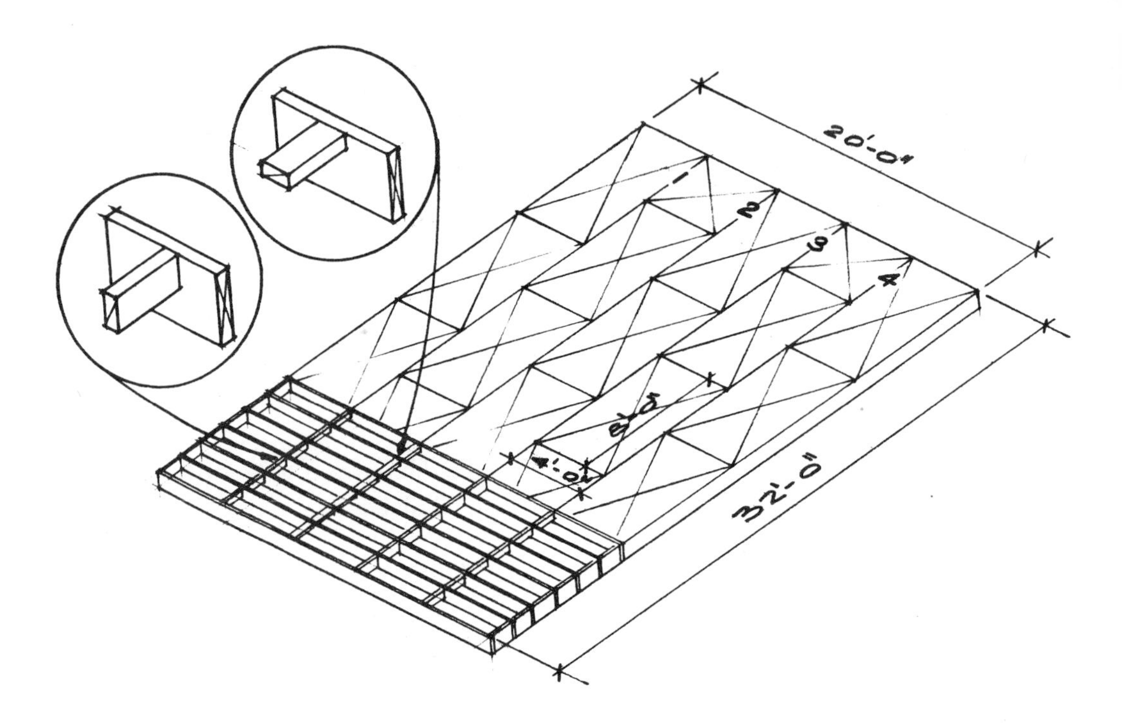

PROCEDURE:

1. Lay out the number of rows of edge blocking needed: 4 rows.
2. Multiply the number of rows (4) by the length of each row (32'-0"): $4 \times 32 =$ 128 lineal feet of 2×4 edge blocking.
3. Order the material in stud length. This length is easy to handle and will allow the carpenter to cut up the crooked pieces for edge blocking.

20

Estimating Floor Sheathing

Subflooring material may consist of 1 × 6 or 1 × 8 smooth or rough boards installed diagonally or at right angles to the joist. When used the boards will have to be covered with hardwood flooring, partical board, or plywood. Plywood will provide a smooth surface for the installation of various types of tile, vinyl sheet material, or carpet.

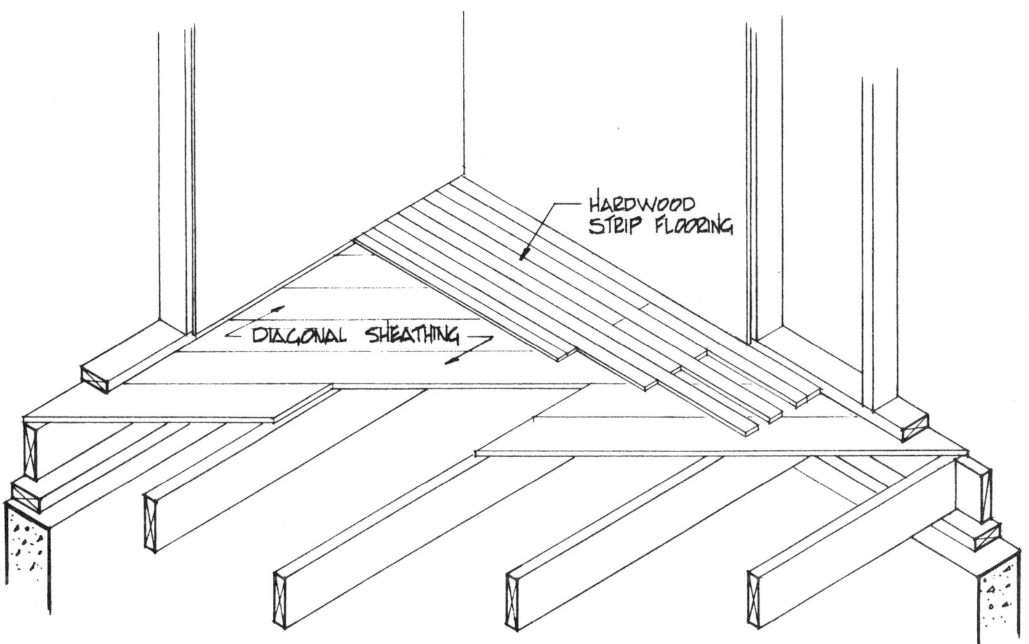

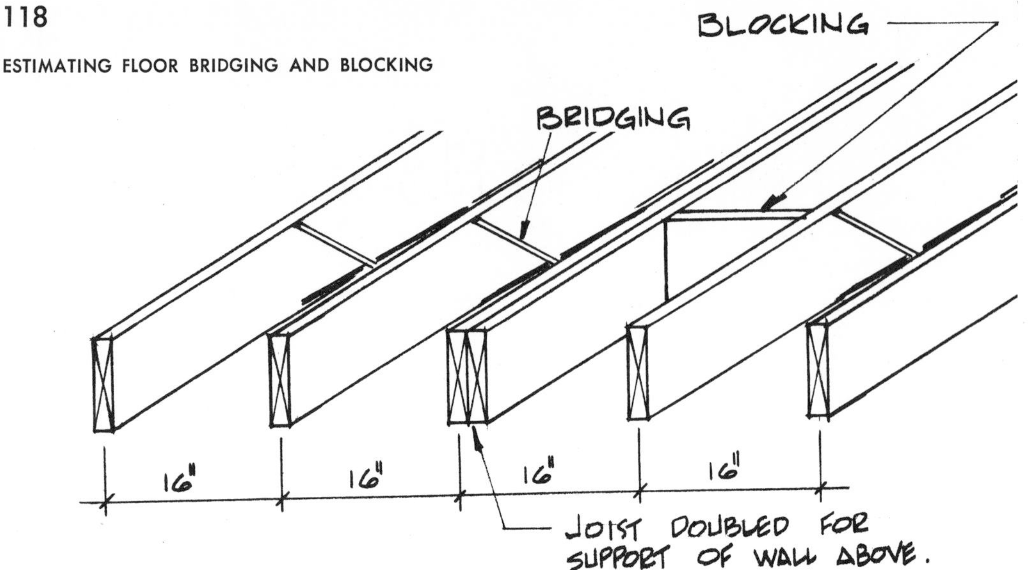

Keep in mind that floor joists that are doubled or not spaced 16″ O.C. will need specially cut cross bridging or solid blocking as illustrated above.

SOLID WOOD BLOCKING

Building with a girder system requires solid wood blocking over each girder and any floor joist span over 8′-0″. The blocking should be the same size as the floor joists.

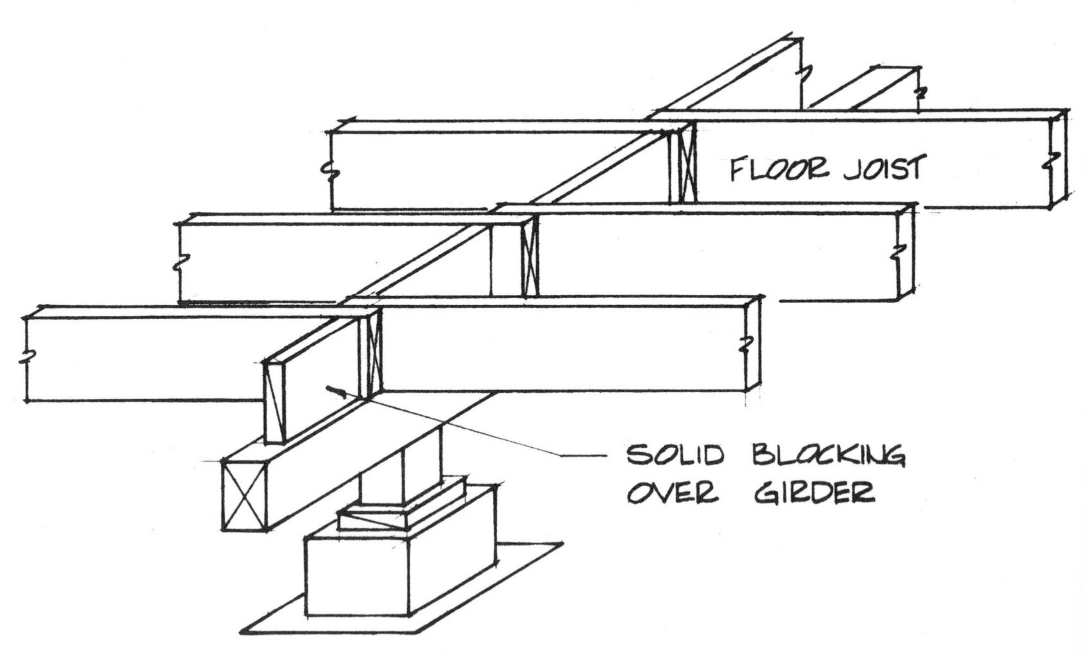

PROCEDURE FOR ESTIMATING SOLID BLOCKING

1. Calculate the lineal feet of each row of blocking and add them. This is the lineal feet of material needed.
2. Order the material in the same lengths as the floor joists. This will allow the cutting of crooked pieces for blocking.

Example Problem:

Estimate the lineal feet of solid blocking needed in the following problem.

PROCEDURE:
Calculate the lineal feet of each row of bridging and add them:

4 rows at 42′ each = 168′
1 row at 18′ each = 18′
Total 186 lin. ft. of 2 × 8 material

PLYWOOD EDGE BLOCKING

When joists are spaced 16″ O.C. and plywood 1/2″ thick or less is used for subflooring material, the code may require 2 × 4 blocks to support the plywood edge as it spans from joist to joist. This blocking may be nailed flat or on edge as illustrated on page 118.

Note: Before spacing and nailing this blocking, the carpenter must know whether the width of the plywood is a full 48 inches or 47-1/2 inches.

PROCEDURE FOR ESTIMATING PLYWOOD EDGE BLOCKING

Example Problem:

Estimate the lineal feet of edge blocking in the illustration on page 118.

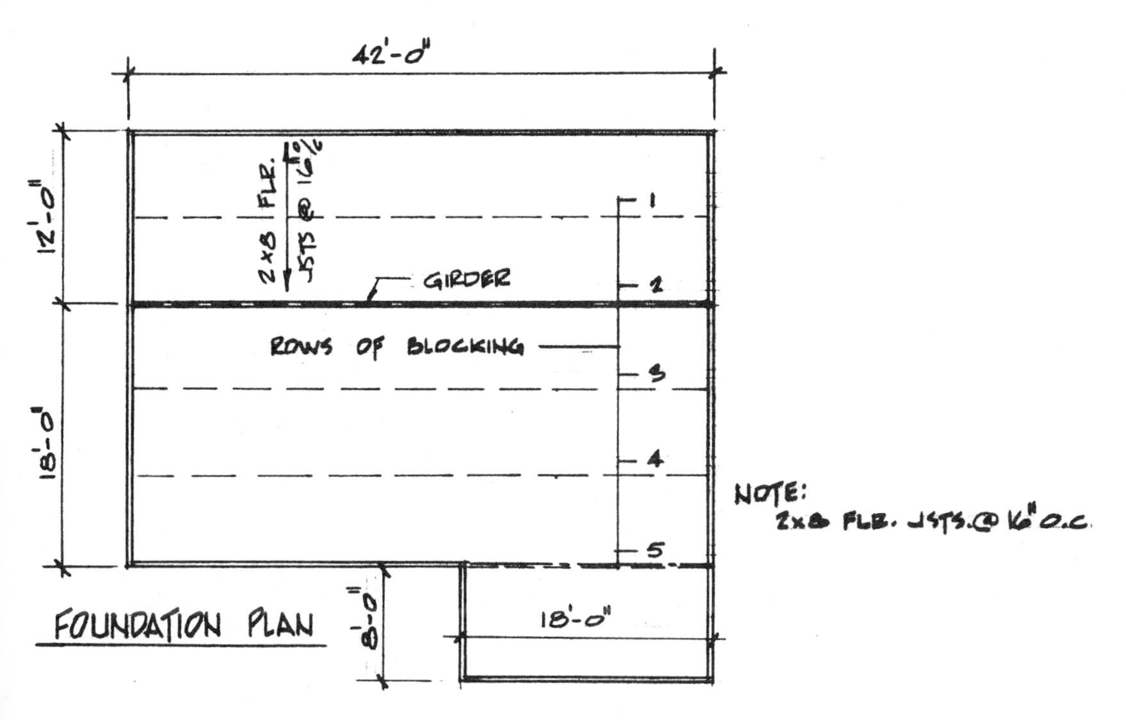

PROCEDURE FOR ESTIMATING BOARD OR STRIP SUBFLOORING

1. Calculate the square foot area the floor sheathing is to cover.
2. Multiply the square foot area to be covered by the waste factor in Table 20-1 for the size and type of boards used.
3. To convert from board feet to lineal feet, do as follows:
   If 1 × 6s are used, multiply the sq. ft. estimated by 2. (There are 2 lin. ft. of 1 × 6 in every bd. ft.)
   If 1 × 8s are used, multiply the sq. ft. estimated by 1.5. (There are 1.5 lin. ft. of 1 × 8 in every bd. ft.)

TABLE 20-1 WASTE ALLOWANCES FOR BOARD SUBFLOORING

SQUARE-EDGE BOARDS			
Nominal Size	*Actual Size*	*Waste percentage when layed at right angles to joist*	*Waste percentage when layed diagonally to joist*
1 × 6	3/4 × 5-1/2	1.22	1.33
1 × 8	3/4 × 7-1/2	1.22	1.33

TONGUE-&-GROOVE AND SHIPLAP BOARDS			
Nominal Size	*Coverage Size for Estimating*	*Waste percentage when layed at right angles to joist*	*Waste percentage when layed diagonally to joist*
1 × 6	3/4 × 5	1.35	1.46
1 × 8	3/4 × 7	1.32	1.43

Note: The factors above take into account the size differential of boards and waste.
Every estimator should adjust the waste factor according to his own needs.

PLYWOOD SUBFLOORING MATERIAL

Plywood used for subflooring may be square edged or tongue and grooved. A common grade for subfloor plywood would be: 5/8″ × 4′ × 8′ CDX T&G 2LE.

This grade is a C grade veneer face with D grade inter cores and back, exterior glue, tongue and grooved two long edges. It is also commonly known as 5/8″ underlayment.

The least amount of plywood waste is acquired by careful layout and alignment of the floor joist. The plywood must be aligned to fall on the center of each joist as illustrated below:

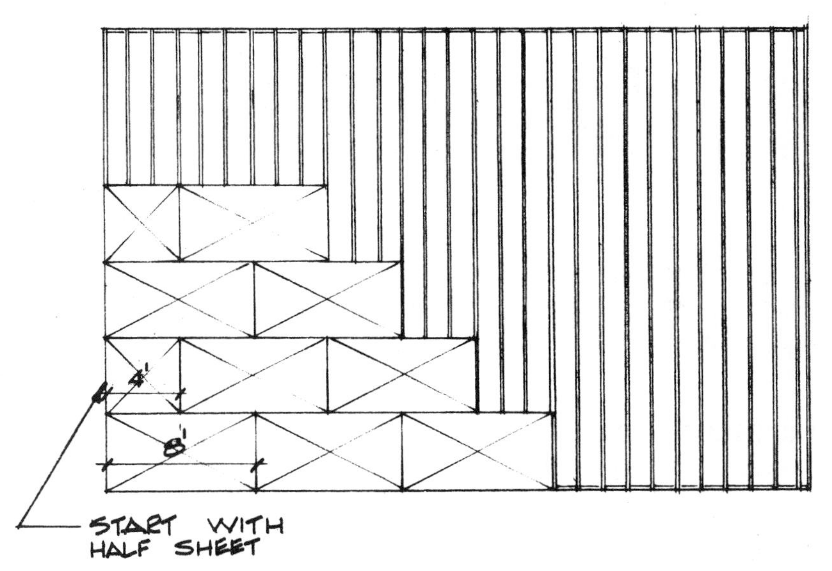

The amount of waste will also depend on the size and shape of the building as shown in the following illustrations.

EXAMPLE 1:

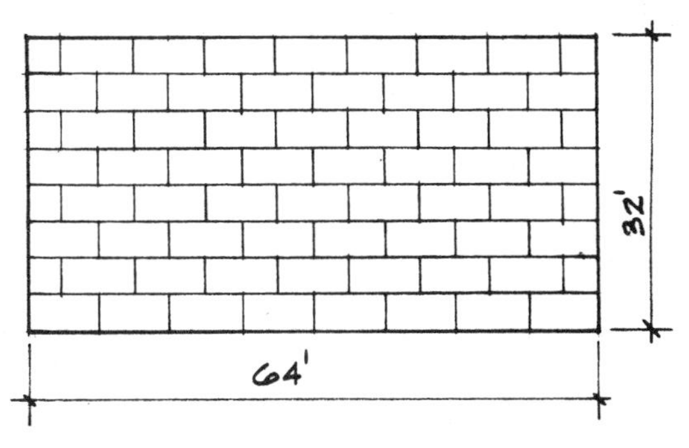

If both the width and length dimensions of a building are on a 4-ft. module, the waste factor will be zero percent.

EXAMPLE 2:

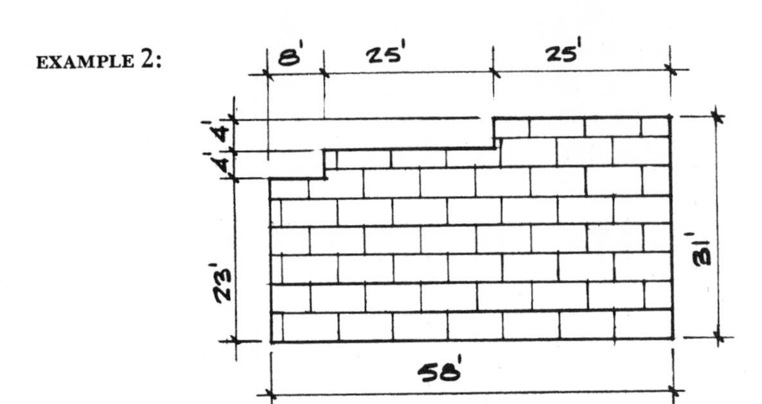

If either the width or length dimension of a building is not on a 4-ft. module, the plywood waste will be approximately 4 percent.

PROCEDURE FOR ESTIMATING PLYWOOD SUBFLOORING

METHOD 1:

The plywood layout can be drawn on the foundation or floor plan, and the number of sheets may be counted.

METHOD 2:

1. Calculate the square footage of floor to be covered.
2. The square foot of floor area is divided by 32 (number of sq. ft. in a 4′ × 8′ sheet of plywood). The result is the number of plywood sheets.
3. If the layout of the building is on a 4-ft. module, very little waste should be added. If the building shape and size are irregular, add 4% waste.

21

Estimating Foundation Vents

Foundation vents are necessary to ventilate and keep dry the crawl space under a conventional wood floor. The size of foundation vents is determined by the size of the floor joist.

Code requirements usually call for 1-1/2 square feet of vent area for every 25 lineal feet of building perimeter. The code also requires that a vent be placed within 3′-0″ of every corner.

For aesthetic reasons, it is not required that foundation vents be placed on the front of a building. The square foot area requirement can be met by placing the vents on the sides and rear of the structure.

Three types of foundation vents that may be used are illustrated below:

1. Stucco Vent

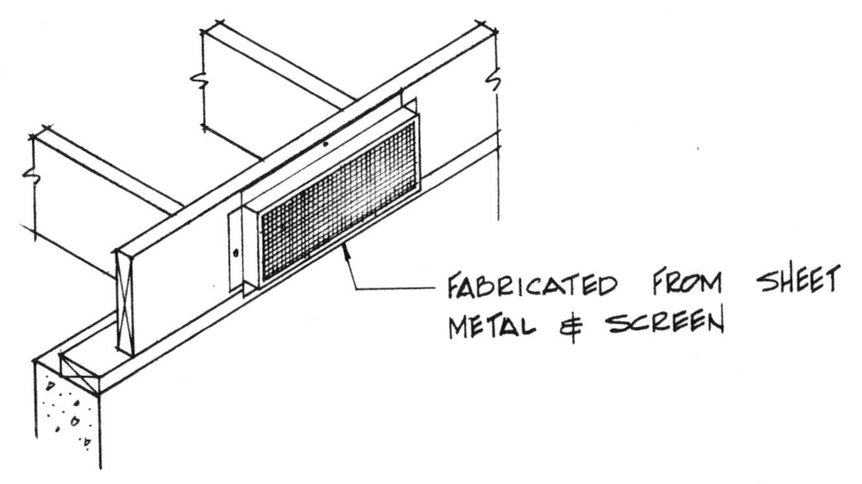

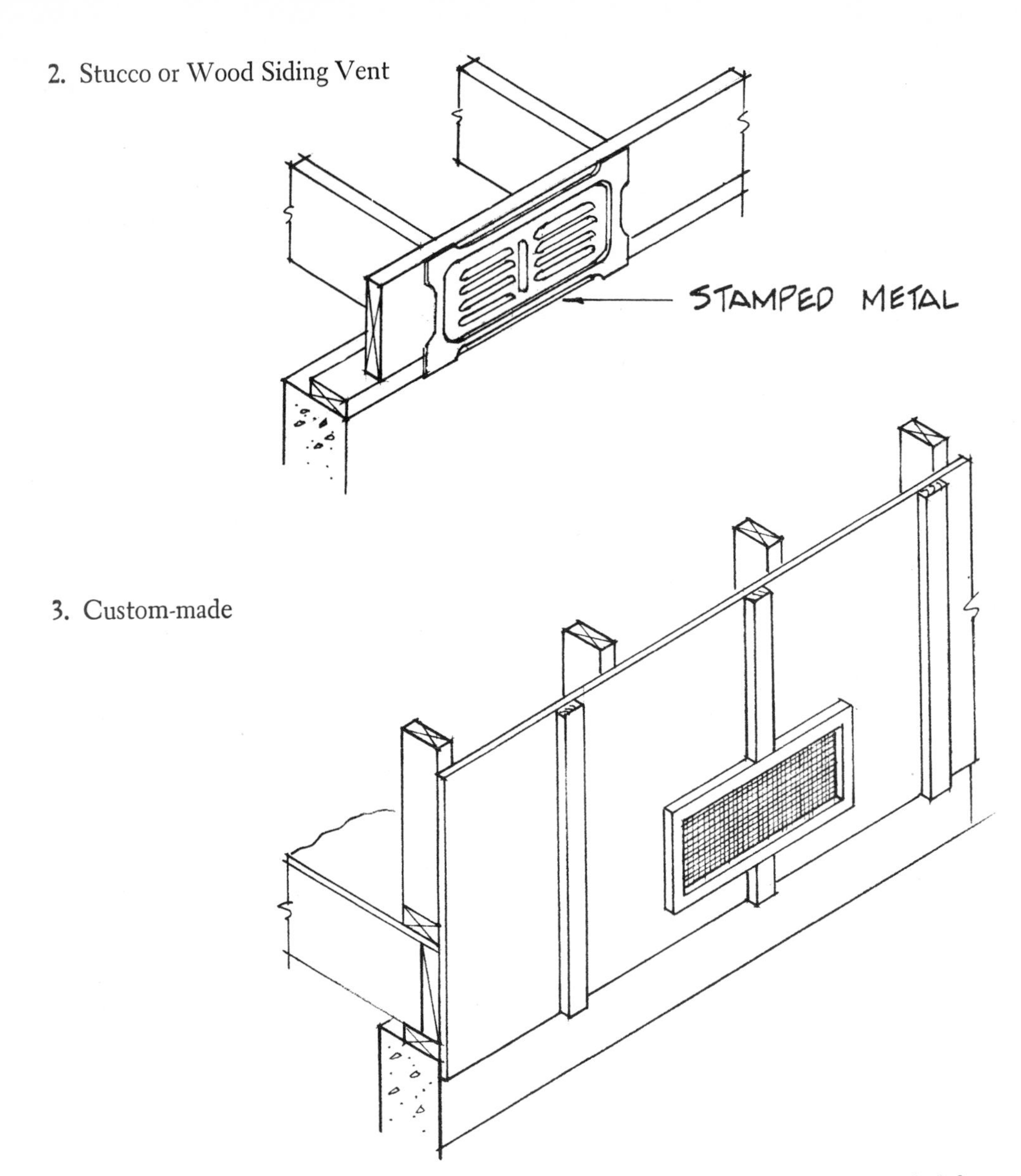

2. Stucco or Wood Siding Vent

3. Custom-made

PROCEDURE FOR ESTIMATING FOUNDATION VENTS

1. Determine the building perimeter.
2. Divide the perimeter by 25.
3. Multiply the answer above by 1-1/2 or the decimal equivalent 1.5, which is the number of sq. ft. needed for every 25 lin. ft. of perimeter. The result is the number of sq. ft. of vent needed.
4. Check the vent size and divide the square footage of vent needed by the sq. ft. or fraction of a foot area of each vent. The result will be the number of vents to order.

22

2-4-1 Floor Framing System

The 2–4–1 floor framing system, commonly called *diaphragm* floor construction, is a modular method using girders 4″ wide spaced 4′-0″ O.C., or floor joists 2″ wide, spaced 32″ O.C. and covered with 1-1/8″ thick tongue-and-groove plywood.

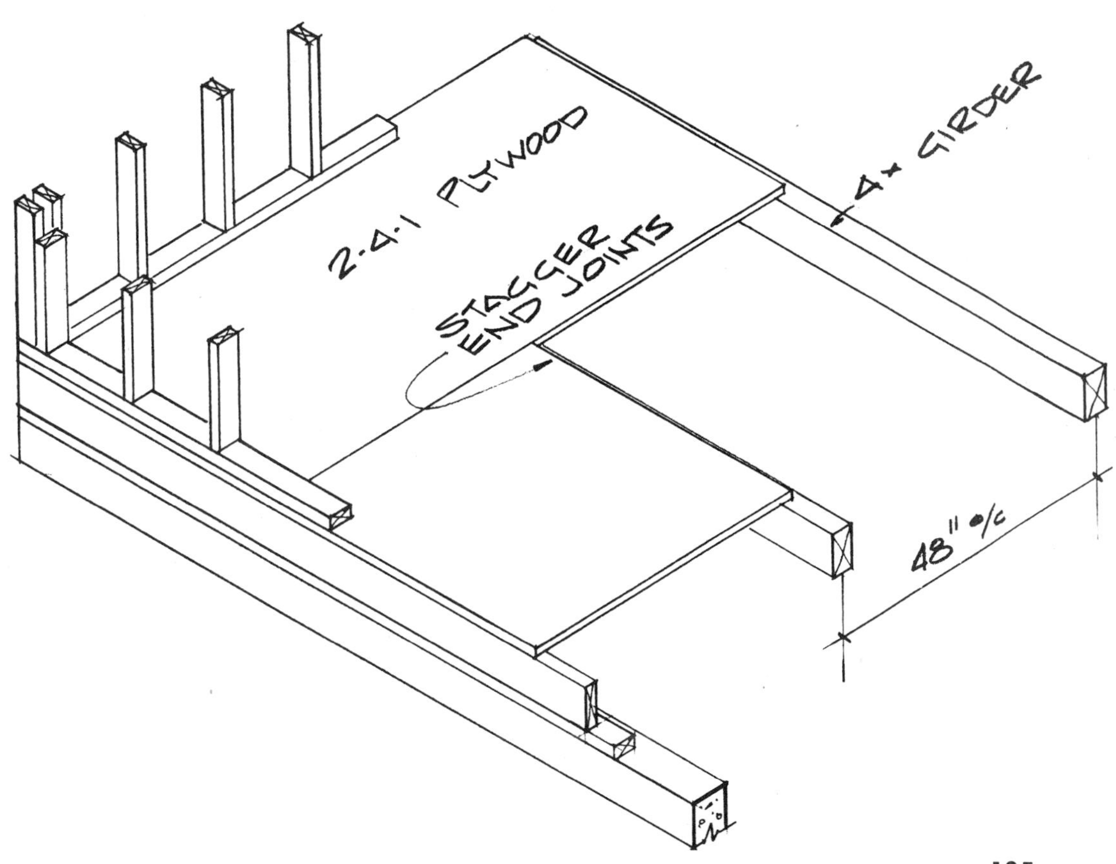

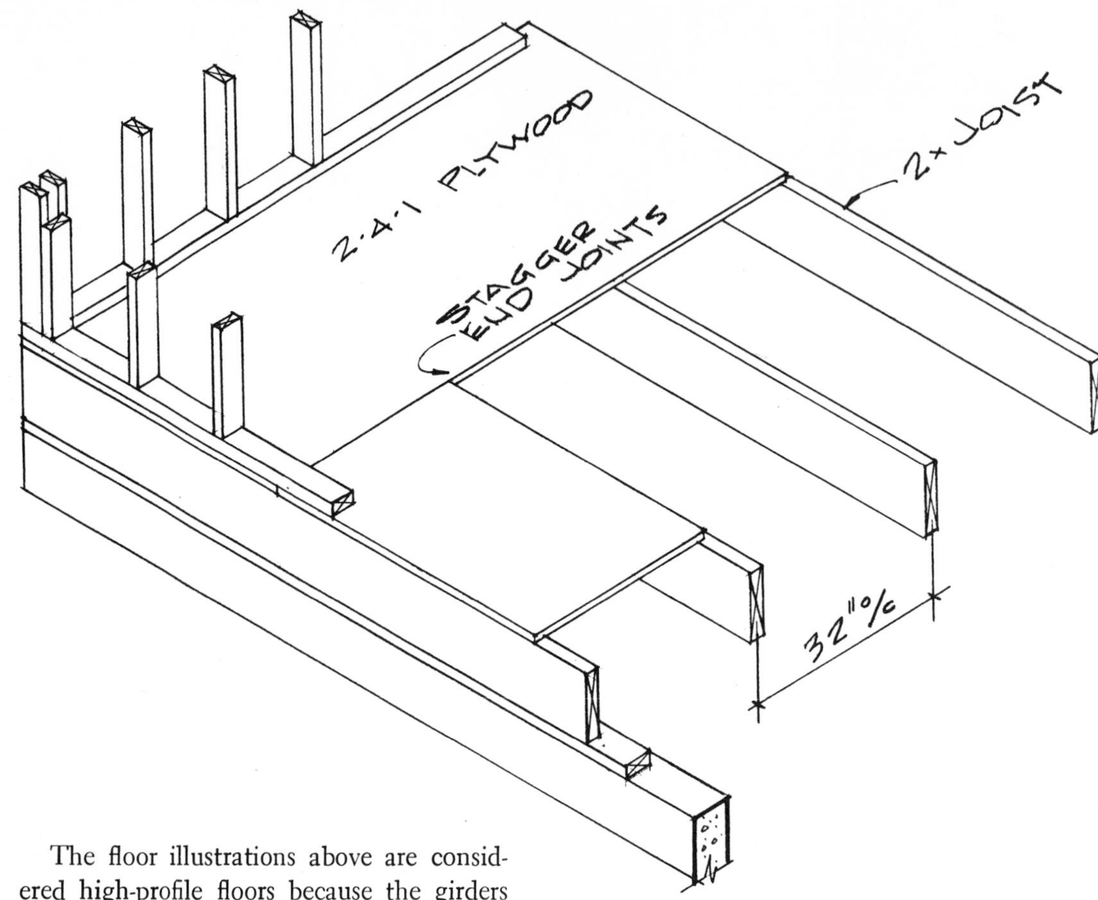

The floor illustrations above are considered high-profile floors because the girders or joists rest on top of the mudsill. For a low-profile building the girders can be set in foundation pockets, or the joist can be lowered with the use of the pier system so that the 2–4–1 plywood bears directly on the mudsill as illustrated in the following details.

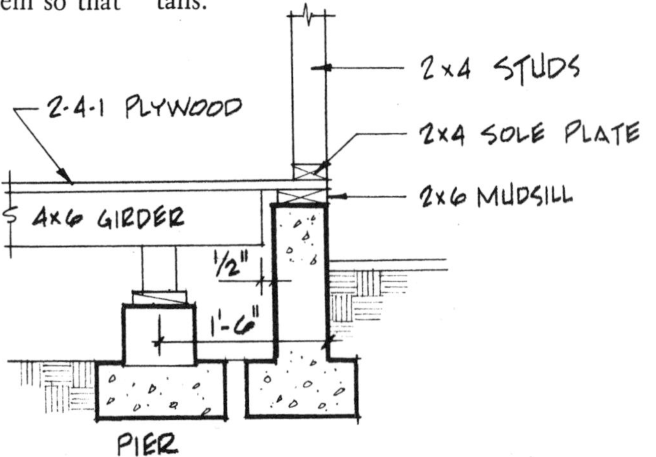

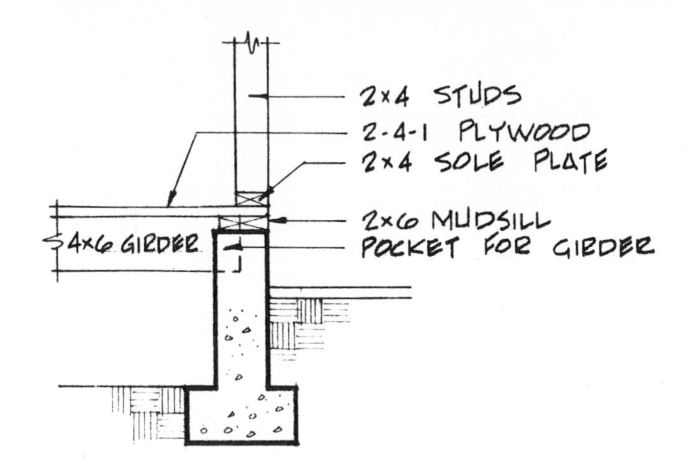

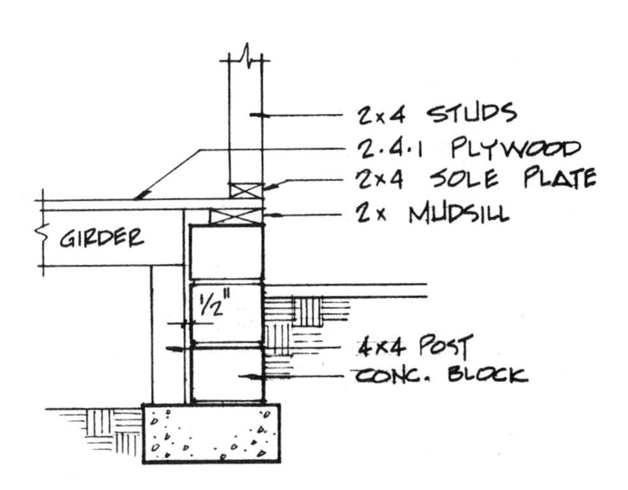

The material to be estimated for the 2–4–1 floor system includes the mudsill, pier posts, girders or floor joists (whichever method is used), blocking (if any is required), and the number of sheets of 2–4–1 plywood. A 2–4–1 floor system foundation plan and a list of material follow.

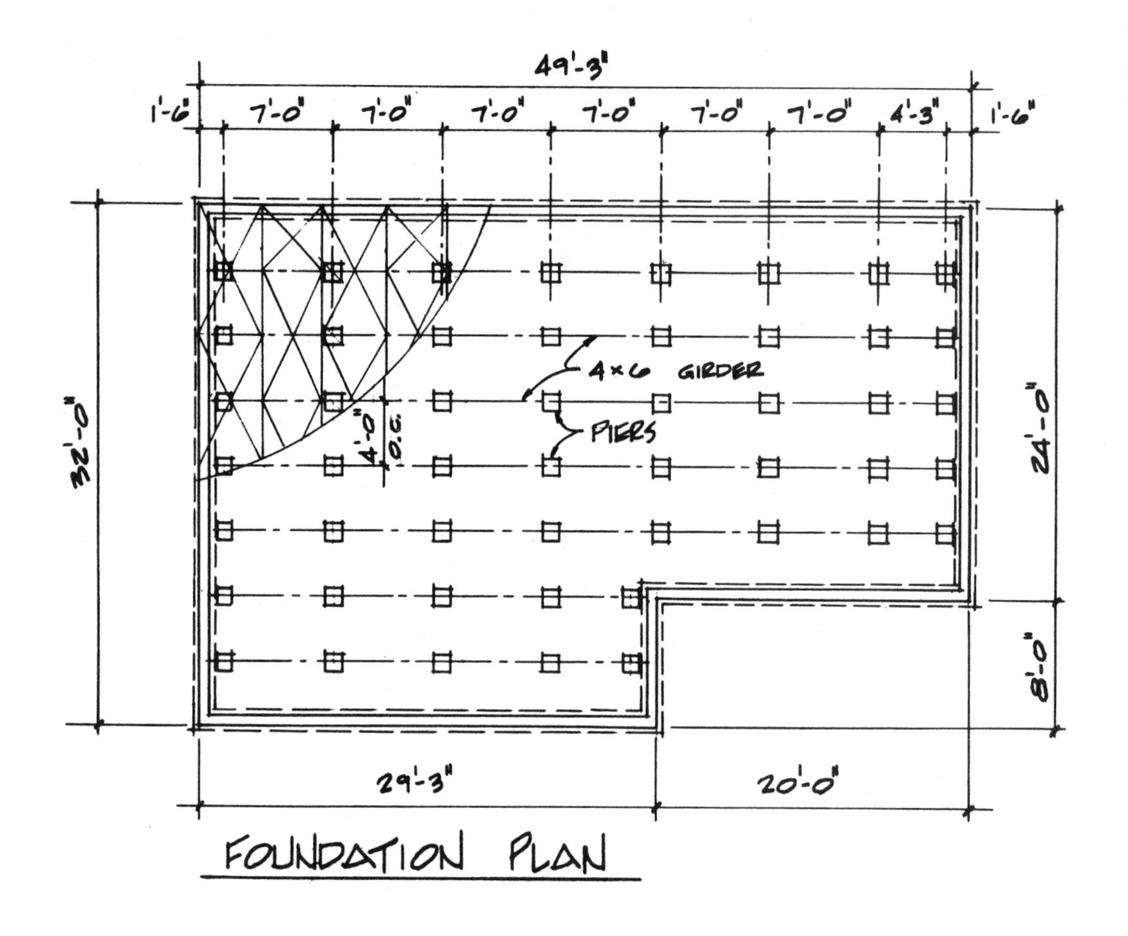

FOUNDATION PLAN

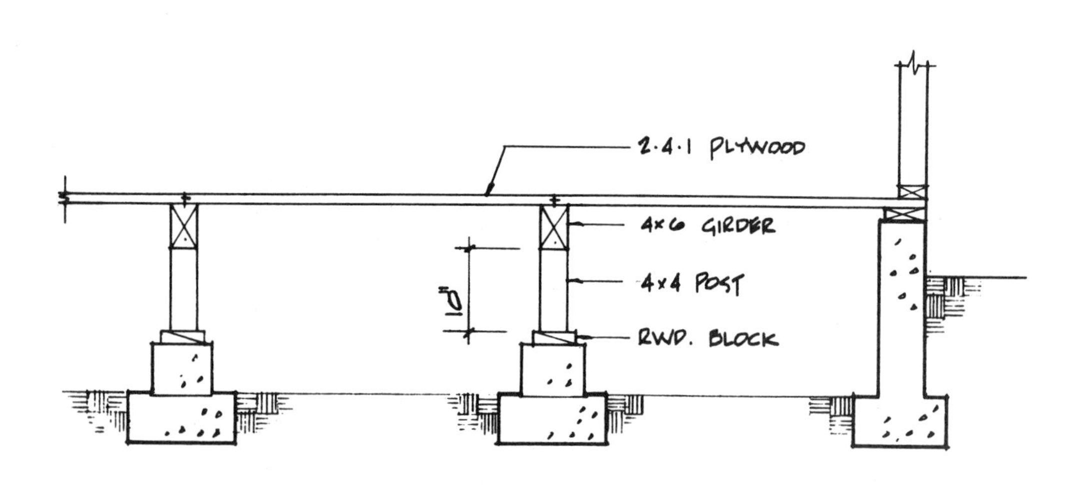

Item No.	*Description*	*Kind & Grade*	*No. of Pieces*	*TXWXL*	*Lin. or Bd. Ft.*	*Cost*
1	Mudsill	Pressure Treated	Random Lengths	2 × 6	164 lin. ft. + 18 waste 10% 182 lin. ft.	65.52
2	Pier Posts	D.F. Const./Std.	3 1	4 × 4 × 12 4 × 4 × 14	50 lin. ft.	28.00
3	Girders	D.F. Const./Std.	7 7 5	4 × 6 × 14 4 × 6 × 16 4 × 6 × 20	310 lin. ft.	260.40
4	Floor Joists					
5	Floor Blocking	D.F. Const./Std.	29	2 × 4 × 12	348 lin. ft.	90.48
6	Plywood	2-4-1 Plywood	45	1-1/8″ × 4′ × 8′	1440 sq. ft.	1,165.50
					TOTAL	$1,609.90

23

Problems—Floor Framing Materials

PROBLEM 1

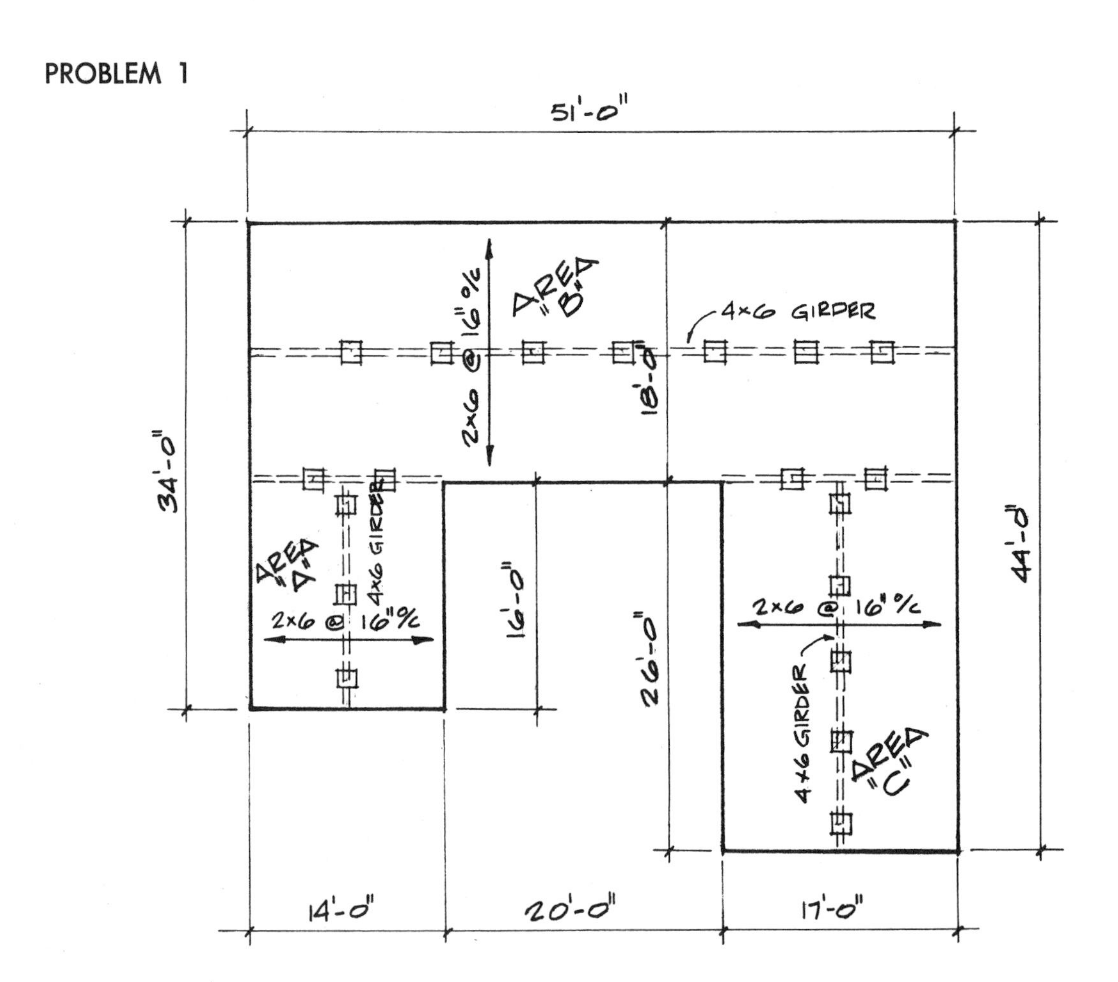

ESTIMATE THE FOLLOWING:

1. Lineal feet of pier post material (average length of pier post 1′-0″) __________
2. Lineal feet of 4 × 6 girder material __________
3. Lineal feet of mudsill material __________
4. Number, size, and length of floor joists for area A __________
   area B __________
   area C __________
5. Lineal feet of header joist material __________
6. Lineal feet of solid blocking material __________
7. Number of 4′ × 8′ sheets of plywood subflooring __________

PROBLEM 2

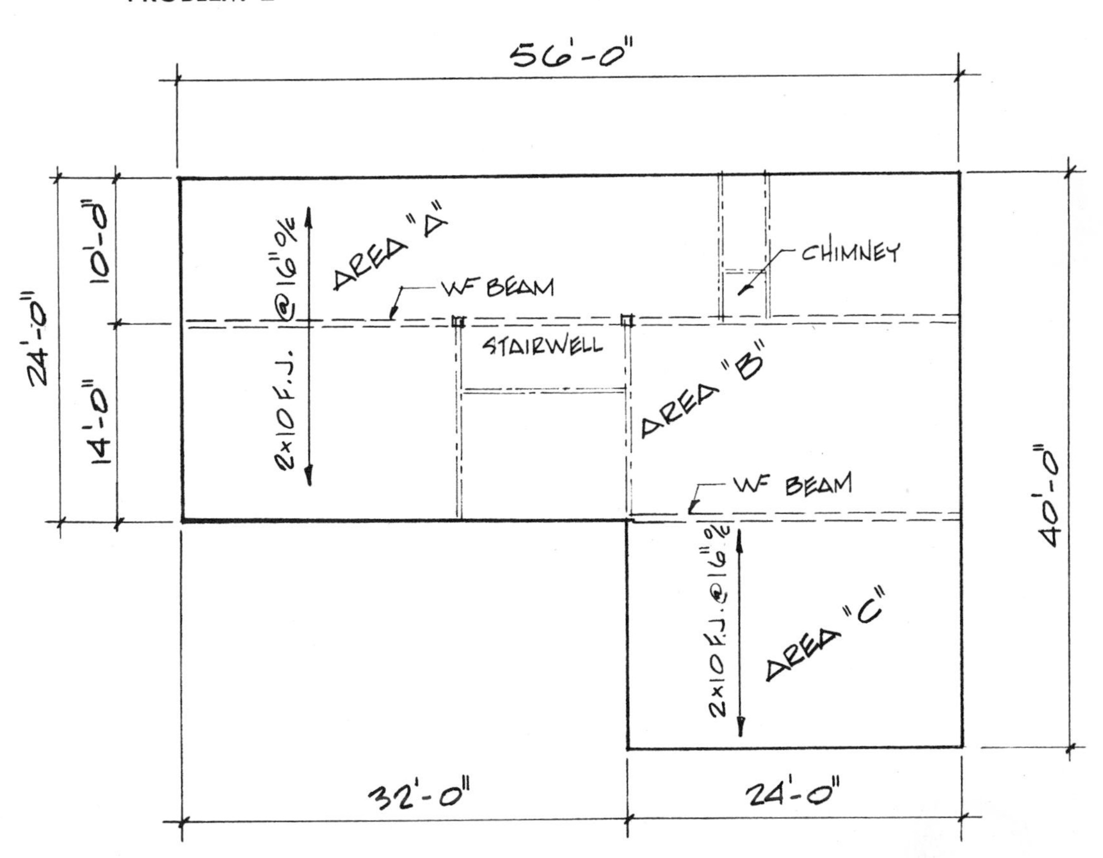

ESTIMATE THE FOLLOWING:

1. Lineal feet of mudsill ____________
2. Number, size, and length of floor joists for area A (including double joists) ____________
   area B ____________
   area C ____________
3. Lineal feet of header joist material ____________
4. Lineal feet of blocking or bridging material ____________
5. Number of 4′ × 8′ sheets of plywood subflooring ____________

PROBLEM 3

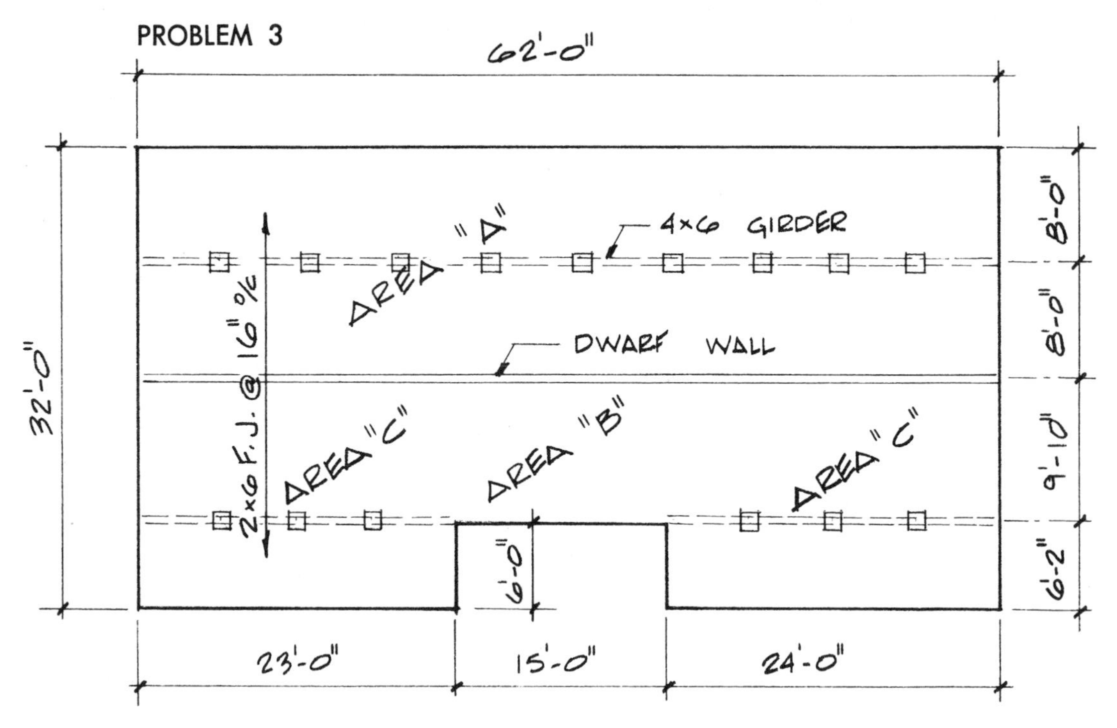

ESTIMATE THE FOLLOWING:

1. Lineal feet of mudsill ____________
2. Number, size, and length of floor joists for area A ____________
   area B ____________
   area C ____________
3. Lineal feet of header joist material ____________
4. Lineal feet of solid blocking material ____________
5. Number of 6″ × 12″ foundation vents (1-1/2 sq. ft. of vent for every 25 lin. ft. of building perimeter) ____________
6. Number of 4′ × 8′ sheets of plywood subflooring ____________

IV

WALL FRAMING

24

Methods of Wall Framing

There are three methods of wall framing with which the estimator should be familiar. The first and most commonly used method is called *conventional stud framing* or *western platform framing*. The distinguishing feature of this type of framing, whether single story or multiple story, is that the wall sections can be assembled quickly on the horizontal platform and tilted into place. Each platform becomes a convenient level on which the carpenters can work. This construction also prevents *flue action* within the walls and eliminates firestops.

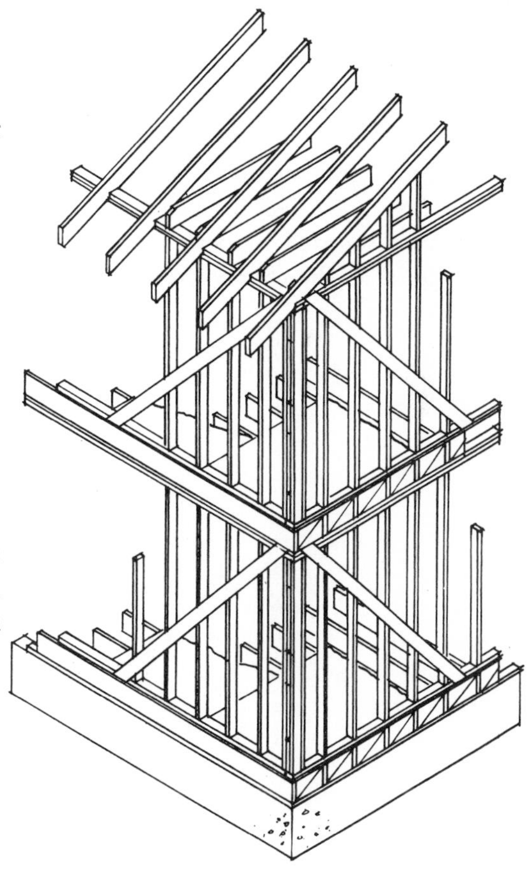

TWO-STORY WESTERN PLATFORM FRAMING

The second method of wall framing deals with both one- and two-story construction. For a two-story building, it is called *balloon framing* and for a single-story building, it is called *semi-balloon framing* or *rake-wall framing*.

The distinguishing feature of balloon framing is that the studs are full height from the mudsill to the second floor top plate. The second-story floor joists are supported on a 1″ × 6″ ribbon and are spiked to the studs. An important advantage of this method is the small amount of shrinkage in the building height.

BALLOON FRAMING

At present, straight balloon framing is very seldom used in the construction industry. But, a combination of platform and balloon framing is used in a two-story building when a stairwell is located on an exterior wall or in the corner of the building. Straight platform framing, in this case, would leave the building with a weakened plane line at the second-story floor joist line. Semi-balloon or rake-wall framing allows some or all of the interior and exterior walls that run parallel to a gable end wall to be framed with the studs running from the

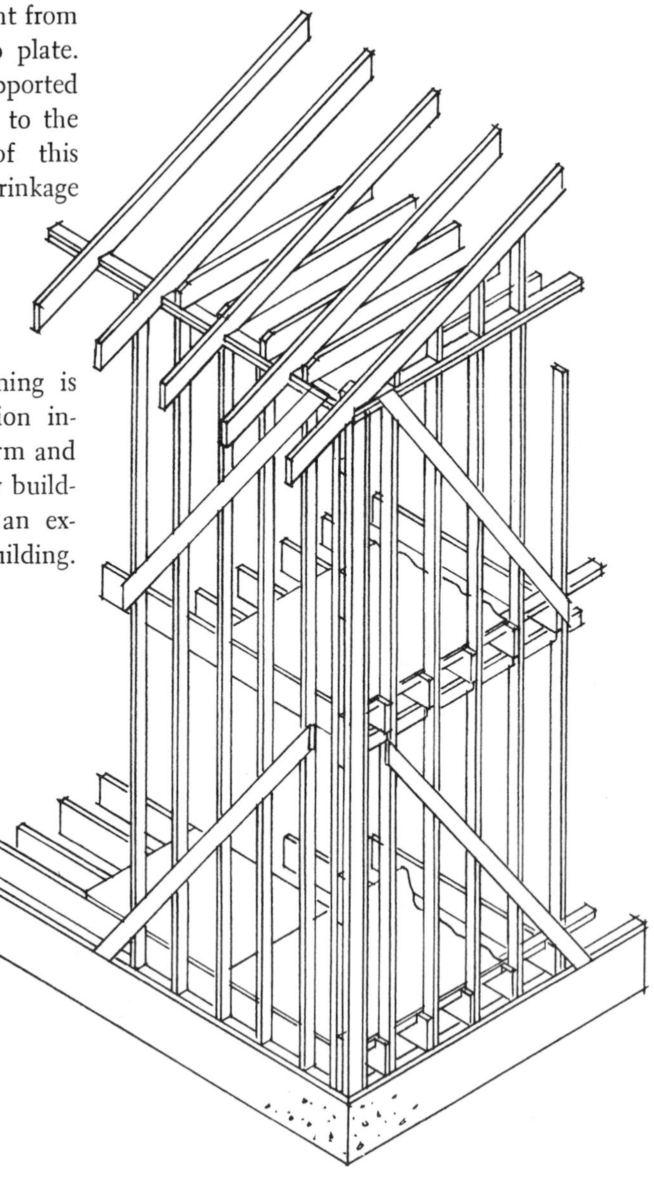

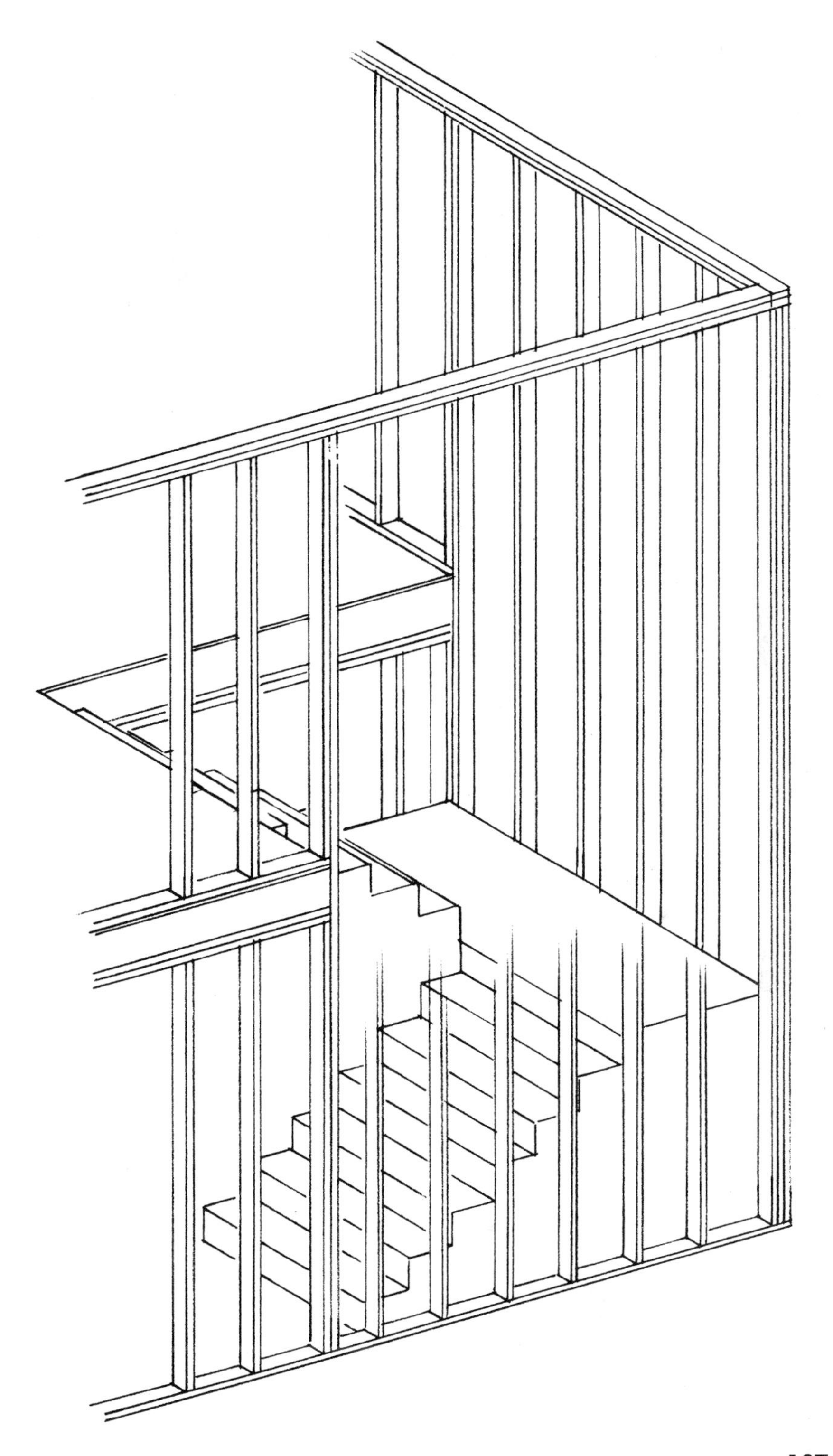

bottom plate to the rafter line as illustrated below:

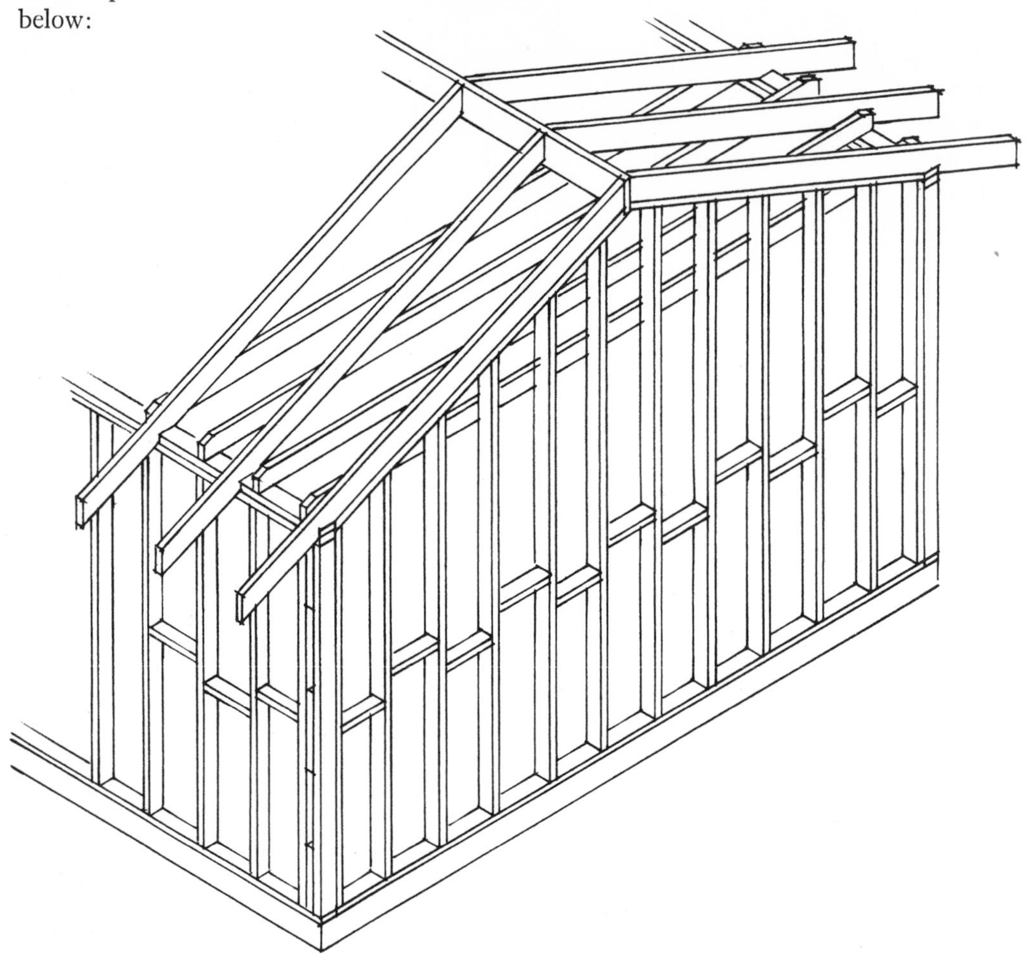

Note: Rake walls may be laid out on the floor deck, built, and tipped into place like any other exterior or interior wall.

Usually the building plans do not specify the use of full-length studs. Code requirements determine at what height a wall will have to be framed in this way. The estimator will have to be familiar with these code requirements in order to do a correct material take-off.

The third kind of wall framing is *post-and-beam* construction. This method uses floor girders, wall posts, and rafters positioned so as to accept standard sizes of building materials. Posts and beams are usually spaced 4′-0″ on center, and intermediate studs are placed between the posts to provide support for interior and exterior wall coverings. The floor deck is usually 2–4–1 plywood, 1-1/8″ thick or 1-1/2″ thick tongue-and-groove decking. The roof sheathing can be made of the same materials as those used for the floor. The underside of the sheathing will be exposed and should be neat in appearance. The roof will have to be covered with batt type or rigid insulation to reduce the heat loss through the roof as detailed in the illustrations on page 252.

POST-AND-BEAM CONSTRUCTION

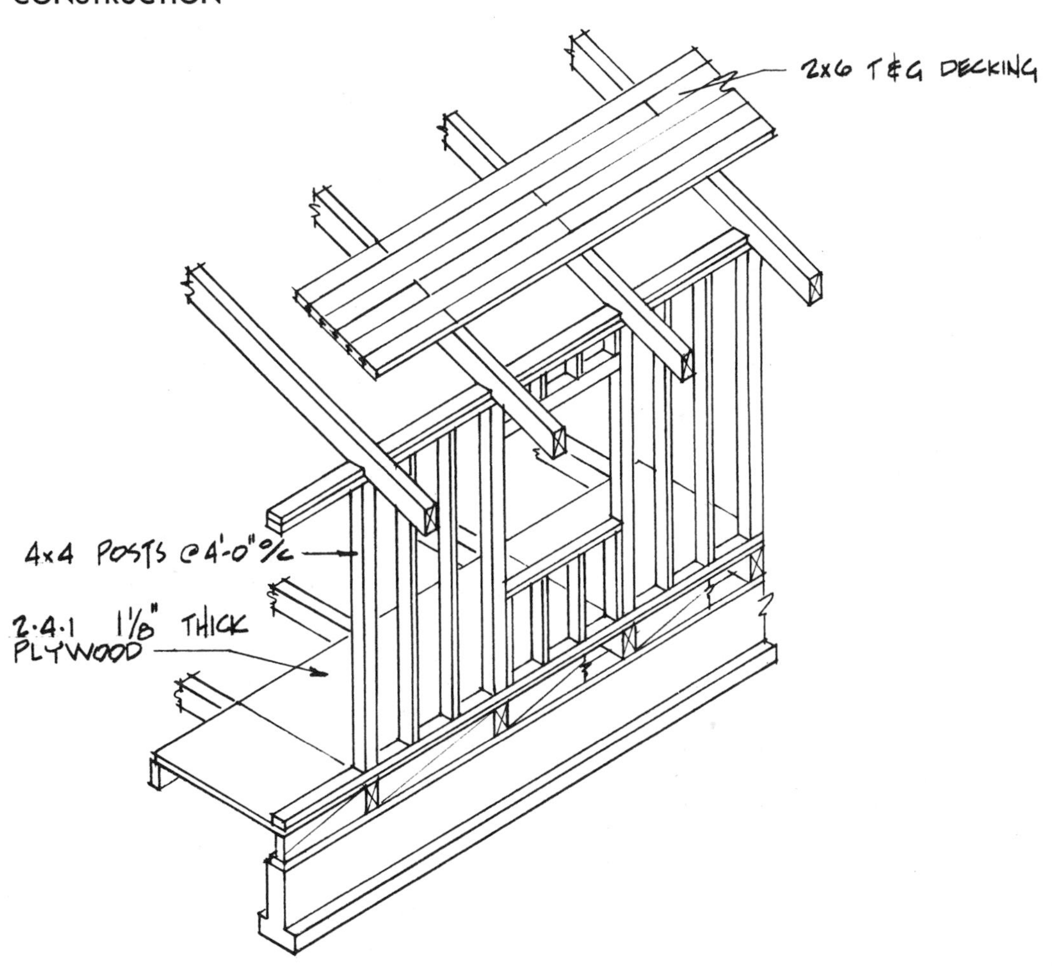

CONVENTIONAL WALL FRAMING

In the following diagram, you will find the basic members that make up a wood-framed wall. Of these members, only the (1) plates, (2) studs, (3) headers, (4) diagonal braces, and (5) wall block material will have to be listed for material take-off.

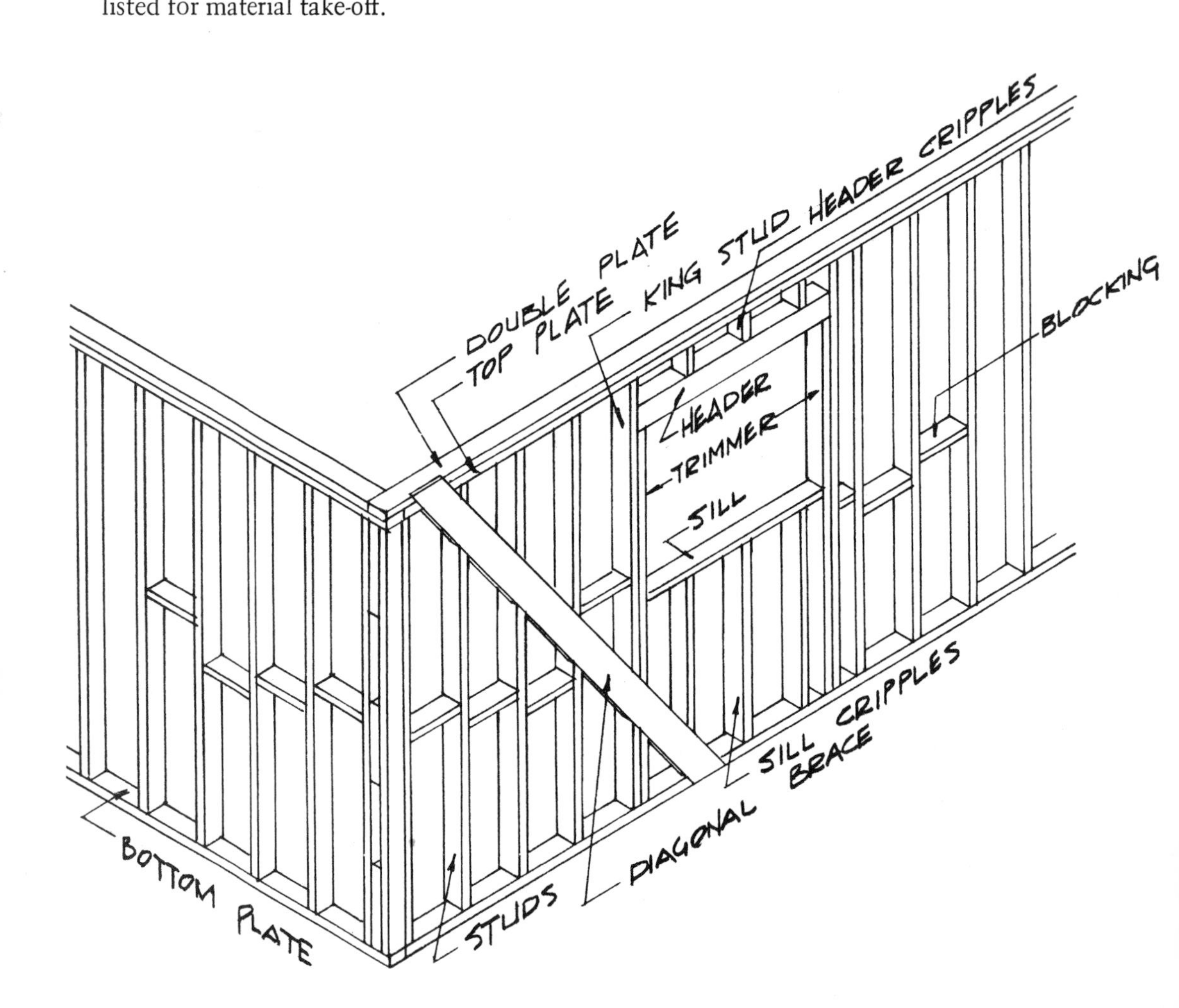

25

Estimating Wall Plates

Wall plates are horizontal members in a wall to which the studs are nailed. The number of plates used in framing a wall will be determined by the weight a wall has to carry.

Some walls are bearing walls and others are nonbearing walls. *Bearing walls* are framed with three plates, a bottom or sole plate, a top plate, and a double plate. *Nonbearing walls* may be framed with two plates, a bottom or sole plate and a top plate, but most builders use two top plates for better construction. This stiffens and strengthens the wall considerably and exposes more wood when nailing the wallboard or plaster lath.

Note: The studs used for a nonbearing single top-plate wall have to be ordered separately and cut 1-1/2″ longer than standard precut studs. Most builders find the added cost for handling and cutting these studs comparable to the cost of material for doubling the top plate. The better construction practice is to use double top plates and standard precut studs.

EXAMPLE:

Bearing walls have single bottom plate and double top plate.

Nonbearing walls have single top plate.

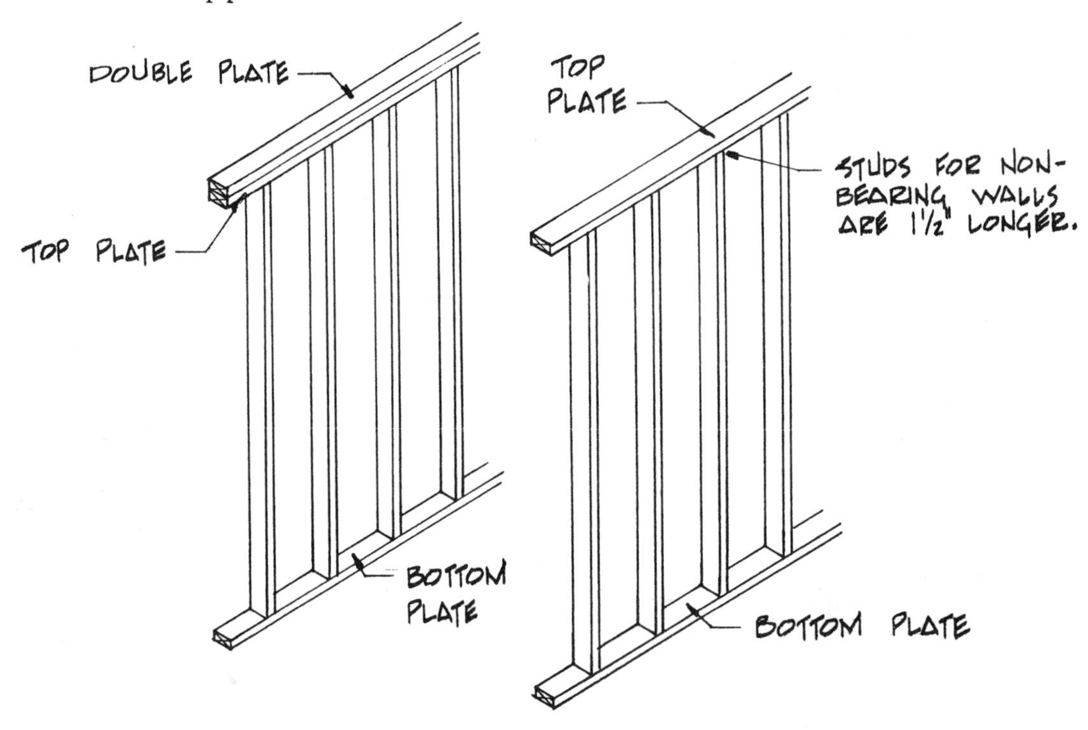

WALL PLATES

The estimator should check the floor plan for any walls that are wider or narrower than a standard 2″ × 4″ frame wall. A list of plates for these walls should be added to the material list. At this time the estimator should also add to the lumber list any special-sized studs for framing these walls. The following example illustrates the different wall sizes and the stud framing method.

EXAMPLE: 2″ wide walls between closets or 6″ wide walls for plumbing pipes.

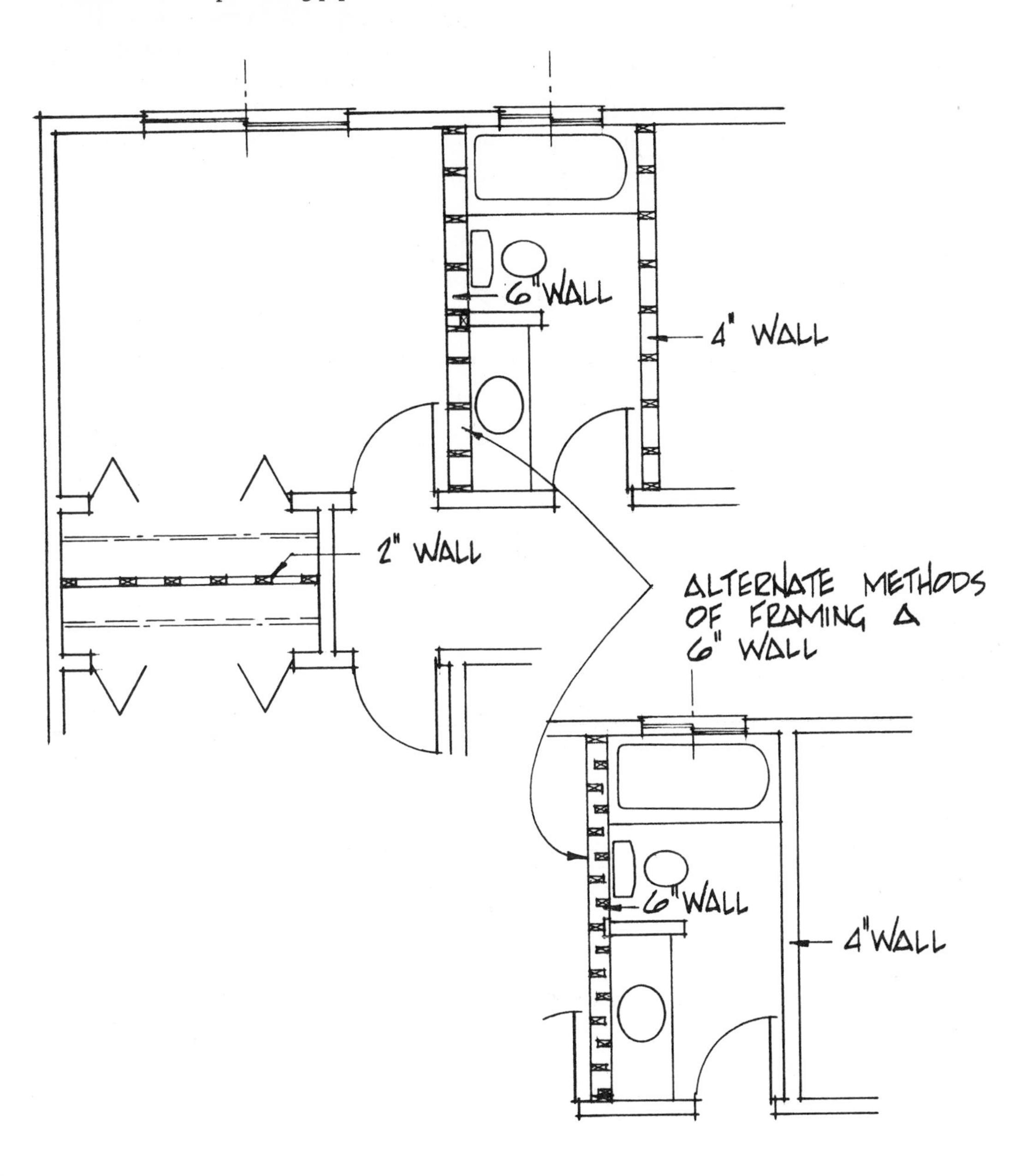

It must be kept in mind that the code usually requires that plates be spliced only in certain places. The least amount of splicing creates the strongest walls. If a building is small, such as a garage, cabin, or shed, the estimator may order the exact lengths needed for each wall. In this case the layout man on the job will have to know the position of each particular length of plate stock.

The following is a floor plan for a small building and a list of plate stock needed for each wall.

EXAMPLE:

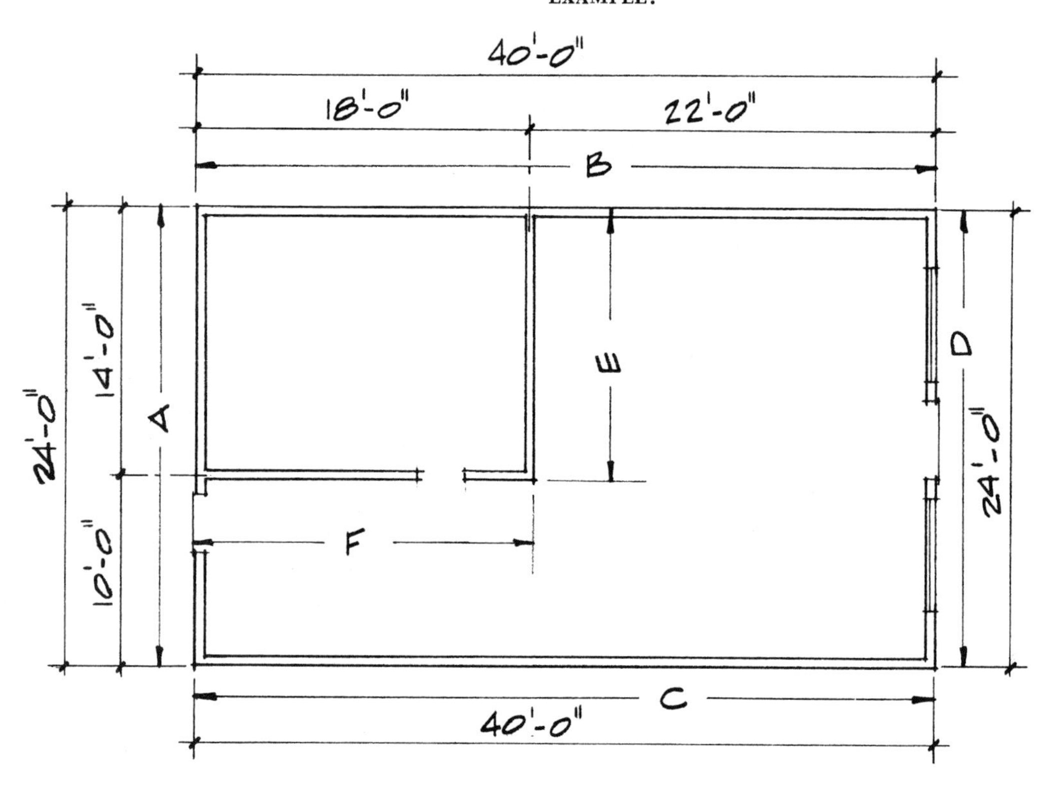

List of Material

Wall Designation	*Number of Pieces*	*Size and Length*	*Wall Designation*	*Number of Pieces*	*Size and Length*
A	3	2 × 4 × 24′	D	3	2 × 4 × 24′
B	3	2 × 4 × 22′	E	3	2 × 4 × 14′
		2 × 4 × 18′	F	3	2 × 4 × 18′
C	6	2 × 4 × 20′			

There is one basic difference between estimating wall plates for conventional wood floor framing and slab floor construction. Both methods of construction have a bottom plate and two top plates, but the bottom plate on a slab floor is in direct contact with the concrete; and a treated plate material, or a type of wood that will not rot or deteriorate when in contact with moisture, should be used.

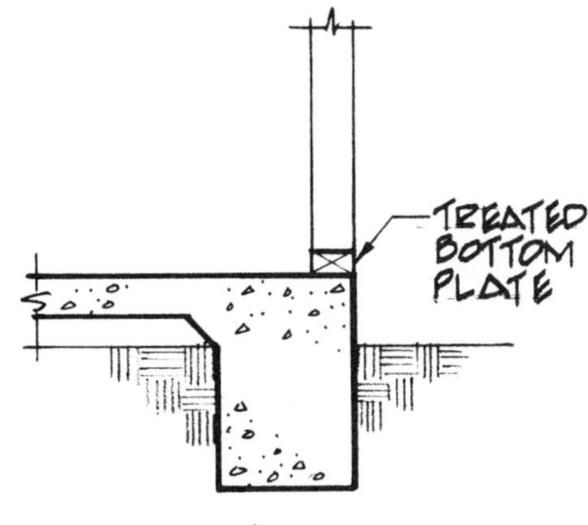

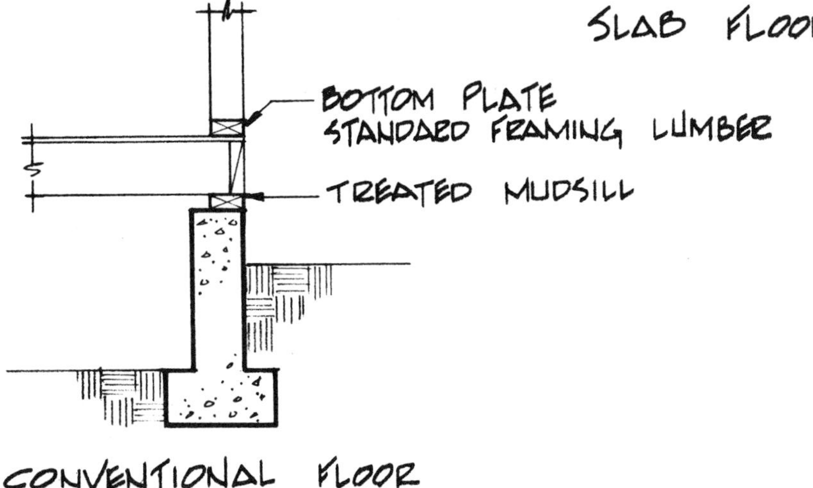

Because of the different wall lengths in a building, an estimator should order long lengths of material for plates, preferably 16′ to 22′ lengths. Plates may all be ordered in one length or a certain number of each length. In either case, they should be cut to allow for the least amount of waste.

WALL PLATES FOR TRUSS ROOF CONSTRUCTION

A truss roof has all the ceiling and roof weight bearing on the exterior walls. The interior walls, whether they run parallel or perpendicular to the ceiling joist, are all nonbearing walls and could be framed with a single top plate. For better construction a double top plate is recommended. If trusses are to be used, the estimator should treat them as a subcontract and obtain cost bids from a "pre-fab" truss manufacturer.

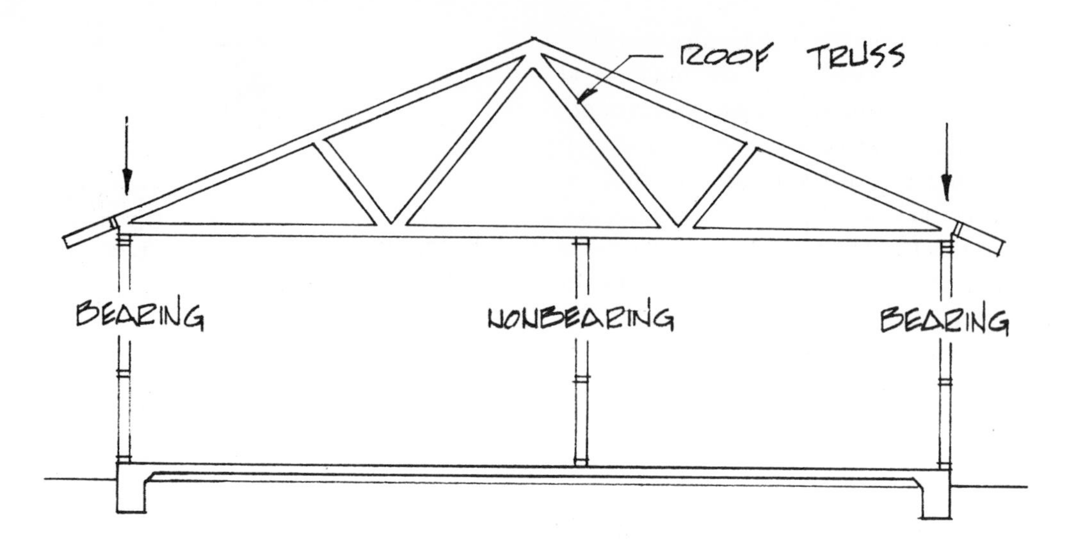

PROCEDURE FOR ESTIMATING WALL PLATES

METHOD 1: Conventional Wood Floor

1. From the floor plan, estimate the lineal feet of the exterior walls. To this add the length of all the walls running horizontally on the plan and then the length of all the walls running vertically on the plan. Thus, when measuring the lineal length of walls, include all window and door opening widths—large or small. When a carpenter frames a wall, he will run the bottom plate stock through the opening; and when the wall is in position, the bottom plate will be cut out of the opening.
2. Multiply the total lineal feet of wall by three (one bottom plate and two top plates).
3. Add 10% for waste. Order wall plates in lengths desired.
4. Check the floor plan for any walls wider or narrower than standard 2″ × 4″ frame wall. The special plates and studs for these walls should be estimated and added to the material list at this time.

Note: If fire or stiffener blocks are required on all interior and exterior walls, the estimator may multiply the total lineal feet of wall by four. This will then include the material for the three plates and the wall blocking. Unit 30 on wall blocking also covers this topic.

METHOD 2: Slab Floor Construction

1. From the floor plan, estimate the lineal feet of the exterior walls. To this add the length of all the walls running horizontally on the plan and then the

length of all the walls running vertically on the plan.

2. Add 10% waste to the total lineal feet of walls. This figure will be the lineal feet of treated bottom plate stock to order.
3. Multiply the lineal feet of treated bottom plate stock by 2 (if a double top plate is used). This material will be common framing lumber. Therefore, the wall plate stock for slab floor construction will contain treated and untreated material.
4. Check the floor plan for any walls wider or narrower than standard 2″ × 4″ frame wall. The special plates and studs for these walls should be estimated and added to the material list at this time.

METHOD 3: Truss Roof Construction

Use the estimating procedure in methods 1 and 2 above, but keep in mind that when using a truss roof the exterior walls bear all the weight and three wall plates will be needed. The interior walls may require only a single top plate.

Note: If fire or stiffener blocks are required on all interior and exterior walls, the estimator may multiply the total lineal feet of treated bottom plate stock by three. This figure will then include the material for the top and double plate and the wall blocking. Unit 30 on wall blocking also covers this subject.

26

Estimating Studs

Studs may be purchased precut to length, or they can be cut to desired length on the job. In residential construction, studs are usually placed 16″ on center (16″ O.C.), but certain structures, such as garages, sheds, and other small buildings, may be placed 24″ O.C. Stud length is determined by the ceiling height of the structure to be estimated.

Normal ceiling height for wallboard construction is 8′-1/2″ from subfloor to finish ceiling or 8′-1″ from subfloor to ceiling joist as illustrated below.

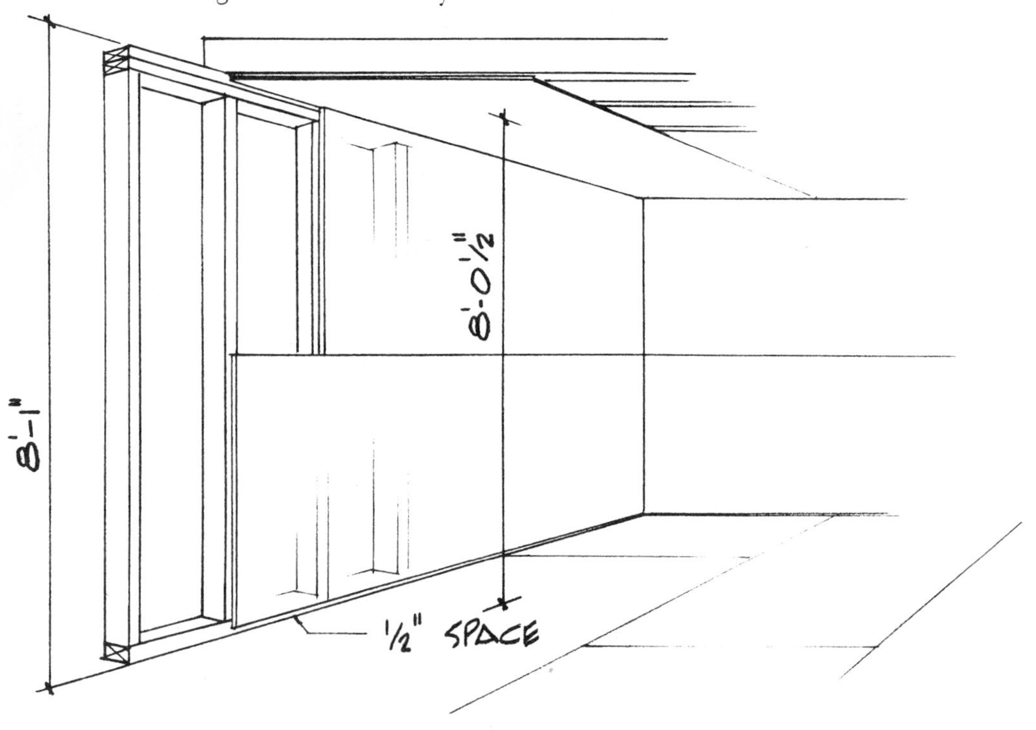

To determine stud length, subtract the thickness of three wall plates from desired subfloor to ceiling joist measurement.

EXAMPLE: 3 × 1-1/2″ (standard dry lumber thickness) = 4-1/2″
8′-1″ − 4-1/2″ = 7′-8-1/2″ or 92-1/2″ stud length

GENERAL RULE FOR ESTIMATING STUDS

When estimating residential structures, it is common practice to order one stud for each lineal foot of wall. This will provide enough extra material for trimmer studs, cripple studs, backing studs, corner posts, and partition channels.

PROCEDURE FOR ESTIMATING STUDS

METHOD 1:

When estimating plate stock, the estimator has already calculated the total lineal feet of interior and exterior bottom plate stock. This lineal footage may be used to short cut step 1 in the procedure below.

1. From the floor plan, estimate the lineal feet of the exterior wall. To this add the length of all the walls running horizontally on the plan, and then add the length of all the walls running vertically on the plan. When measuring the lineal length of walls, include the window and door opening widths, whether large or small. When a carpenter frames a wall, he will run the bottom plate stock through the opening; and when the wall is in position, the bottom plate will be cut out of the opening.
2. Using the general rule for estimating studs, an estimator can determine the number of studs needed from the total lineal feet calculated above.
3. If not already estimated with the plate stock, special size stud material for walls wider than 4 inches should be estimated at this time.

METHOD 2:

When estimating studs for small structures that have very few wall openings—such as garages, sheds, room additions, etc.—the estimator may use the center-to-center constant table as in the procedure below.

TABLE 26-1 CENTER-TO-CENTER STUD SPACING

Spacing	*Constant*	*Decimal Equivalent*
12″ O.C.	1	1.00
* 16″ O.C.	3/4	.75
18″ O.C.	2/3	.6667
20″ O.C.	3/5	.60
24″ O.C.	1/2	.50
32″ O.C.	3/8	.375
36″ O.C.	1/3	.3333

* Usual stud spacing

1. Calculate the lineal feet of exterior and interior walls.
2. Multiply the lineal footage above by the constant for the stud spacing from Table 26-1.
3. To the answer above, add 4 studs for every room or closet. The result will be the number of studs needed.

METHOD 3:

For small structures an estimator might want to take the time to lay out and count the exact number of studs on the building plans.

27

Estimating Gable Studs

Gable studs are the studs that run between the top plate of a wall and the pair of gable rafters at the end of a building.

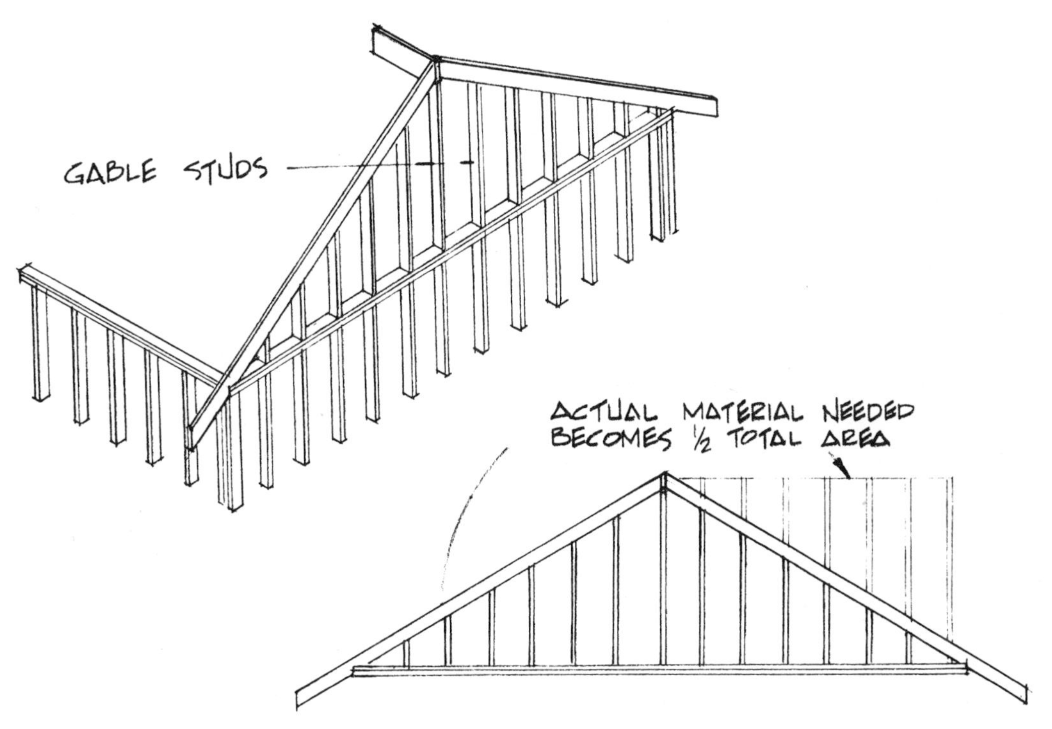

PROCEDURE FOR ESTIMATING GABLE STUDS

1. Multiply the width of the building by the center-to-center spacing constant. This will provide the number of gable studs needed.
2. Multiply the number of studs by the length of the longest gable stud increased to the nearest foot. The length of the longest gable stud may be scaled from a section or elevation view.
3. Divide the lineal footage from above by 2. The result will be the lineal length of material needed on one gable end.
4. Multiply the lineal feet of material by the number of gable ends of equal size.

Note: If gable ends of different widths are used, the material for each gable should be estimated separately.

28

Estimating Headers and Beams

Headers are horizontal members that carry the weight above an opening. The header's size varies according to the width of the opening and the amount of weight to be carried. Headers may be of solid stock or of two or more pieces of material nailed together.

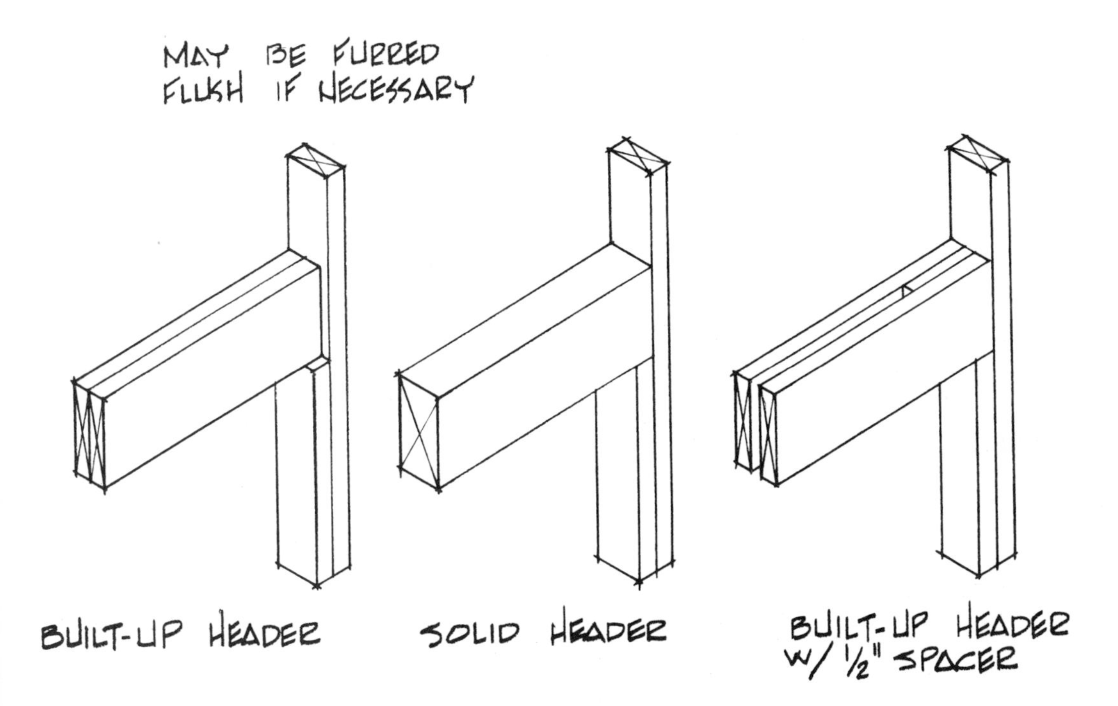

When estimating header material, the estimator has a twofold job. First he has to determine the size of material to use over each opening, and second he has to determine the length of the material needed.

Headers fall under two categories, bearing headers and nonbearing headers. Bearing headers carry the weight of the structure above, and nonbearing headers do not carry this weight.

The following illustration gives examples of bearing and nonbearing headers. It should be kept in mind that nonbearing headers are normally used in walls that run parallel to the ceiling joist but they could also be used in walls that run perpendicular to the ceiling joist as illustrated below.

Following are the recommended header sizes for various opening spans according to most of the building codes used throughout the United States.

Material Size	*Span*
4″ × 4″	up to 4′-0″
4″ × 6″	from 4′-0″ to 6′-0″
4″ × 8″	from 6′-0″ to 8′-0″
4″ × 10″	from 8′-0″ to 10′-0″
4″ × 12″	from 10′-0″ to 12′-0″
4″ × 14″	from 12′-0″ to 14′-0″

If the building has two stories, the first floor interior and exterior bearing headers are increased in width by 2 inches. For example, if a 4 × 4 header is required for an opening in a one-story building, the same opening on the first floor of a two-story building would require a 4 × 6 header. The second-story headers will be the same as a single-story building.

EXAMPLE:

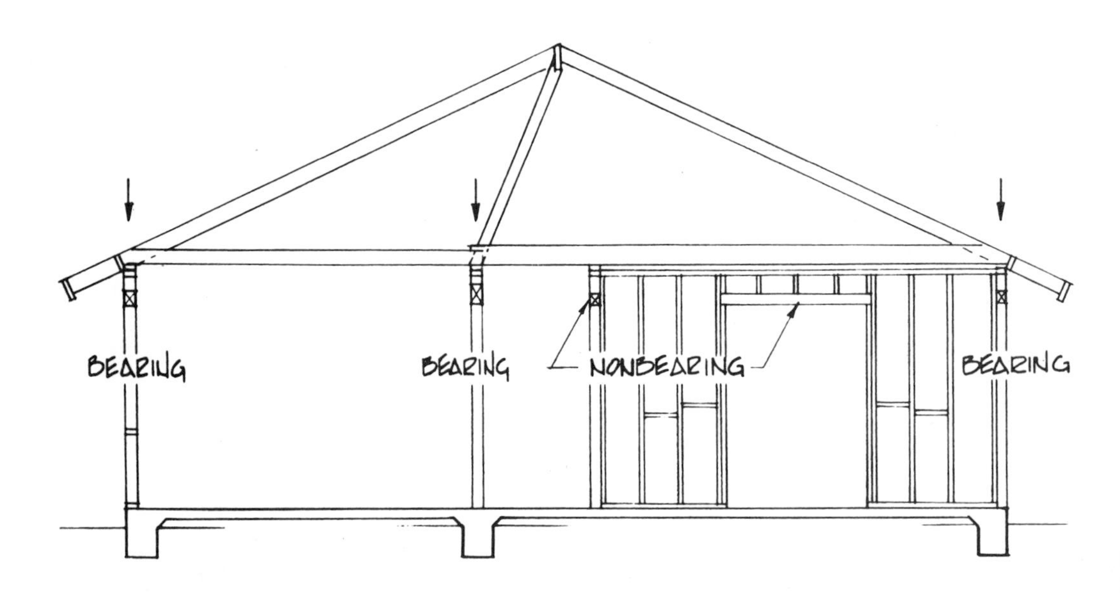

HEADERS FOR WINDOW AND SLIDING DOOR OPENINGS

Window and door manufacturers will usually provide rough-in sizes on a specification sheet.

EXAMPLE:

Window Size	*Type*	*Rough-in Size*
60 30	Aluminum Slider	71-5/8 × 35-1/2

Note: The width of the window is listed first and then the height:

60 meaning 6′-0″ wide
30 meaning 3′-0″ high

From the rough-in size, the length of header material can be determined.

Note: The header extends 1-1/2″ over each trimmer stud so actual header length will be 3″ wider than rough opening size. In the illustration above, a 4″ × 6″ × 7′ header is needed.

HEADERS FOR INTERIOR DOORS

Interior door sizes may be found on the floor plan and are usually designated as such: 2868 meaning a door 2′-8″ wide by 6′-8″ high.

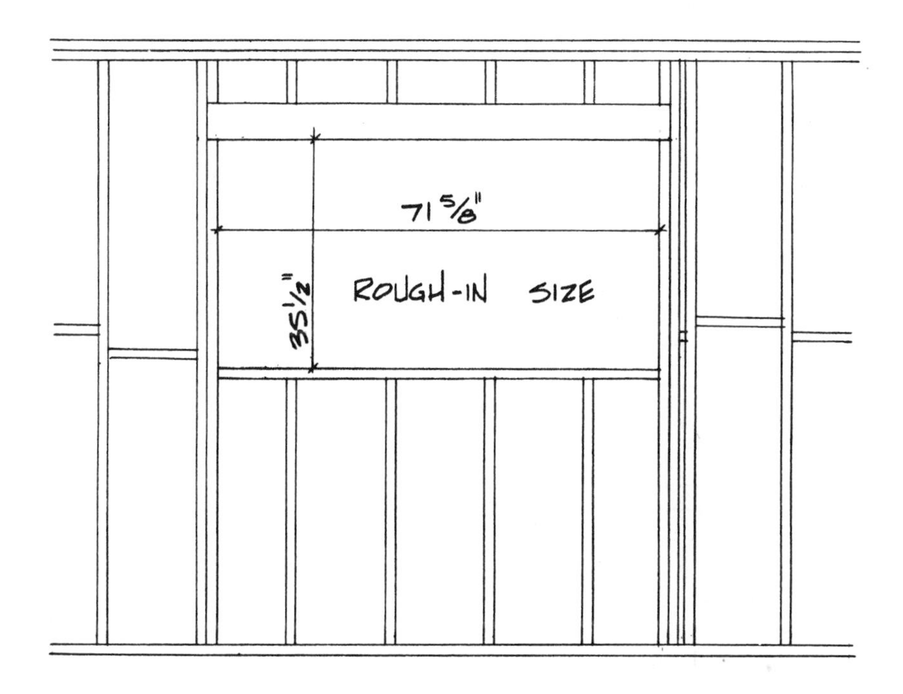

EXAMPLE:

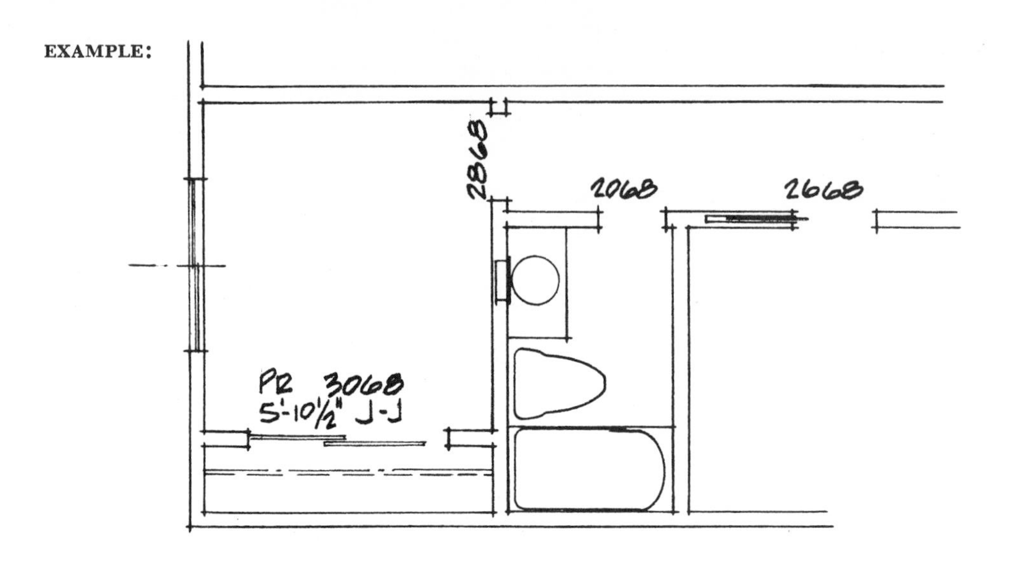

J–J means "jamb to jamb." This allows the sliding doors to lap approximately 1-1/2" at center.

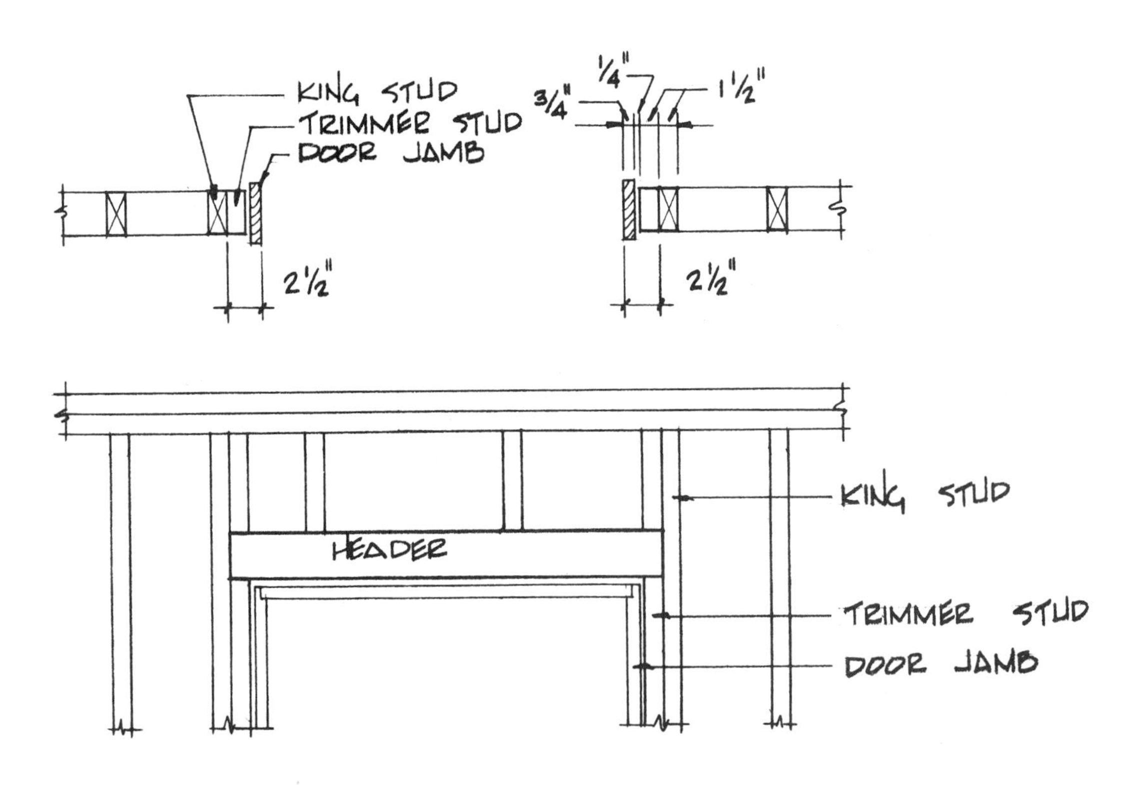

TABLE 28-1 Calculating Interior Door-Header Lengths

Interior Door Size	*Allowance*	*Actual Length*
1′-8″	+5 =	2′-1″
1′-10″	+5 =	2′-3″
2′-0″	+5 =	2′-5″
2′-2″	+5 =	2′-7″
* 2′-4″	+5 =	2′-9″
* 2′-6″	+5 =	2′-11″
* 2′-8″	+5 =	3′-1″
2′-10″	+5 =	3′-3″
3′-0″	+5 =	3′-5″
3′-2″	+5 =	3′-7″
3′-4″	+5 =	3′-9″
3′-6″	+5 =	3′-11″

* Most common interior door sizes

In this case, the rule for interior doors would be the door size plus 5 inches. This would be the exact header length.

A list of all the header lengths should be made and combined into convenient lumber lengths for the least amount of waste. Table 28-1 will help you determine the interior door-header lengths.

Some builders estimate their headers to fit just between 16″ O.C. studs and float the trimmer studs to the window rough-in width. This is illustrated in the following drawing.

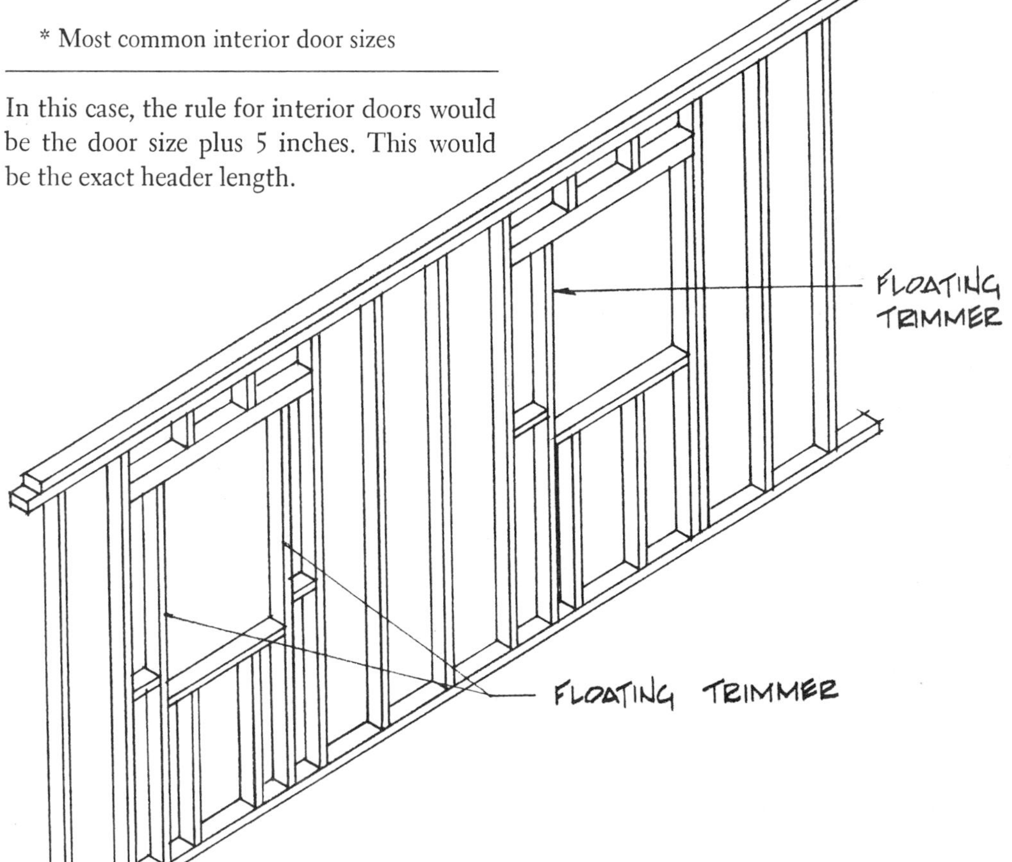

In this case, the estimator will have to figure the length of material that will fit between the studs at 16″ centers and be wide enough to span the rough-in width of the required opening.

HEADER BEAMS FLUSH

Header beams are members that support ceiling and roof loads over openings between rooms where doors are not used. These beams should be estimated along with the rest of the headers. They are designated on the floor plan as illustrated below.

Sometimes the architect may choose to designate the header sizes next to the exterior window openings as illustrated below. In this case the estimator must simply estimate the header length.

Flush beams usually carry large ceiling loads and should have good support bearing on each end as illustrated on page 159. Code requirements call for 2″ minimum bearing for beam ends, but carpenter's rule of thumb for beam bearing is full thickness of wall or 3-1/2″ as in the illustration on page 159.

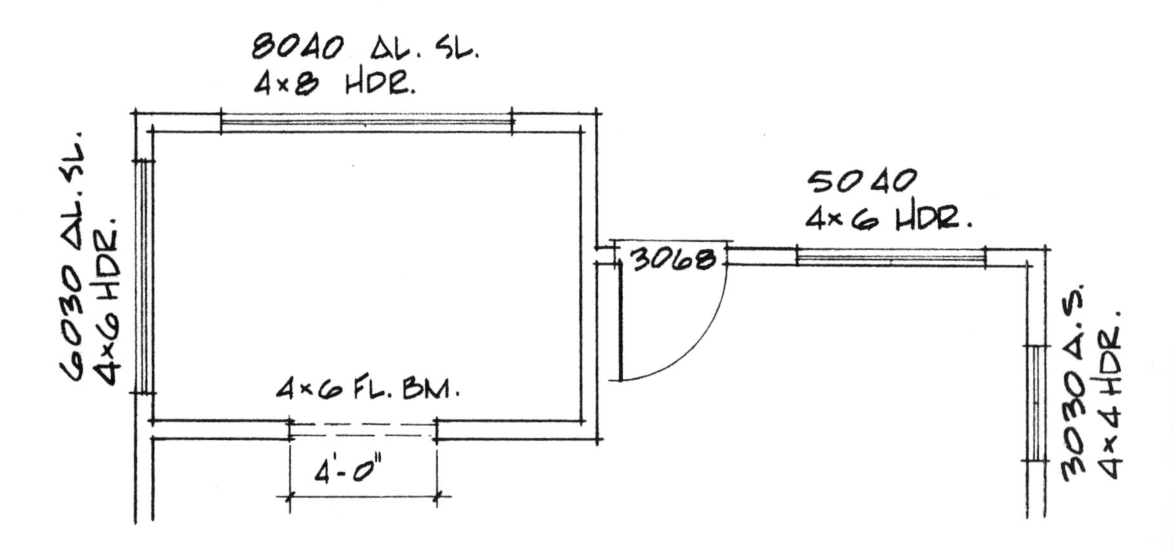

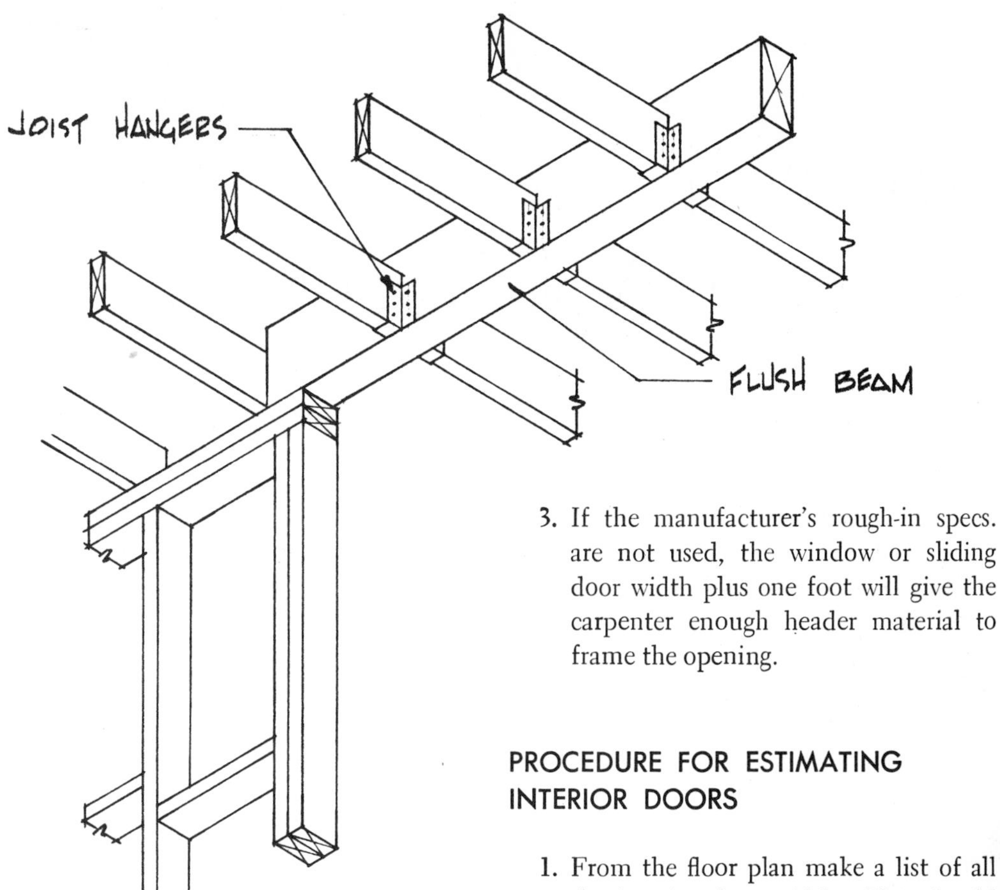

PROCEDURE FOR ESTIMATING HEADERS FOR EXTERIOR WINDOW AND SLIDING DOOR UNITS

1. From the floor plan make a list of all the exterior door and window units.
2. If using the manufacturer's rough-in specs., bear in mind that the rough-in width plus 3″ equals the actual header length.
3. If the manufacturer's rough-in specs. are not used, the window or sliding door width plus one foot will give the carpenter enough header material to frame the opening.

PROCEDURE FOR ESTIMATING INTERIOR DOORS

1. From the floor plan make a list of all the interior door widths. This should also include closet openings.
2. Actual interior door-header lengths are found by adding 5″ to the door widths as illustrated in Table 28-1.

   Note: Some estimators prefer to add one foot to each header width for the material needed. This will create more waste than the rule above.
3. All the exterior window and door headers plus interior door headers should be listed and combined into convenient lumber lengths to ensure the least amount of waste.
4. Estimate and add to the list above any header beams needed.

29

Estimating Diagonal Wall Braces

A diagonal wall brace is a stiffener brace that is placed as close to a 45° angle as possible. The general rule for diagonal wall braces is one for every exterior corner of the building and one for every 25 lineal feet of exterior wall between the corners. (In some localities diagonal braces may also be required on interior walls.) Three types of diagonal bracing methods and one alternative method are illustrated below.

1. ***1 × 4 or 1 × 6 Let-in Braces.*** The let-in brace is cut into the studs, so as to be flush with the stud edge.

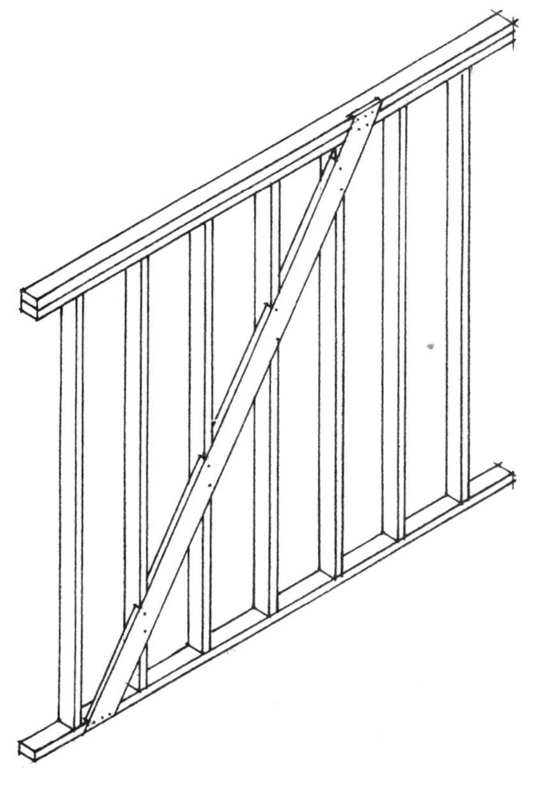

2. 2×4 *Diagonal Brace.* In some localities the code will not allow 2×4 diagonal bracing.

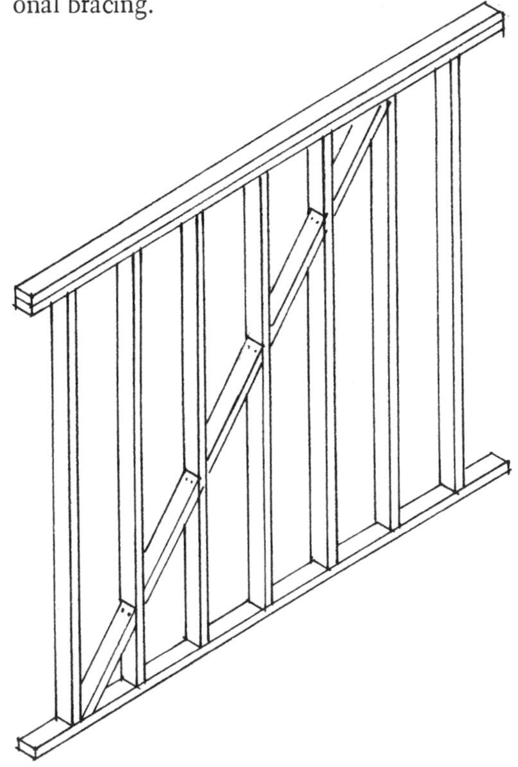

EXAMPLE 1:

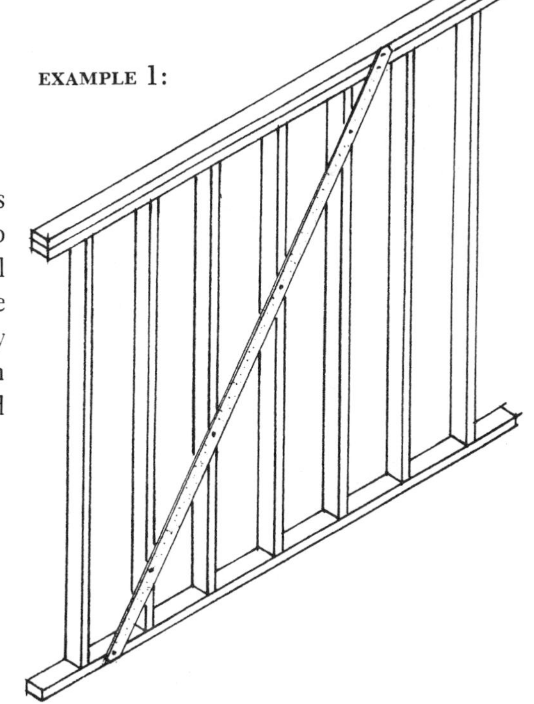

3. *Metal Diagonal Braces.* Metal straps with predrilled holes are nailed directly to the studs. In some localities one metal brace will suffice for every diagonal brace needed. In other localities the code may require that two metal braces be used in place of one wood let-in brace, as illustrated in Example 2.

EXAMPLE 2:

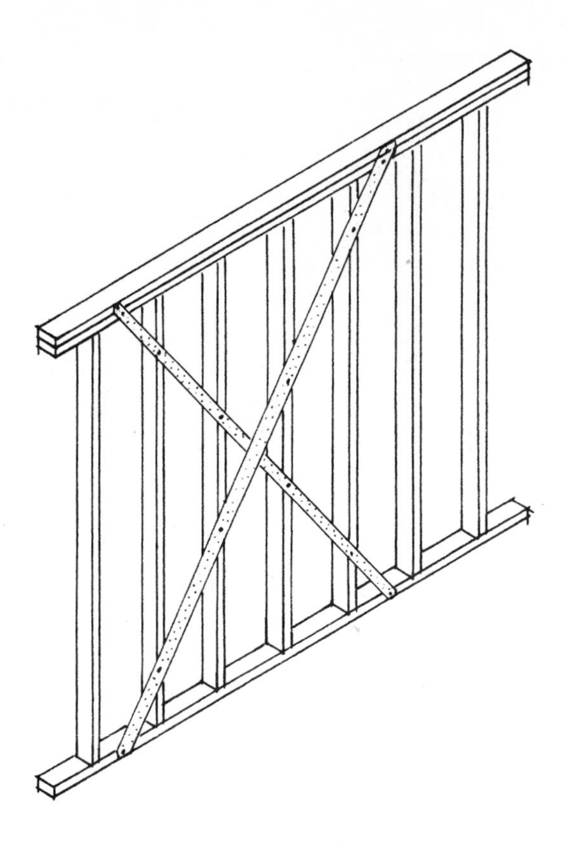

4. *Alternative System—Shear Panel.* In some cases, a diagonal brace may not be able to be placed according to code requirements. If this is so, a shear panel may be used. Shear panels are made from plywood, usually 5/16″ or thicker, and will be designated on the elevation views as illustrated below.

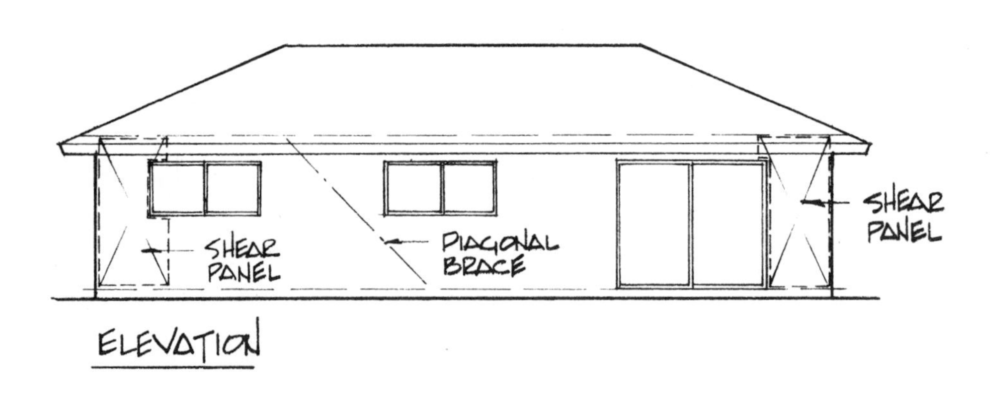

Shear Panel Detail:

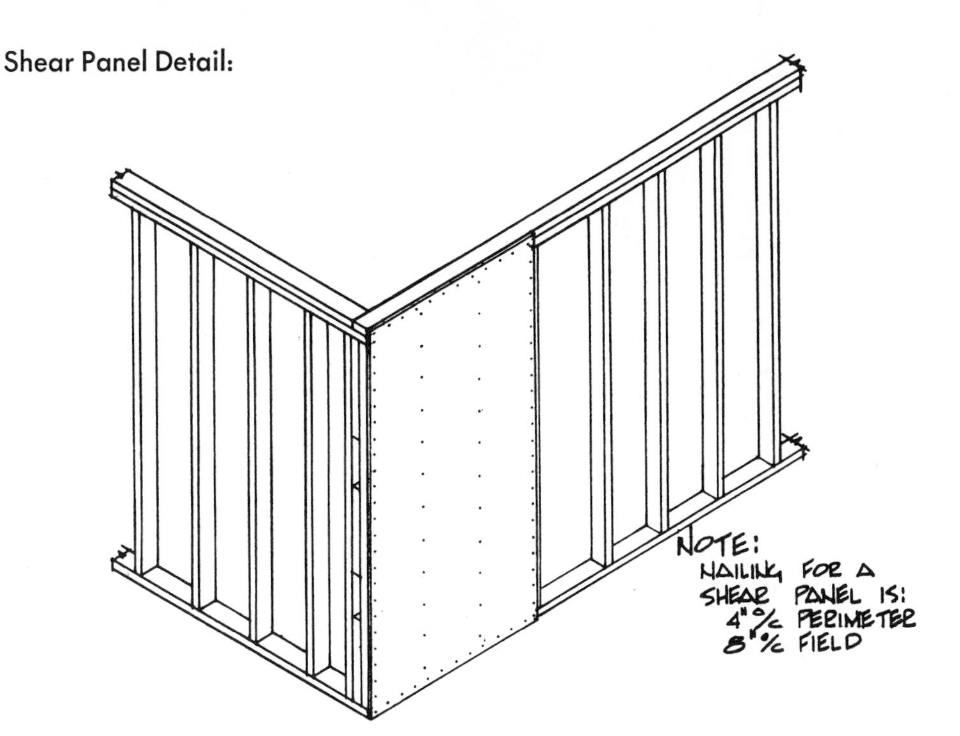

Note: Nailing for a shear panel is 4″ O.C. perimeter, 8″ O.C. field.

PROCEDURE FOR ESTIMATING WALL BRACES

Determining the length of diagonal braces is common to both methods of estimating below. The length of the brace is the diagonal measurement of a right triangle whose sides are equal to the height of the wall. This should be increased to a standard lumber length. The approximate length can also be scaled on a piece of scratch paper. For 8′-0″ high walls, a 12′-0″ piece of material is needed.

METHOD 1:

1. Many times the architect draws in the braces on the elevation views as in the previous illustration. If this is the case the number of braces may be counted from the elevation views.
2. If wall braces are also required on the interior walls, the estimator must study the floor plan to determine the number needed.
3. Check the building plans for any shear panel material needed.

METHOD 2:

1. If the diagonal braces are not designated on the elevation views, the estimator must take one brace for every wall corner and one brace for every 25 lineal feet between the corners.
2. If wall braces are also required on the interior walls, the estimator must study the floor plan to determine the number needed.
3. Check the building plans for any shear panel material needed.

30

Estimating Wallblocking

In certain areas the code will require that 2 × 4 blocks be nailed between the studs to stiffen and strengthen corners, doors, and window openings. Areas that require blocking will usually specify 3 blocks from each corner and partition and 2 blocks from a door or window opening on exterior walls. Some localities may also require blocks on interior walls. In this case, 2 blocks are usually required from door openings, cor-

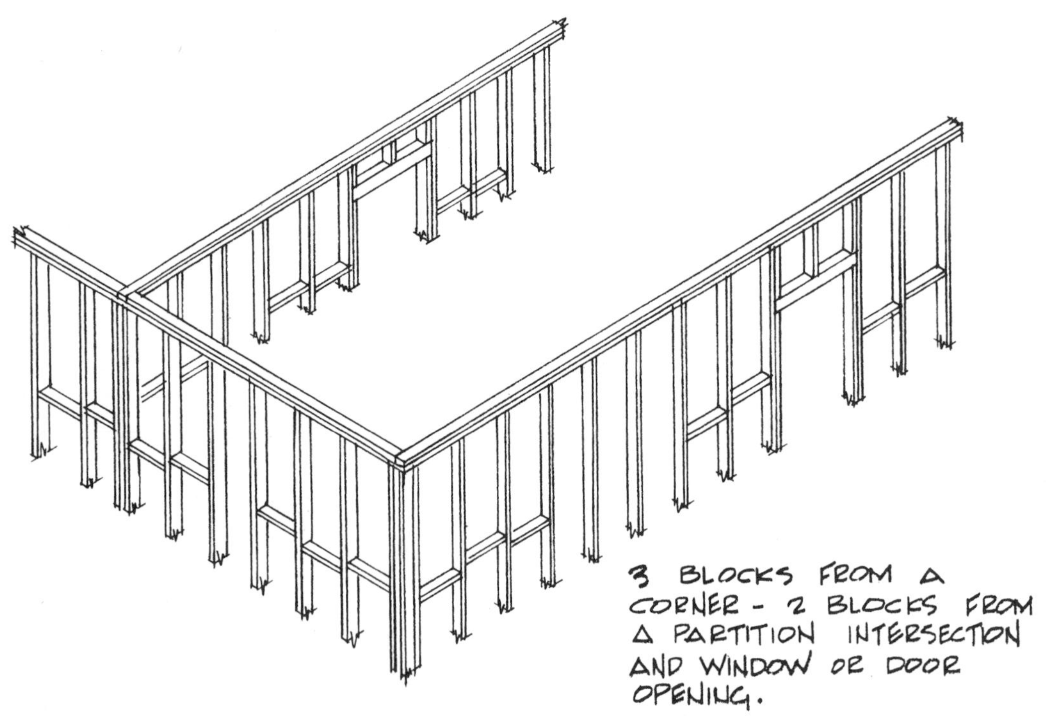

ners, and partition intersections. Code restrictions may also require continuous blocking on the lower floor of two-story construction.

PROCEDURE FOR ESTIMATING WALLBLOCKING

One method for estimating wallblocking was previously discussed in Unit 25 (wall plates). If continuous blocking is required, the estimator must multiply the lineal feet of bottom plate stock by 4. This provides material for the 3 wall plates and a row of wallblocking.

If wallblocking is only needed on exterior walls, the lineal feet of material equaling the building perimeter must be added to the material list. If wood board siding is to be used on the building exterior and applied in a vertical direction, the code will require that backing blocks be placed 2′-0″ O.C. for nailing.

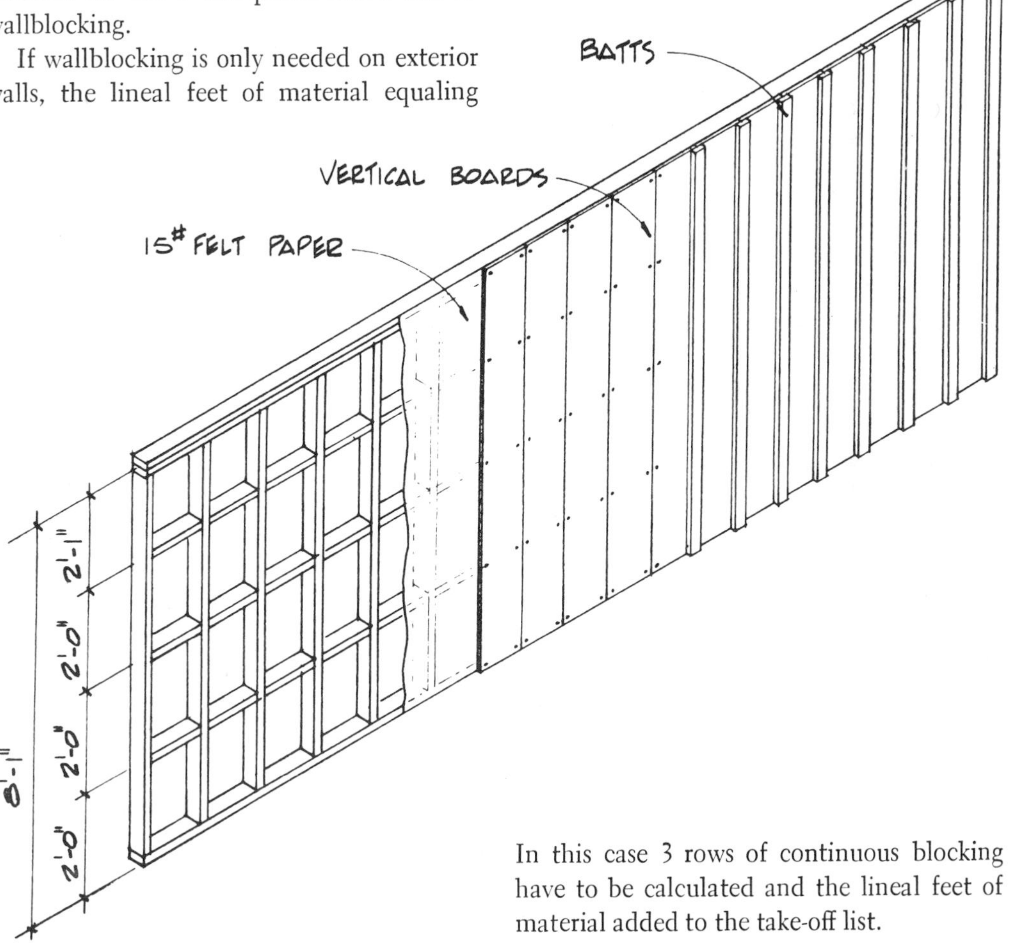

In this case 3 rows of continuous blocking have to be calculated and the lineal feet of material added to the take-off list.

31

Estimating Wall Sheathing

Exterior wall sheathing, depending both on the type and on the proper nailing of same, helps to strengthen and add rigidity to the exterior walls of a structure. It also provides the solid backing needed for applying certain types of siding materials. In colder geographical areas of the country, exterior wall sheathing is required by code and is sometimes called *wind* or *storm sheathing*. In milder climates exterior sheathing may not be required.

Following are various types of exterior wall sheathing, their applications, and estimating procedures.

BOARD WALL SHEATHING

Usually 1 × 6 or 1 × 8 size boards are used for wall sheathing. They may be square edged, or if a better wind seal is needed, shiplap or tongue-and-groove is used. They may be applied horizontally or diagonally, usually at a 45° angle as illustrated below.

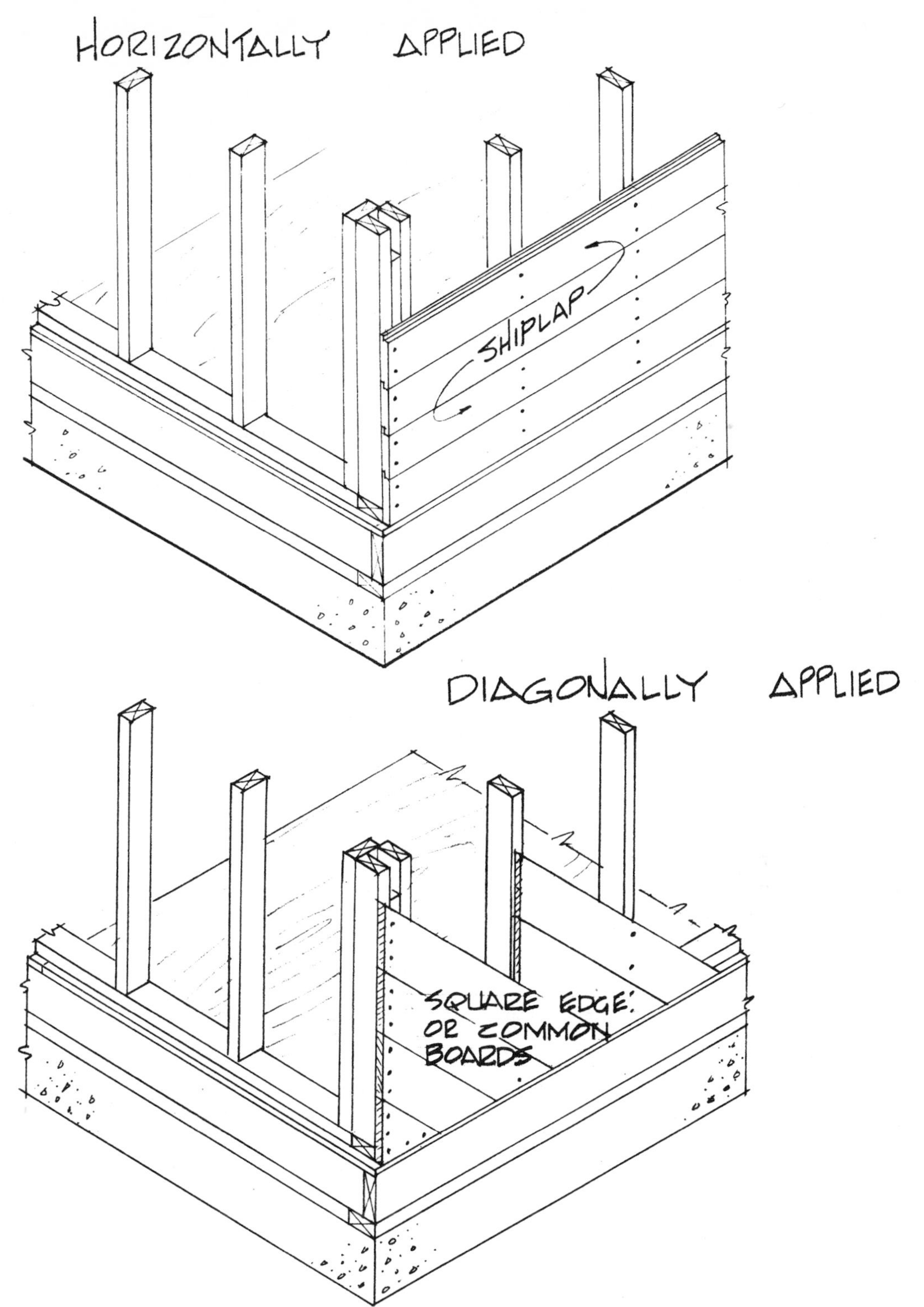
HORIZONTALLY APPLIED
SHIPLAP
DIAGONALLY APPLIED
SQUARE EDGE OR COMMON BOARDS

PROCEDURE FOR ESTIMATING WALL SHEATHING

1. Calculate the square foot area of the exterior wall to be sheathed. This may be done by multiplying the wall height by the wall width, or if all sides of the building are to be sheathed, multiply the building perimeter by the wall height and subtract the window and door areas. The result is the number of square feet of wall to be covered. (For these openings the estimator rounds off to the nearest foot. For example, a 3068 door is considered 3′ × 7′ or 21 sq. ft.)

2. Multiply the square foot area to be covered by the waste factor in Table 31-1. The result will be the number of board feet needed.

Note: The factors below take into account the size differential of boards and waste. Every estimator should adjust the waste factor according to his own framing needs.

3. To convert from board feet to lineal feet, do as follows:

   If 1 × 6 is used multiply the sq. ft. estimated by 2. (There are 2 lin. ft. of 1 × 6 in every bd. ft.) If 1 × 8 is used multiply the sq. ft. estimated by 1.5. (There are 1.5 lin. ft. of 1 × 8 in every bd. ft.)

TABLE 31-1 Estimating Wall Sheathing

Square-Edge Boards

Nominal Size	*Actual Size*	*Waste Factor: Applied Horizontally*	*Waste Factor: Applied Diagonally*
1 × 6	3/4 × 5-1/2	1.22	1.33
1 × 8	3/4 × 7-1/2	1.22	1.33

Tongue-&-Groove and Shiplap Boards

Nominal Size	*Coverage Size for Estimating*	*Waste Factor: Applied Horizontally*	*Waste Factor: Applied Diagonally*
1 × 6	3/4 × 5	1.35	1.46
1 × 8	3/4 × 7	1.32	1.43

Example Problem:

How many lineal feet of 1×8 tongue-and-groove sheathing boards applied diagonally are needed to cover 1,840 sq. ft. of wall area?

PROCEDURE:

1. Multiply the wall area (1,840) by the factor from the table for diagonally applied 1×8 tongue-and-groove boards (1.43): $1{,}840 \times 1.43 = 2{,}632$ bd. ft.
2. Multiply the bd. ft. area above (2,632) by the number of lin. ft. of 1×8 in 1 bd. ft. (1.5): $2632 \times 1.5 = 3948$ lin. ft. of 1×8 tongue-and-groove sheathing.

PLYWOOD SHEATHING

Plywood for wall sheathing may be 5/16″, 3/8″ or 1/2″ thick, and it comes in sheet sizes of 4×8, 4×9, and 4×10. It may be applied horizontally or vertically and will butt together over a stud or blocking depending on the direction applied. If properly nailed, plywood acts as shear paneling and eliminates the need for diagonal bracing. Plywood makes a good nailing base for wall shingles and similar types of wall siding.

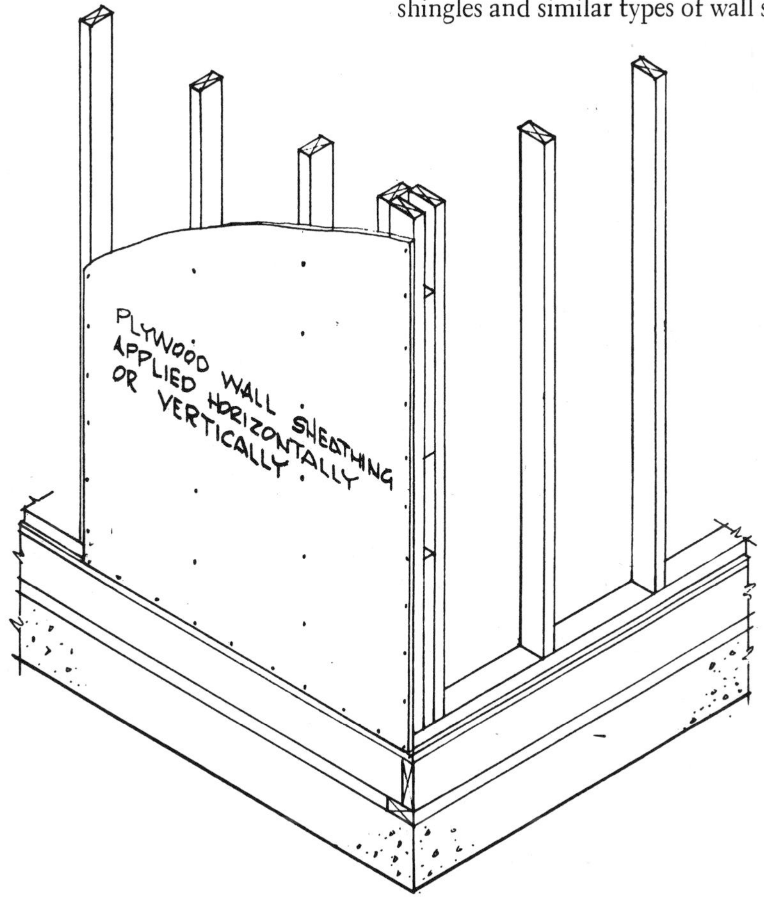

FIBER INSULATING BOARD

Fiber insulating board for wall sheathing comes in 1/2″ or 25/32″ thicknesses in either 2′ × 8′ or 4′ × 8′ sheets. The 2 × 8 sheets are applied horizontally with staggered joints. The 4 × 8 sheets are usually applied vertically, but they may also be applied horizontally. Exterior fiberboard is asphalt impregnated or coated for water resistance. This same material without the coating is sometimes used on certain interior walls and is commonly called *sound board.*

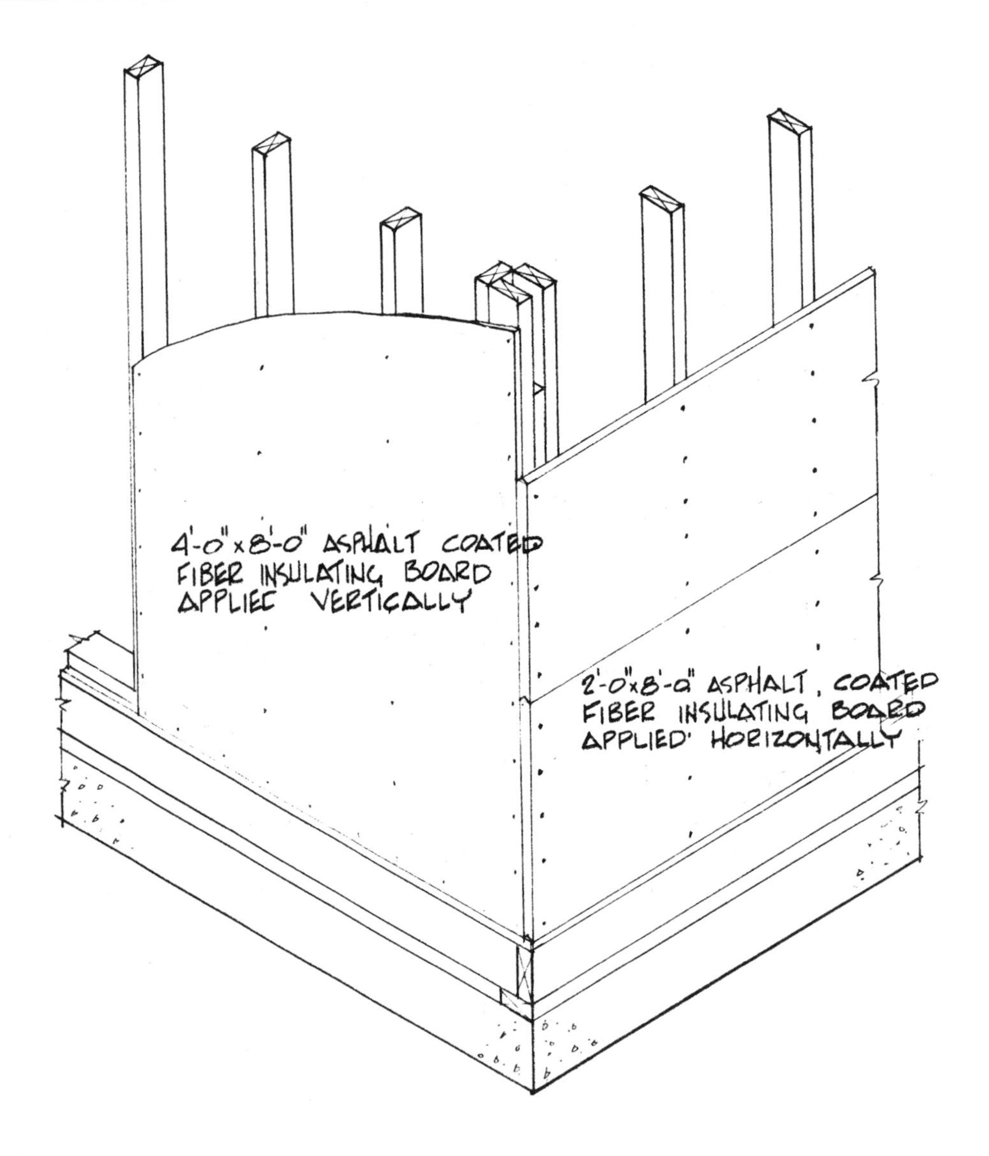

GYPSUM BOARD SHEATHING

Gypsum board comes 1/2″ thick in 2′ × 8′ sheets. It is applied horizontally with staggered joints. Gypsum or fiberboard cannot be used as a nailing base for certain siding materials. If shingles are used, nailing strips or skip sheathing must be applied over the existing sheathing as illustrated on page 325.

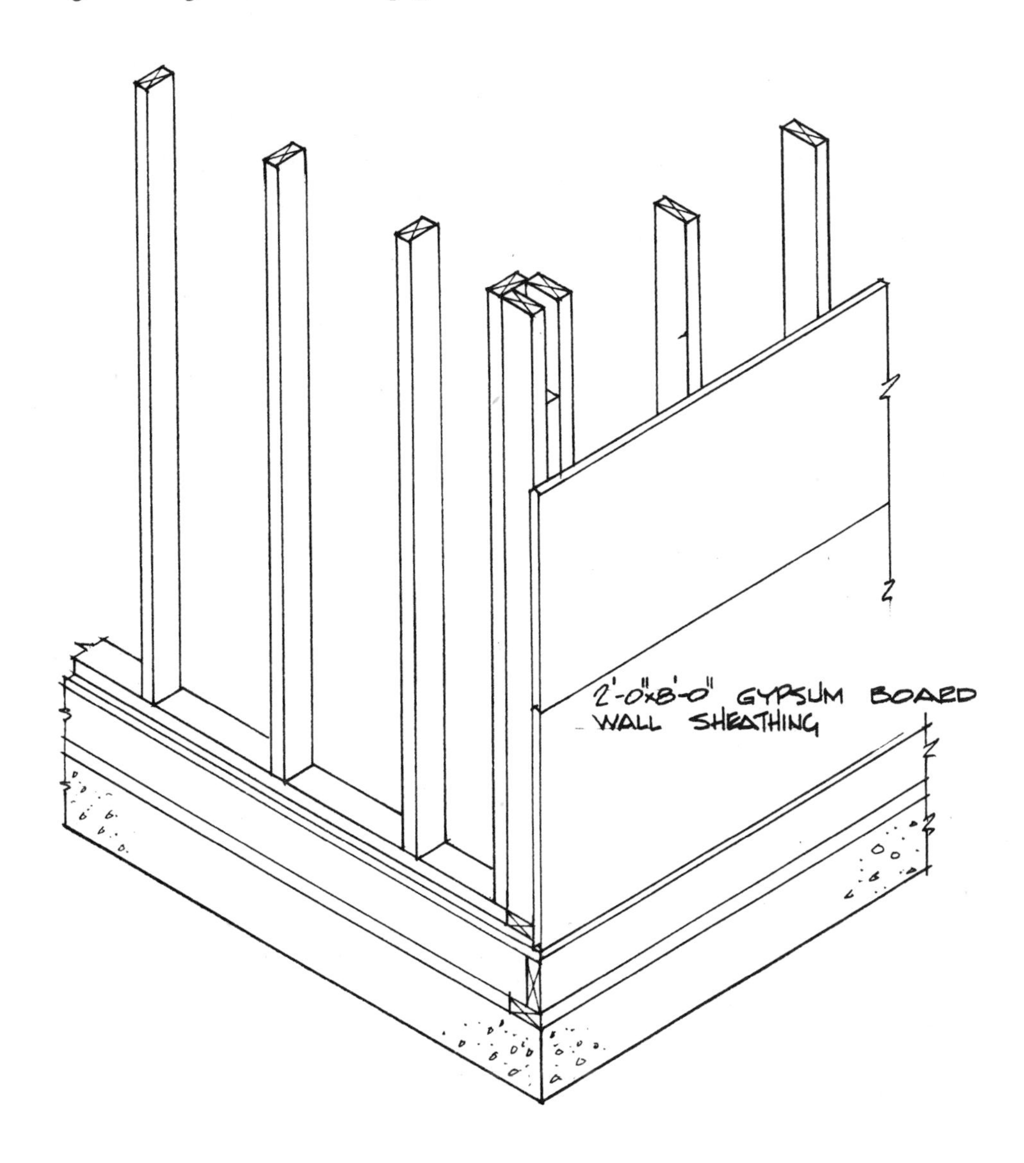

ESTIMATING PROCEDURE FOR PLYWOOD, FIBERBOARD, AND GYPSUM BOARD

1. Calculate the square foot area the exterior wall sheathing is to cover. *Do not* subtract for the window and door openings.
2. Divide the total wall area by the number of square feet in the size of materials used. The result is the number of sheets needed. (A 2 × 8 sheet is 16 sq. ft.; a 4 × 8 sheet is 32 sq. ft.)

Example Problem:

How many 2 × 8 sheets of fiberboard are needed to cover 1,620 sq. ft. of wall area?

PBOCEDURE:

1. In this problem the square foot area needed for the wall sheathing has been calculated (1,620 sq. ft.).
2. Divide the total wall area (1,620) by the sq. ft. in a 2 × 8 sheet (16).
   1620 ÷ 16 = 101.2 or 102 pieces of 1/2 × 2 × 8 fiberboard

32

Estimating Temporary Wall Bracing Material

Some material has to be added to your estimate for temporary braces needed to plumb and align the interior and exterior walls. These braces are usually 2 × 4s in 12- or 14-foot lengths. They are used to hold the walls straight and plumb until the ceiling joist and rafters are nailed. This material may be used later in the project for drop ceilings, blocking, stud backing, trimmers, etc.

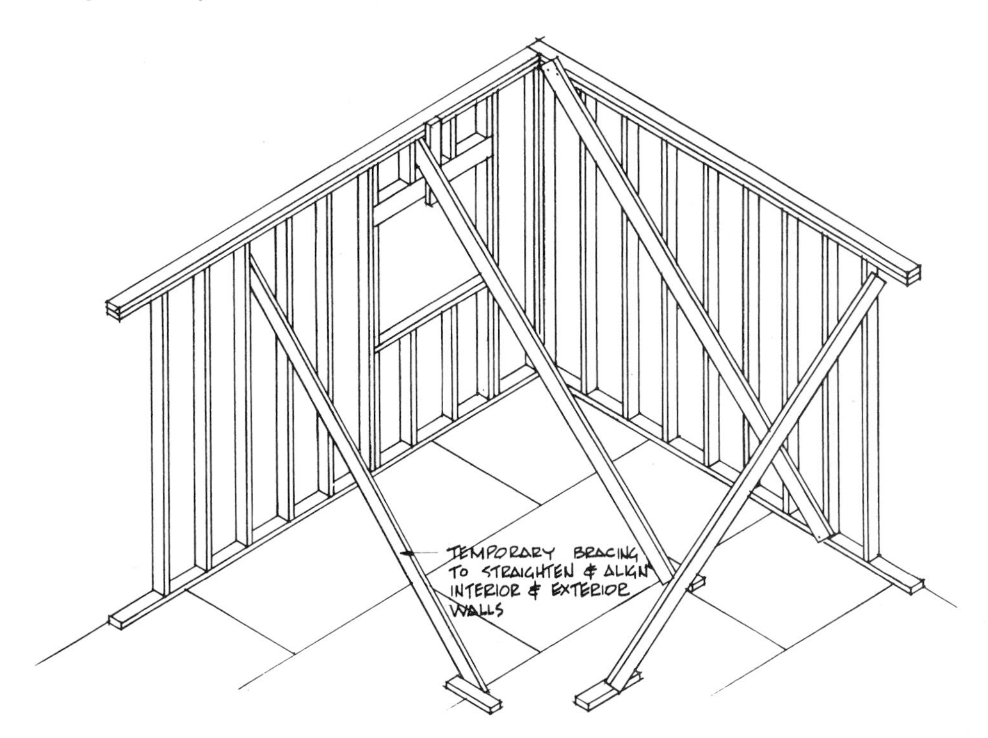

The amount of material needed for bracing is hard to estimate for a beginner, for the amount of material may vary depending on how straight and plumb the walls are. If the walls are plumb, the following floor plan has the usual placement of braces and amount of material needed:

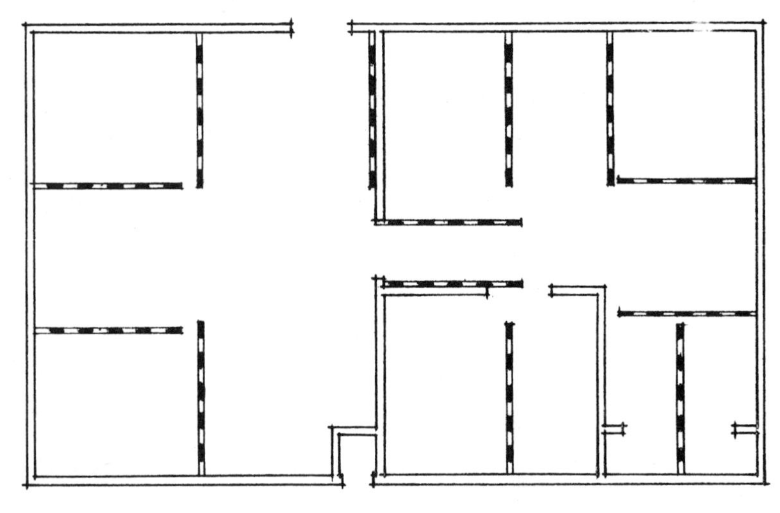

MATERIAL:
13 PCS. 2x4x12'

33

Problems—Estimating Wall Framing Material

PROBLEM 1: SLAB FLOOR

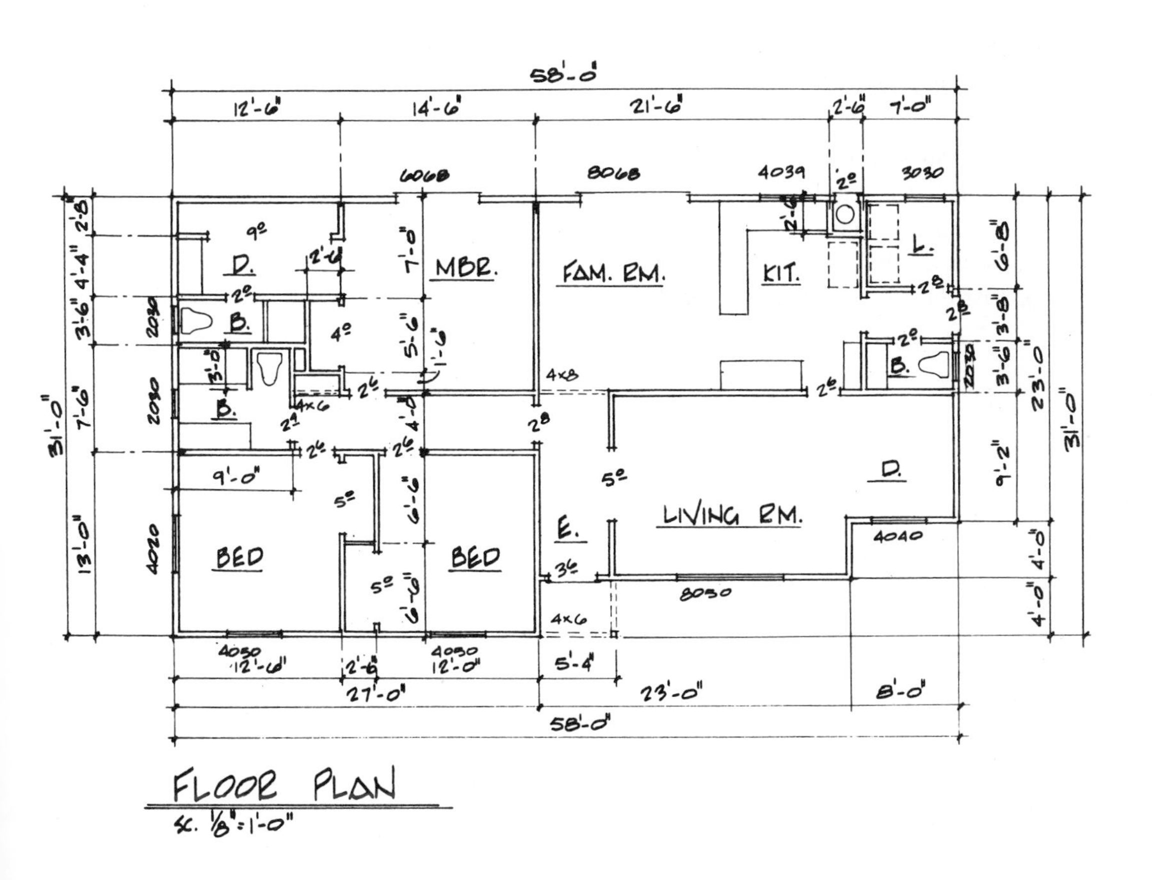

ESTIMATE THE FOLLOWING:

1. Lineal feet of treated bottom plate material ____________
2. Lineal feet of top plate material ____________
3. Number of studs ____________
4. Number of 1 × 6 × 12′ diagonal braces for exterior walls ____________
5. List the size and length of headers and beams ____________

____
____
____
____
____

PROBLEM 2: WOOD FLOOR CONSTRUCTION

ESTIMATE THE FOLLOWING:

1. Lineal feet of wall plate material ____________
2. Number of studs ____________
3. Number of 1 × 6 × 12′ diagonal braces for exterior walls ____________
4. List the size and length of headers and beams ____________

____
____
____
____
____

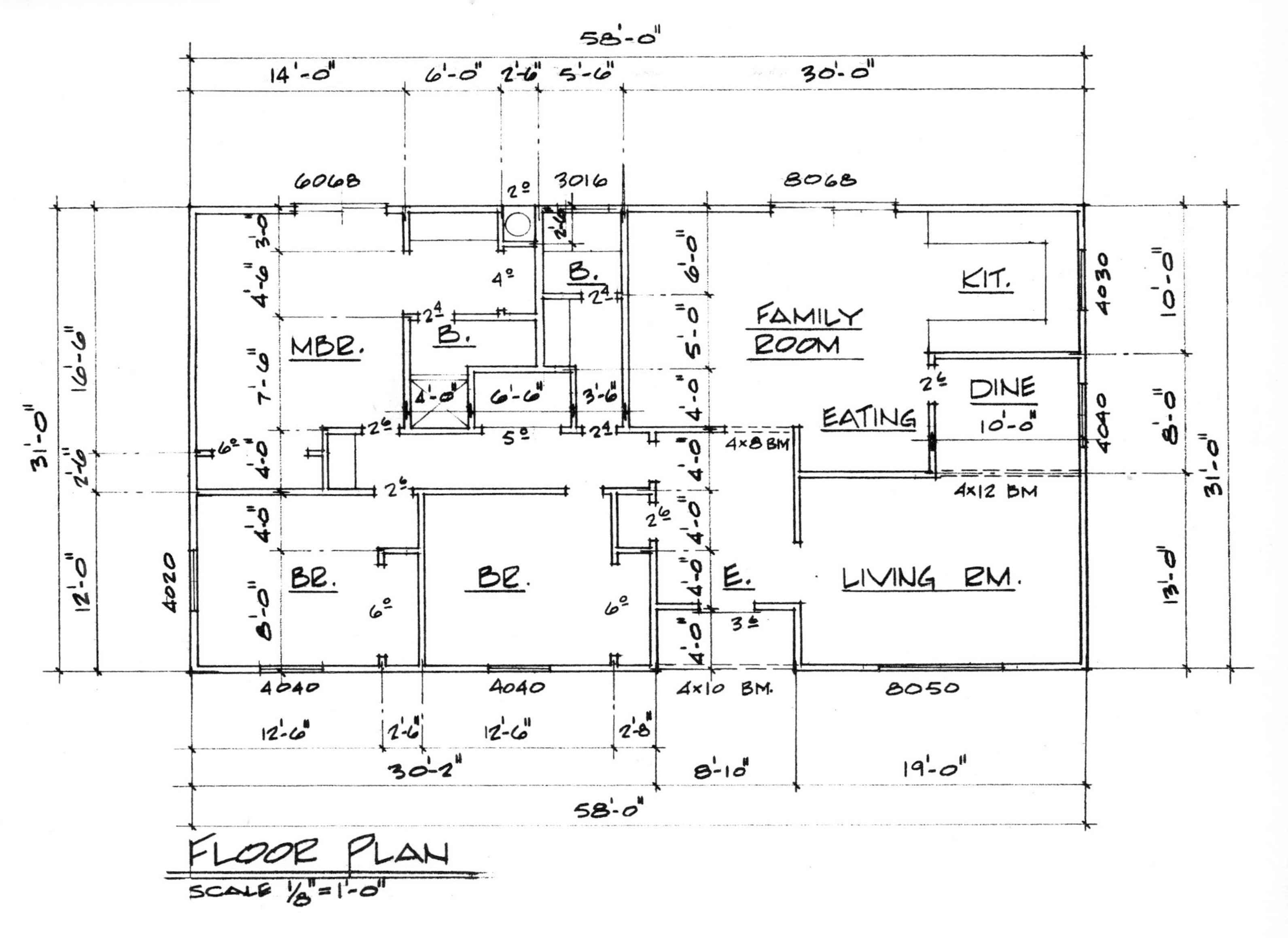
58'-0"
14'-0"
6'-0"
2'-6"
5'-6"
30'-0"
6068
3016
8068
MBR.
B.
B.
FAMILY ROOM
KIT.
DINE
10'-0"
EATING
4x8 BM
4x12 BM
BR.
BR.
E.
LIVING RM.
4020
4030
4040
31'-0"
16'-6"
2'-6"
12'-0"
10'-0"
8'-0"
13'-0"
31'-0"
4040
4040
4x10 BM.
8050
12'-6"
2'-6"
12'-6"
2'-8"
30'-2"
8'-10"
19'-0"
58'-0"
FLOOR PLAN
SCALE 1/8" = 1'-0"

PROBLEM 3: SLAB FLOOR

ESTIMATE THE FOLLOWING:

1. Lineal feet of treated bottom plate material ____________
2. Lineal feet of top plate material ____________
3. Number of studs ____________
4. Number of 1 × 6 × 12′ diagonal braces for exterior walls ____________
5. Lineal feet of 2 × 4 wall stiffener block (exterior walls only) ____________
6. List the size and length of headers and beams ____________

____

____

____

____

____

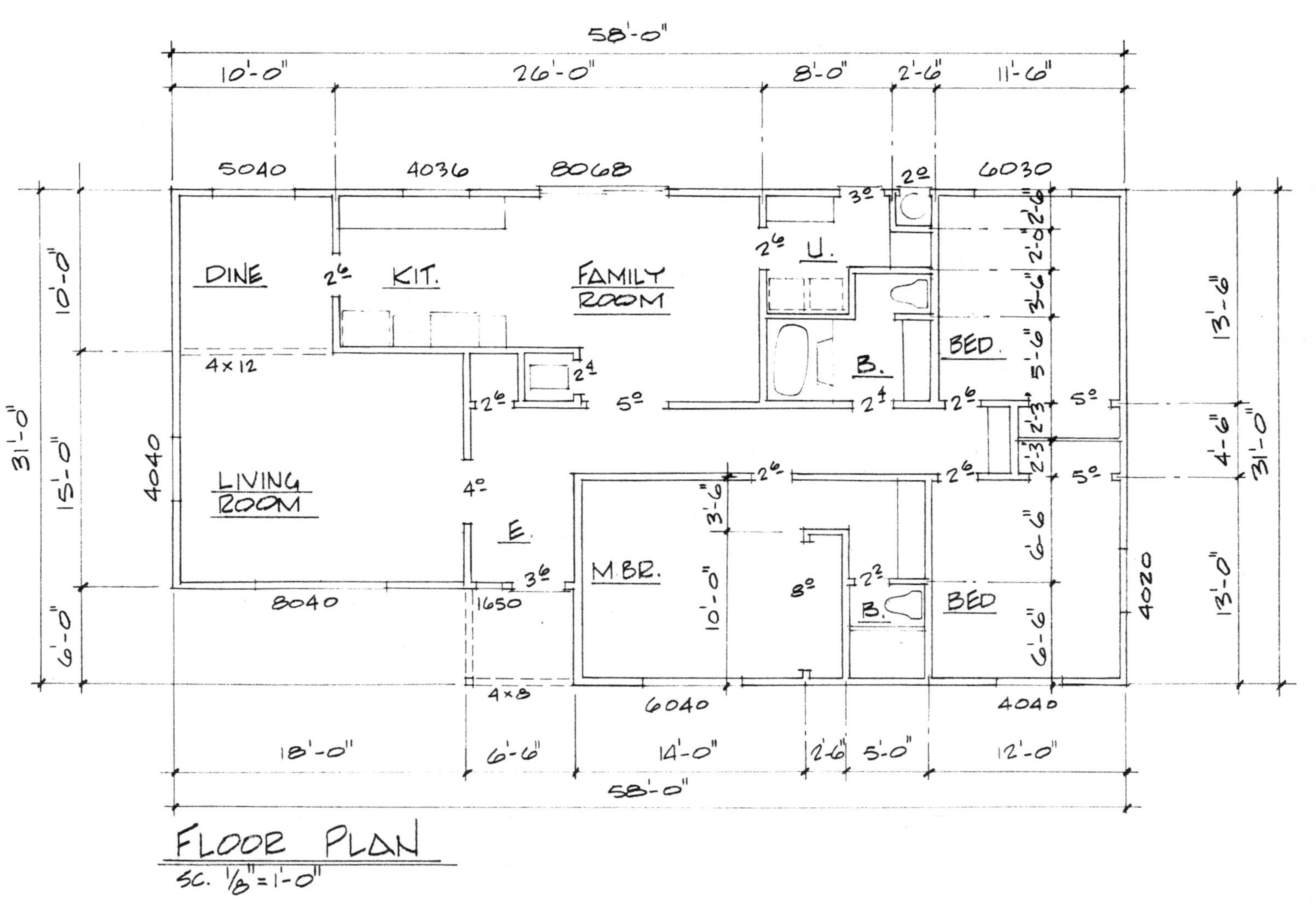

FLOOR PLAN

SC. 1/8"=1'-0"

PROBLEM 4: WOOD FLOOR CONSTRUCTION

ESTIMATE THE FOLLOWING:

1. Lineal feet of wall plate material ______________
2. Number of studs ______________
3. Number of 1 × 6 × 12′ diagonal braces for exterior and interior walls ______________
4. Lineal feet of 2 × 4 wall stiffener blocks for exterior and interior walls ______________
5. List the size and length of headers and beams ______________

____

____

____

____

____

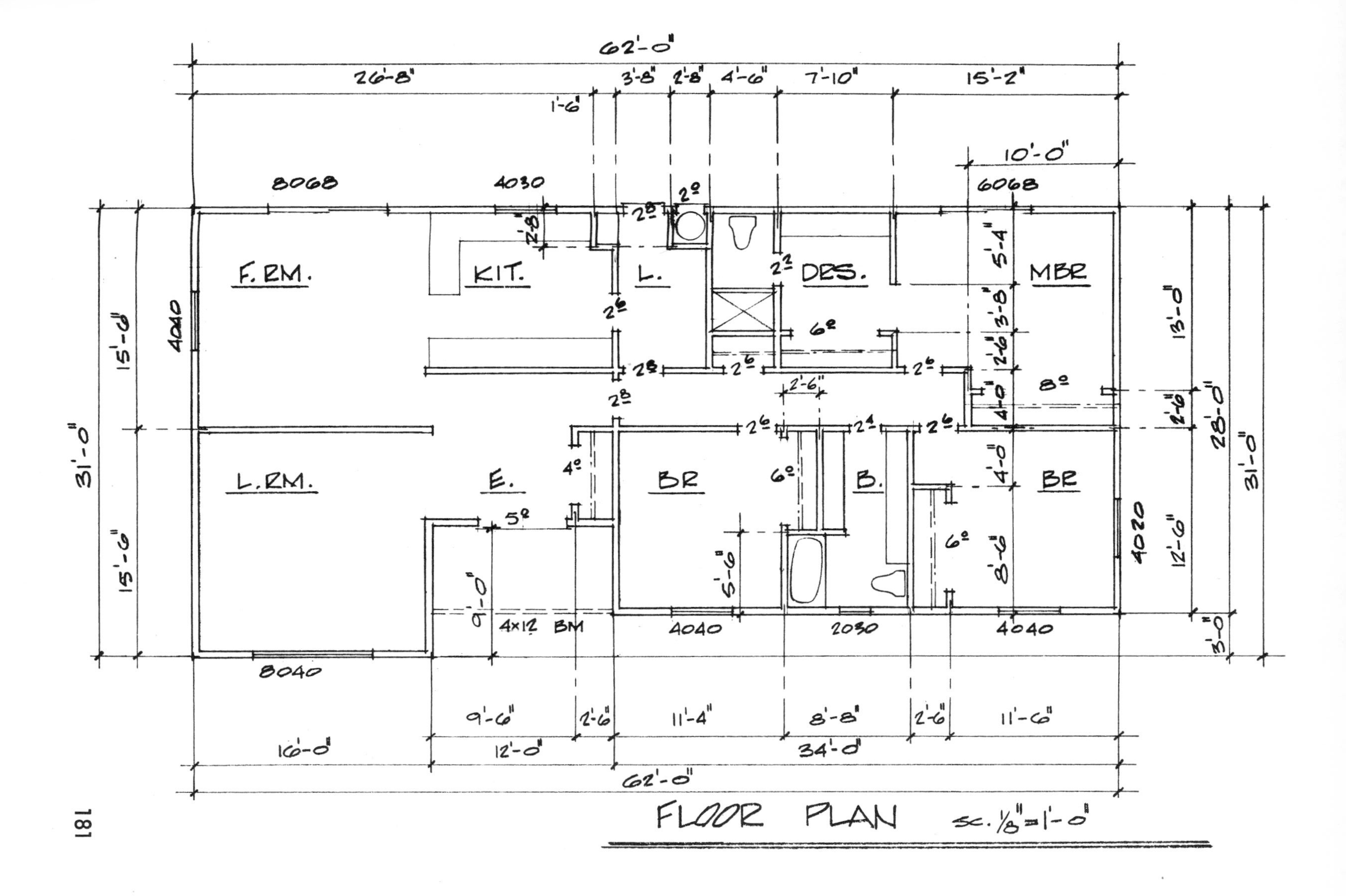
F. RM.
KIT.
L.
DES.
MBR
L. RM.
E.
BR
B.
BR
4x12 BM
FLOOR PLAN
SC. 1/8"=1'-0"

V
CEILING FRAMING

34 Estimating Ceiling Joists

The size and spacing of the ceiling joist may be found on the floor plan. It is usually designated the same as the floor joist:

2 × 6 16″ O.C.

⟵⟶

Ceiling Joist

The length of the ceiling joist will be determined by checking the floor plan. On small buildings, the joist may span from one exterior wall to another. On larger buildings it may be spliced or lapped once or twice on interior bearing walls as illustrated below.

Note: To find the bearing walls, check the foundation plan to find out under which wall the girder or support beams run.

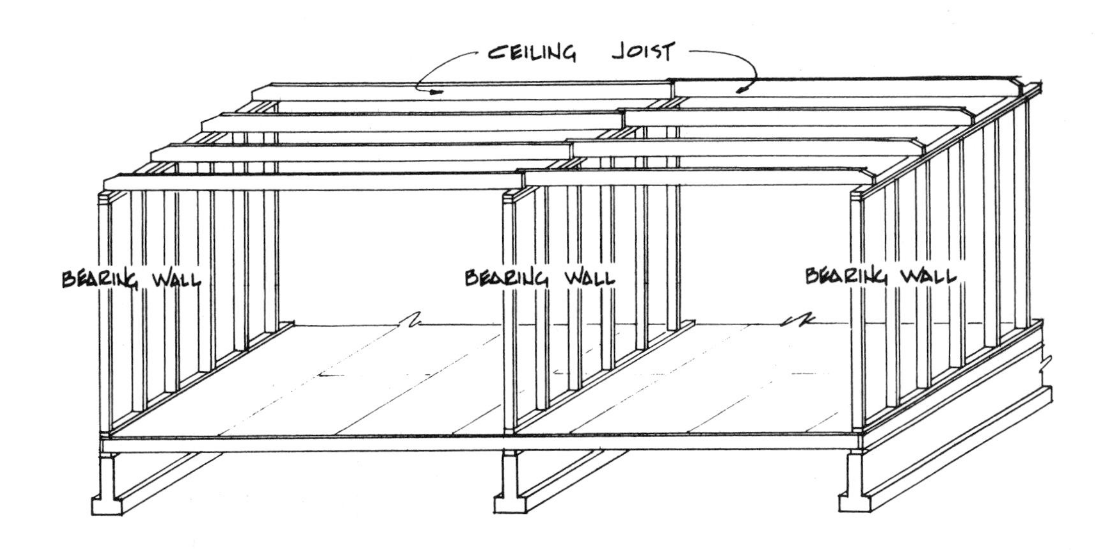

The procedure for estimating the material for ceiling joist is the same whether the building has a gable or hip roof. The estimator estimates the ceiling joist material for a hip roof as if it were a gable roof. In other words, the amount of ceiling joist material needed for both roofs is the same, but the joist layout of the two roofs is different as illustrated below:

CEILING JOIST LAYOUT FOR GABLE ROOF

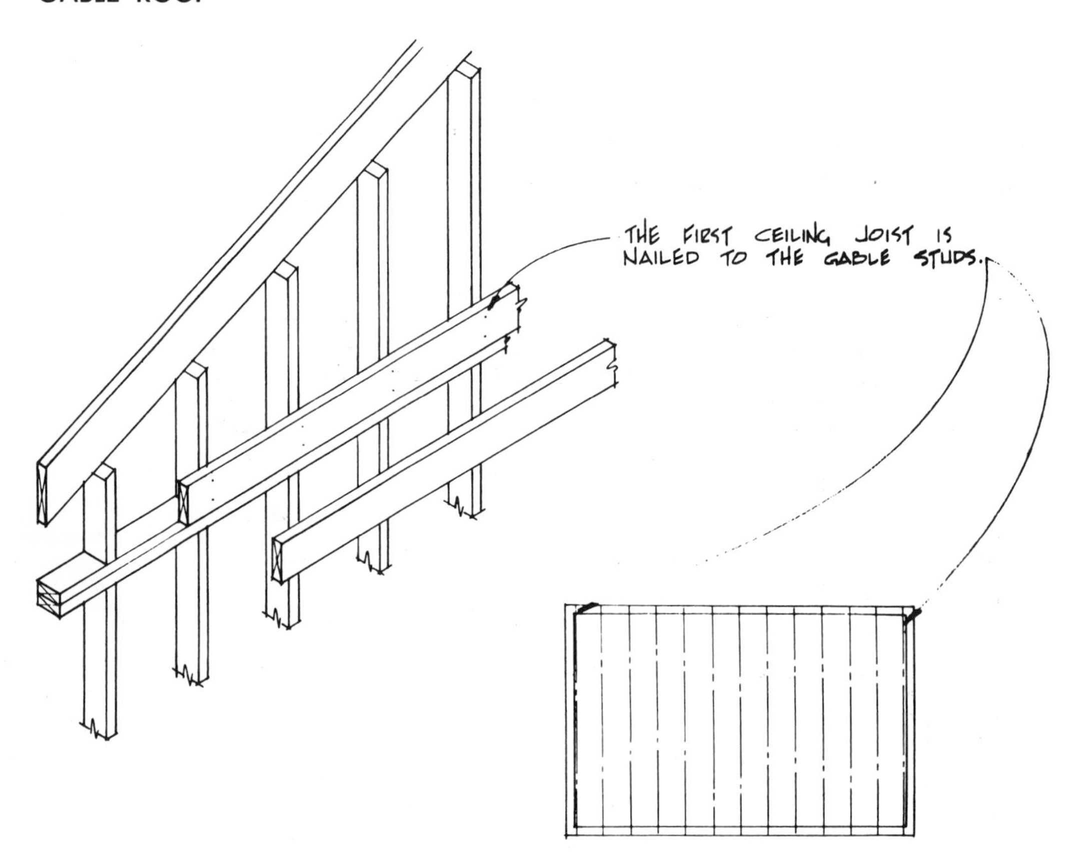

CEILING JOIST LAYOUT FOR HIP ROOF

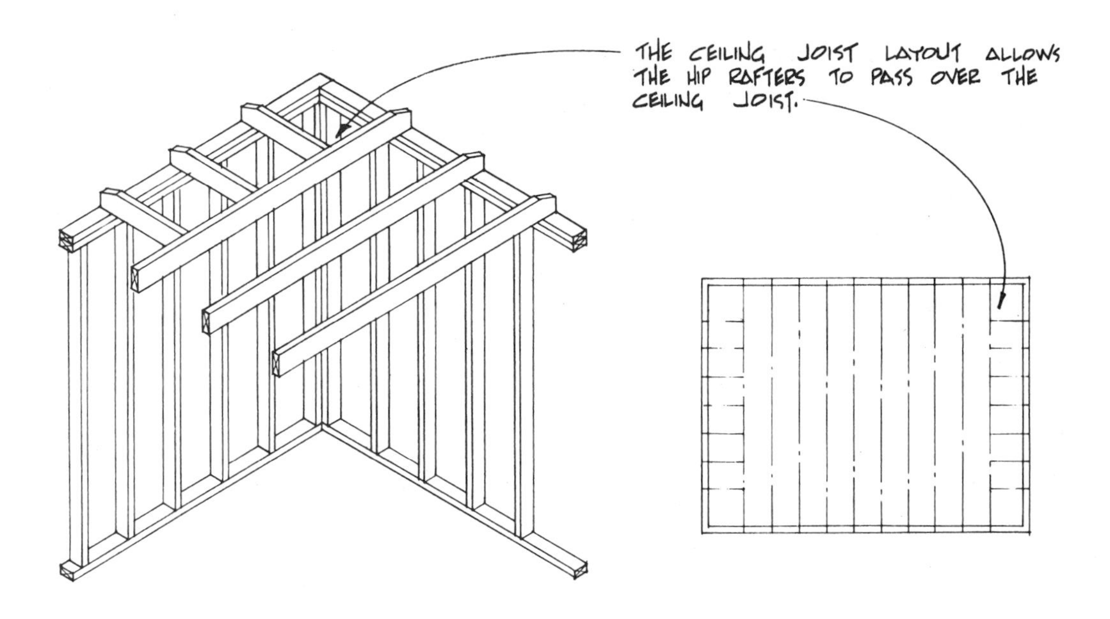

Once the size and spacing of the ceiling joist has been established, the center-to-center spacing table may be used to estimate the number of joists needed.

TABLE 34-1 CENTER-TO-CENTER CEILING JOISTS

Joist Spacing	*Constant*	*Decimal Equivalent*
12″ O.C.	1	1.00
16″ O.C.	3/4	.75
18″ O.C.	2/3	.6667
20″ O.C.	3/5	.60
24″ O.C.	1/2	.50
32″ O.C.	3/8	.375
36″ O.C.	1/3	.3333

PROCEDURE FOR ESTIMATING CEILING JOIST

1. Determine from the building plans the length of wall, in lineal feet, on which the ceiling joists rest.
2. Multiply the length by the spacing constant from the table above. Then add one.
3. If the joist spans the full width of the building, the figure from the calculation above will result in the number of ceiling joists needed. If the joists are lapped on a bearing wall and the wall falls in the center of the building, the number of ceiling joists estimated in step two above should be multiplied by two.

4. If the ceiling joists are lapped on a bearing wall and the bearing wall does not fall directly in the center of the building, the size of material, the length of material, and the number of joists per section will have to be calculated separately as illustrated in the joist plan below:

CEILING JOIST LAYOUT

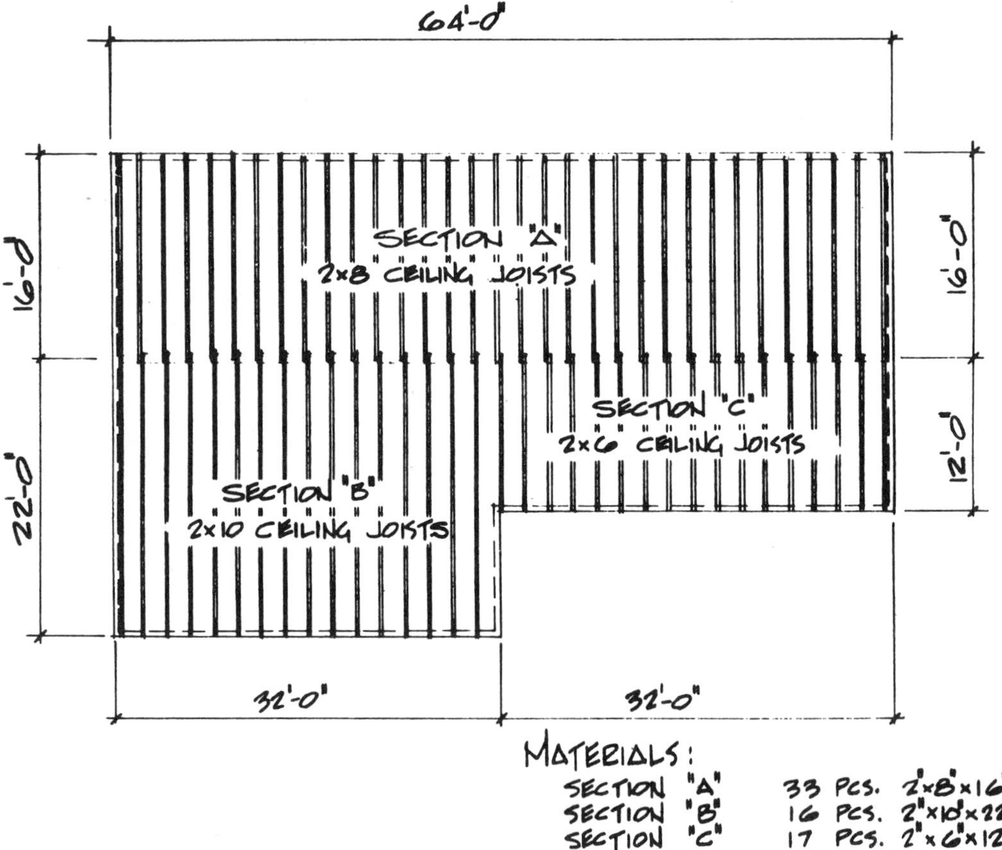

Note: Check for any additional ceiling joists that may appear on the plan.

35

Estimating Ceiling Backing

Ceiling backing provides nailing support for wallboard or plaster lath next to walls that run parallel to the ceiling joist.

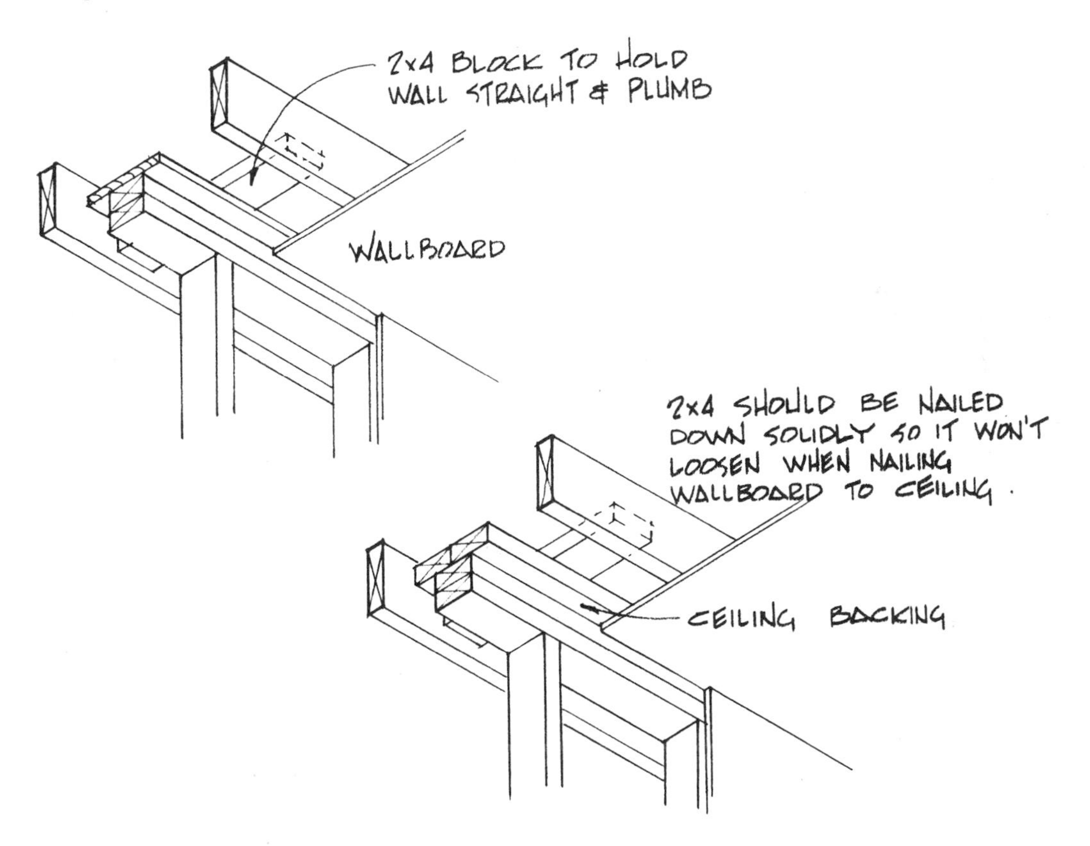

PROCEDURE FOR ESTIMATING CEILING BACKING

1. From the floor plan, calculate the lineal footage of all the walls running parallel to the ceiling joist.
2. If 1 × 8 or 2 × 8 material is being used for ceiling backing, the lineal feet calculated above will be the amount needed. If 2 × 4 material is used, the lineal feet calculated above will have to be multiplied by two to determine the material needed.

Note: Some lineal footage may be subtracted if walls running parallel to the ceiling joists are soffited. The carpenter will also try to save backing material by moving ceiling joists that happen to fall directly on top of parallel running walls to one edge of the wall as illustrated below.

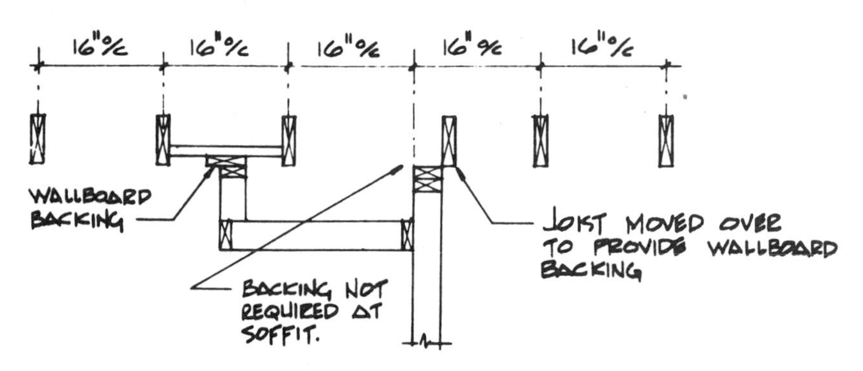

Following is a floor plan and a list of 2 × 4 backing material needed for those walls running parallel to the ceiling joist.

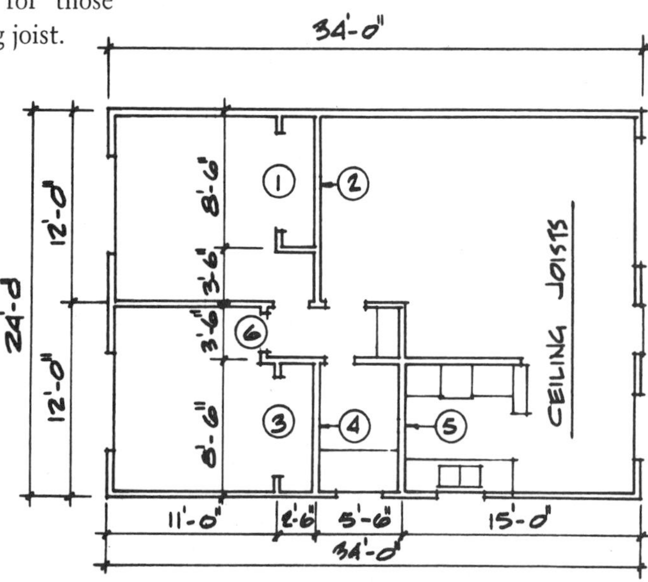

Ceiling backing material:

Walls 1, 3, 4	= 3 pcs. 2 × 4 × 18′
Walls 2, 5	= 4 pcs. 2 × 4 × 12′
Wall 6	= 1 pc. 2 × 4 × 8′

36

Estimating Stay Lath

A stay lath is a piece of lumber, usually 1 × 6″ or 2 × 4″ material, that is used to space, align, or stiffen the ceiling joist. It spaces the ceiling joists on the prescribed centers (for example, 16″ O.C., 24″ O.C.); it aligns the ceiling joists—not only holding them in a vertical position, but also straightening them from end to end; and it stiffens the joist system by fastening each joist together at mid-span as illustrated below.

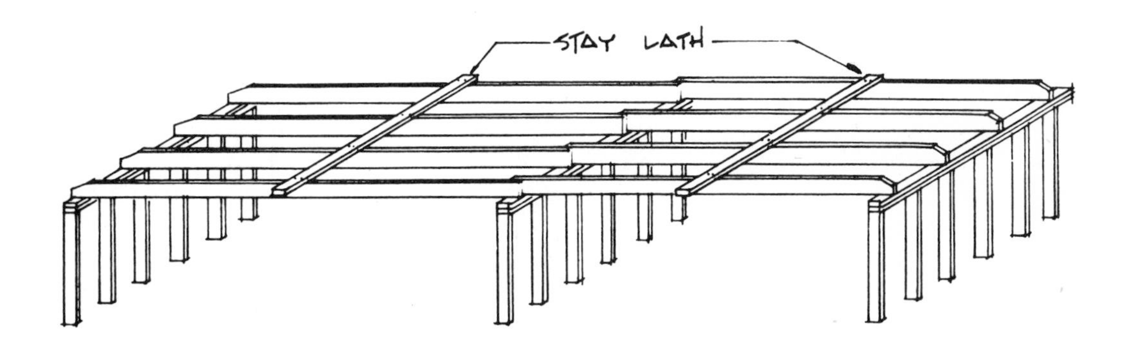

Two-by-four material is used for best results. It is stiff enough when nailed to the ceiling joist to draw the crowned joist down and raise the sagging joist up, giving the ceiling a straighter line before receiving wallboard or lath.

PROCEDURE FOR ESTIMATING STAY LATH

For a straight gable or hip roof, the lineal feet of stay lath material needed will usually be two times the length of the building. For more complicated roofs, the approximate amount may be estimated by studying the floor plan.

37

Estimating Purlins, Ridge Props, and Collar Ties

Local code restrictions may require the use of any one or a combination of purlins, ridge props, and collar ties.

PURLINS

Purlins are used to provide support for rafters. The use of a purlin cuts down the rafter span, allowing the use of smaller size rafter material, greater center-to-center spacings, or both. The number of 2″ × 4″ purlin braces needed to support a purlin depends on the size of the purlin. Following are the usual code requirements.

2″ × 4″ purlin	purlin braces	4′-0″ O.C.
2″ × 6″ purlin	purlin braces	6′-0″ O.C.
2″ × 8″ purlin	purlin braces	8′-0″ O.C.

The following diagram illustrates the method and use of a purlin.

Note: Purlins are placed at a 45° angle from the ceiling joist.

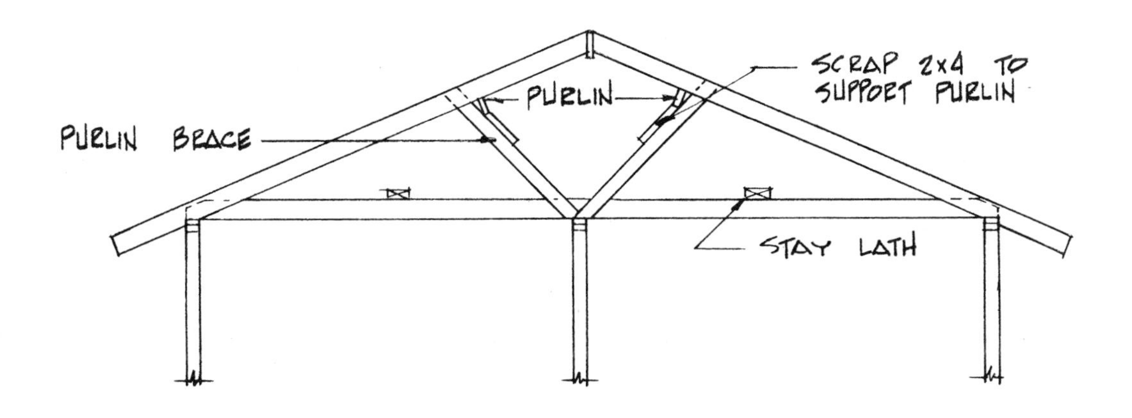

PROCEDURE FOR ESTIMATING PURLIN MATERIAL

1. Determine the size of the purlin material. This is usually found on a section view of the building plans.
2. Calculate the lineal feet of purlin material needed. For a straight gable roof, the lineal footage needed will be two times the length of the building (one length for each side of the roof). Not as much purlin material will be needed for a hip roof because of the slope on the ends of the roof. In such cases the estimator will have to approximate the amount needed.
3. Calculate the lineal feet of 2 × 4 purlin brace material. The approximate length of the purlin brace may be scaled from the section view or an elevation view. The number of these braces needed may be found by dividing the length of the purlin material by the purlin brace spacing.

RIDGE PROP

A ridge prop is a 2″ × 4″ piece of material running from the top plate of a bearing wall to the bottom edge of the ridge board. The ridge prop helps support the ridge and is usually placed 8′-0″ O.C. or less.

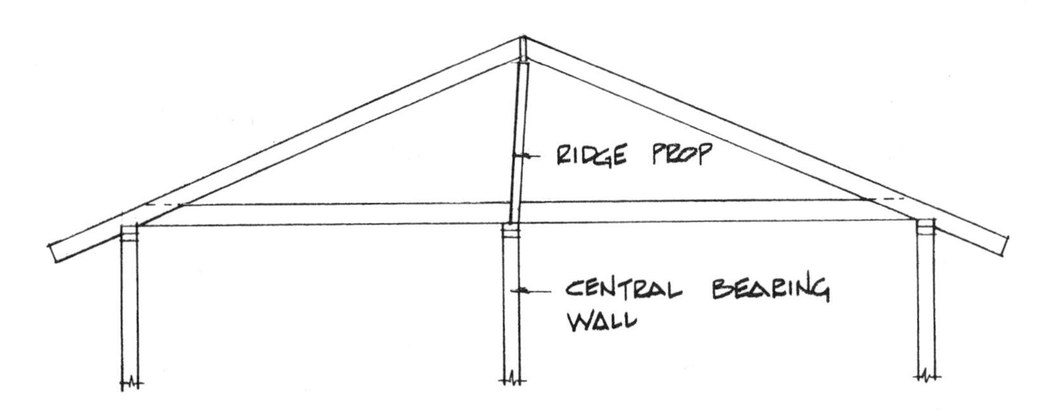

COLLAR BEAMS OR COLLAR TIES

Collar beams run in a horizontal direction between a set of common rafters and are usually 1″ × 6″ or 2″ × 4″ material. Collar ties help stiffen and hold the ridge and rafters from spreading under roof loads. The architect may refer to the size and spacing of collar ties in a note or on a section view of the plans. He may also choose to leave them off the plans and rely on the contractor to provide them according to local code requirements.

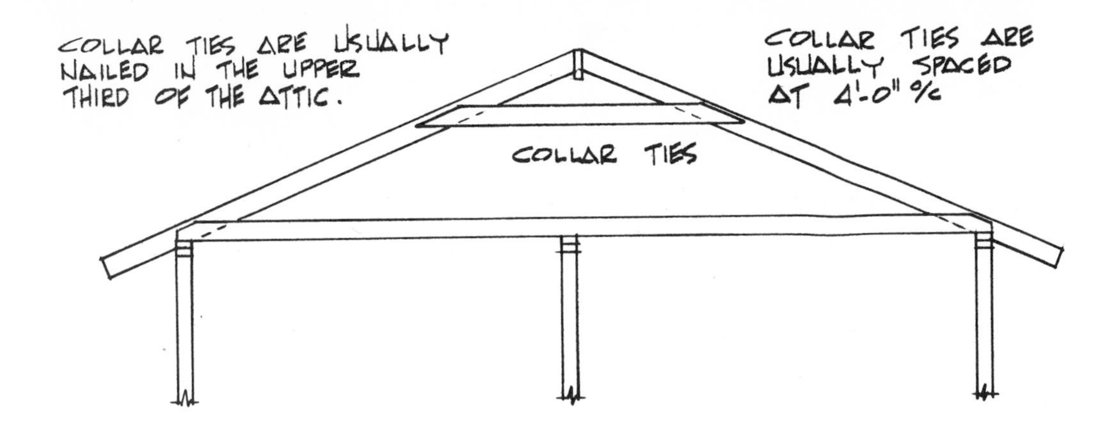

The amount of material needed for purlins, ridge props, and collar ties will depend on local code requirements, and the estimator should not forget them on the material list.

PROCEDURE FOR ESTIMATING RIDGE PROPS

1. Determine the length of the ridge prop by scaling a section or elevation view of the building plans.
2. To calculate the number of props, divide the length of the ridge by the ridge prop spacing. The result will be the number needed.

PROCEDURE FOR ESTIMATING COLLAR TIES

1. The approximate length of a collar tie may be scaled from a section or elevation view of the building plans.
2. To calculate the number of collar ties, divide the length of the ridge by the collar tie spacing. The result will be the number needed.

Note: For a hip roof the estimator may add a few extra lengths of collar tie material. These extra pieces may be used beyond the ends of the ridge board and will connect a pair of jack rafters.

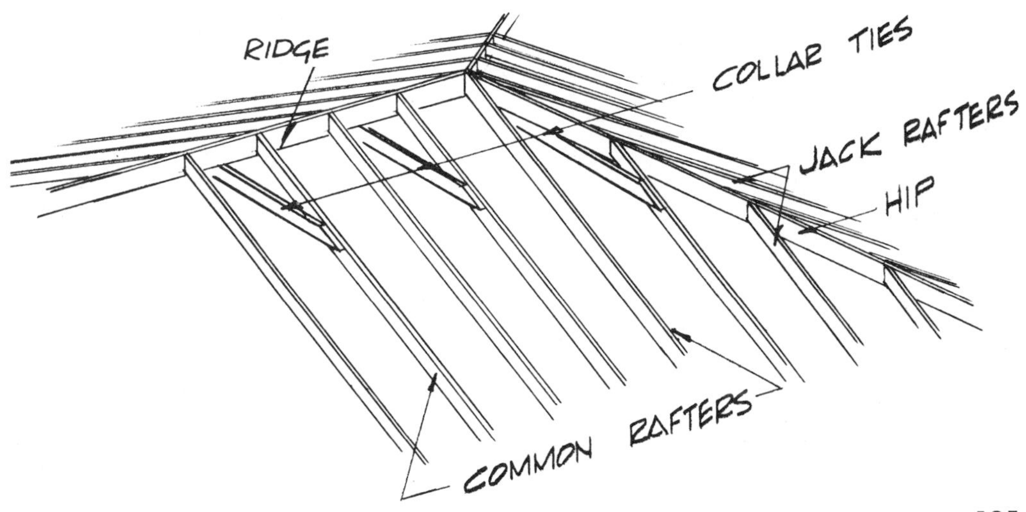

38

Estimating Dropped or Soffited Ceilings

A dropped or soffited ceiling is one that has been lowered for aesthetic or psychological reasons. Some people just like the way dropped ceilings look; to others lowered ceilings may convey a close or warm feeling. Dropped ceilings also make rooms and hallways appear wider. Owing to code requirements, most dropped ceilings are lowered 6″ to a ceiling height of 7′-6″. Lowered ceilings may be shown on the plans in the following way.

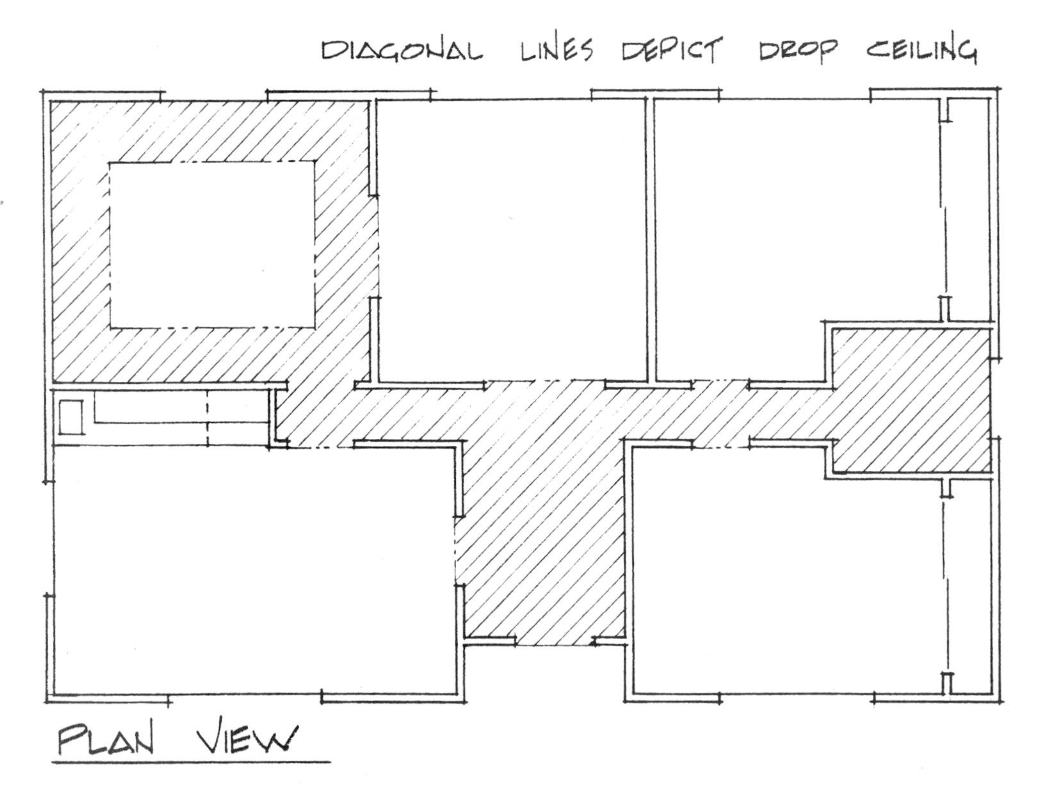

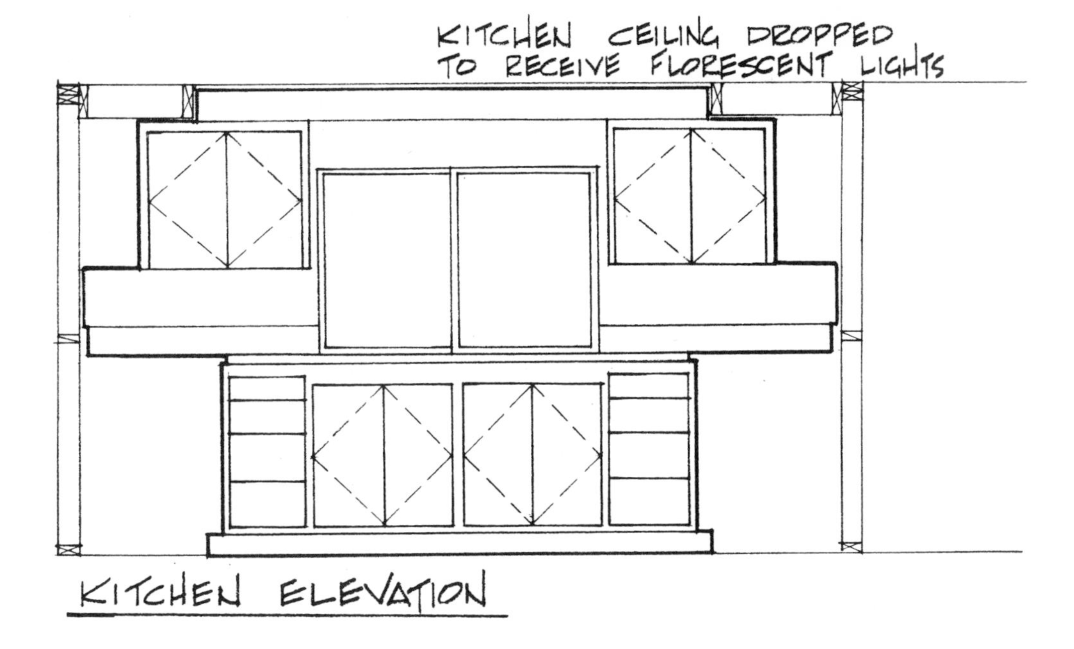

Some of the basic methods of framing dropped or soffited ceilings are detailed below.

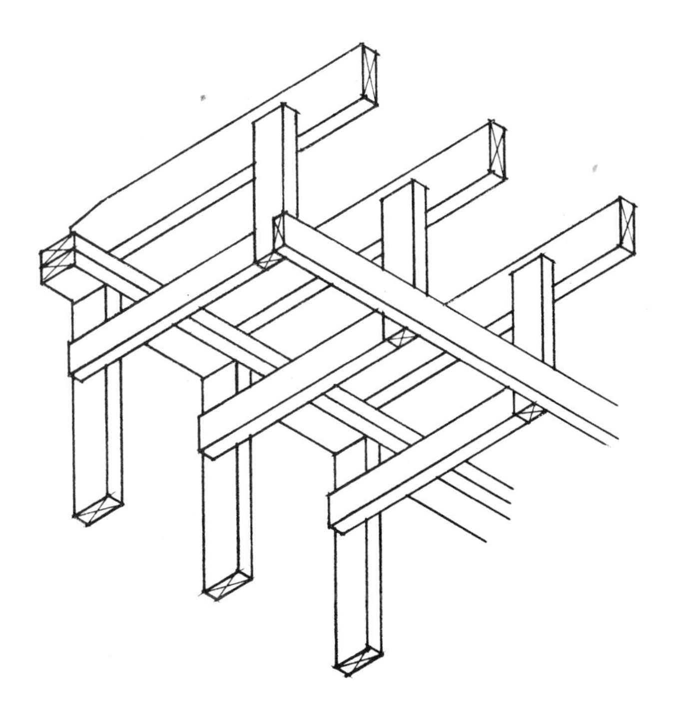

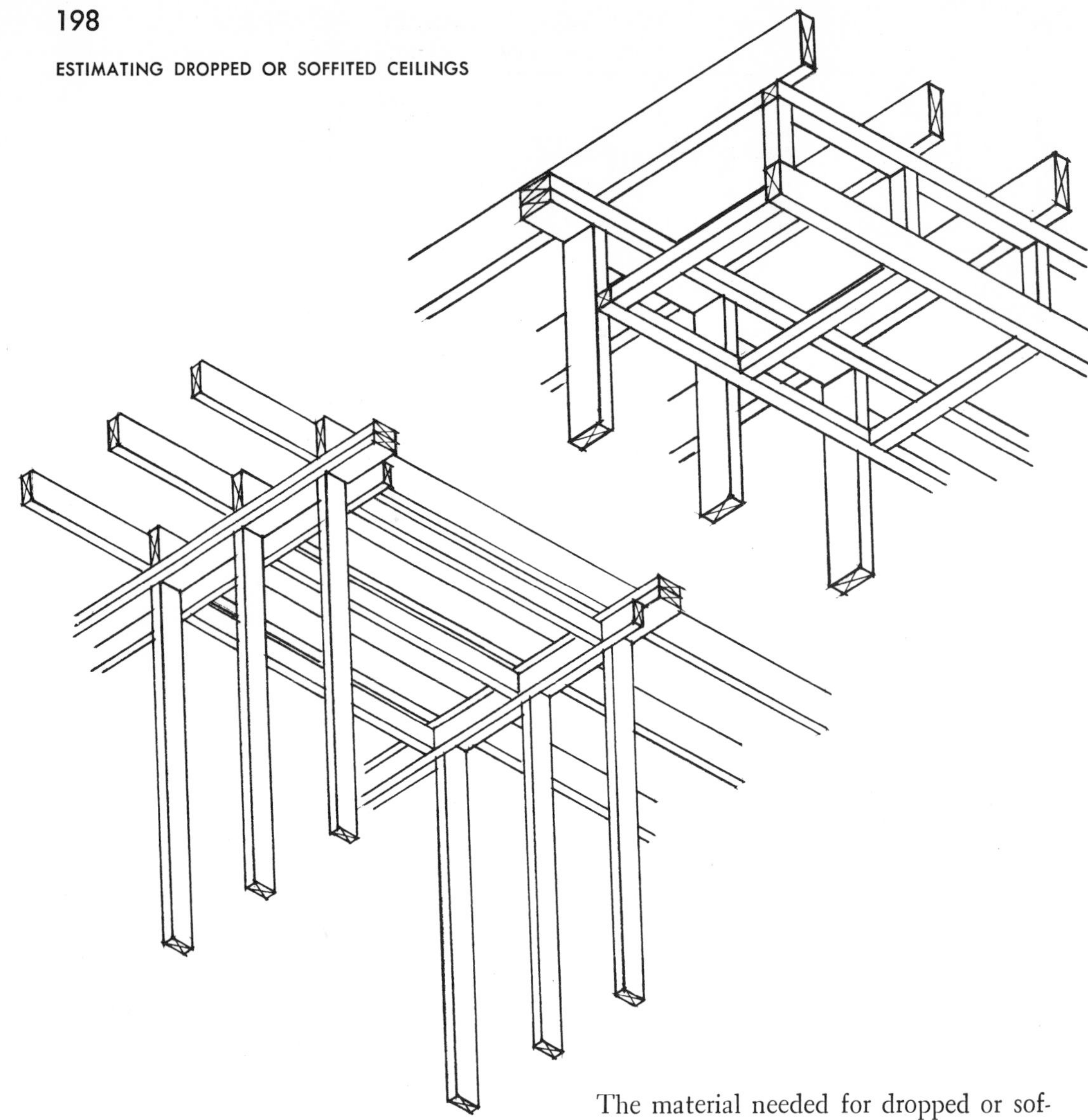

The material needed for dropped or soffited ceilings is acquired from the scrap and leftover lumber from the various members estimated in the building. As previously mentioned in Unit 30, once the ceiling joist and rafters are nailed to the top plate, the wall bracing material can be removed and also used for this purpose.

VI

ROOF FRAMING

39

Types of Roofs

In order to estimate the various framing members in a roof structure, the estimator must be familiar with the various types of roofs being built today. The following diagrams illustrate these roofs, starting with those most commonly built.

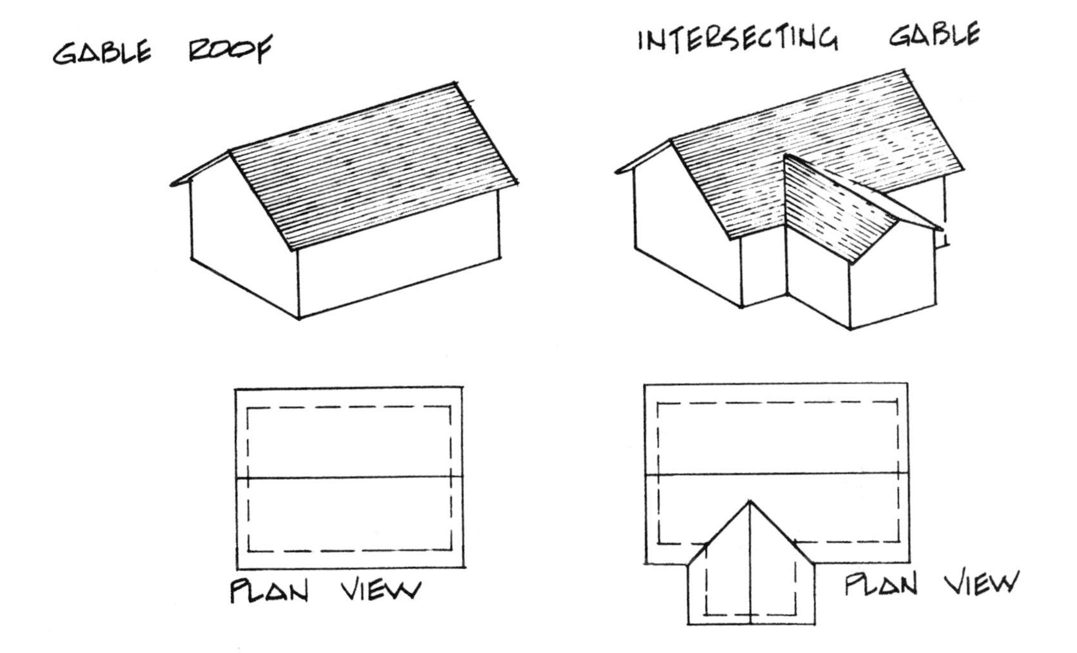

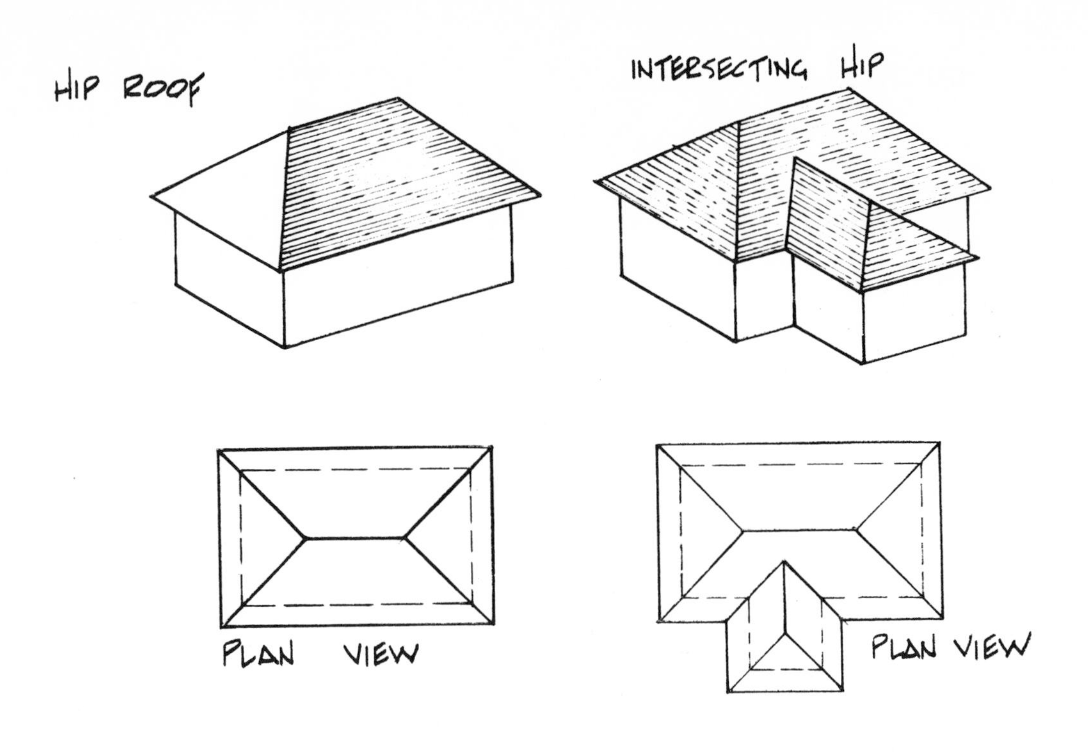
HIP ROOF
INTERSECTING HIP
PLAN VIEW
PLAN VIEW

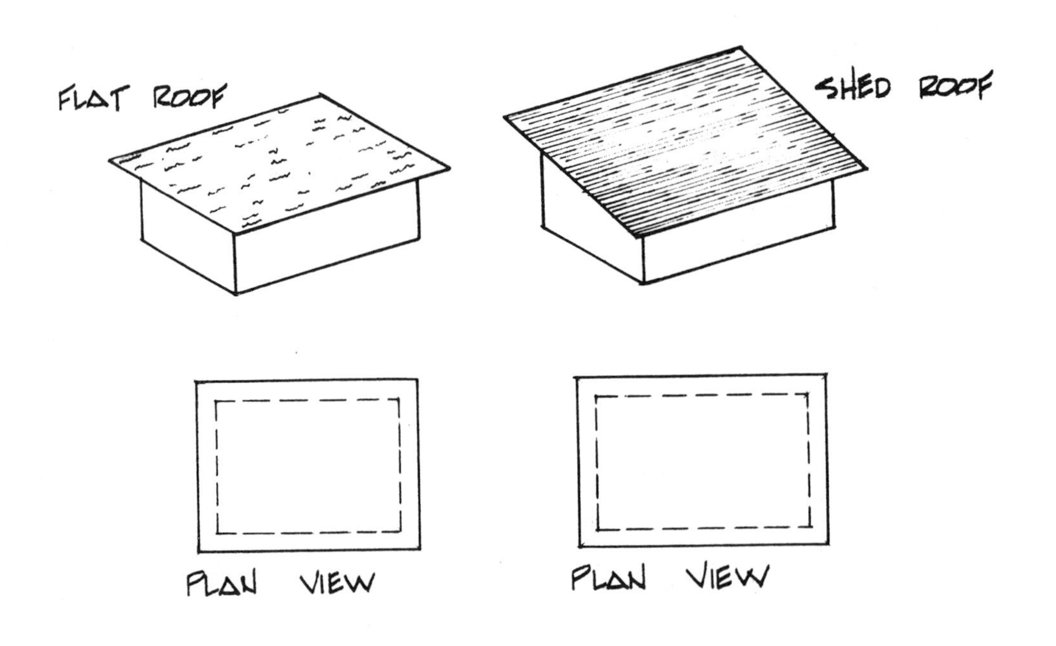
FLAT ROOF
SHED ROOF
PLAN VIEW
PLAN VIEW

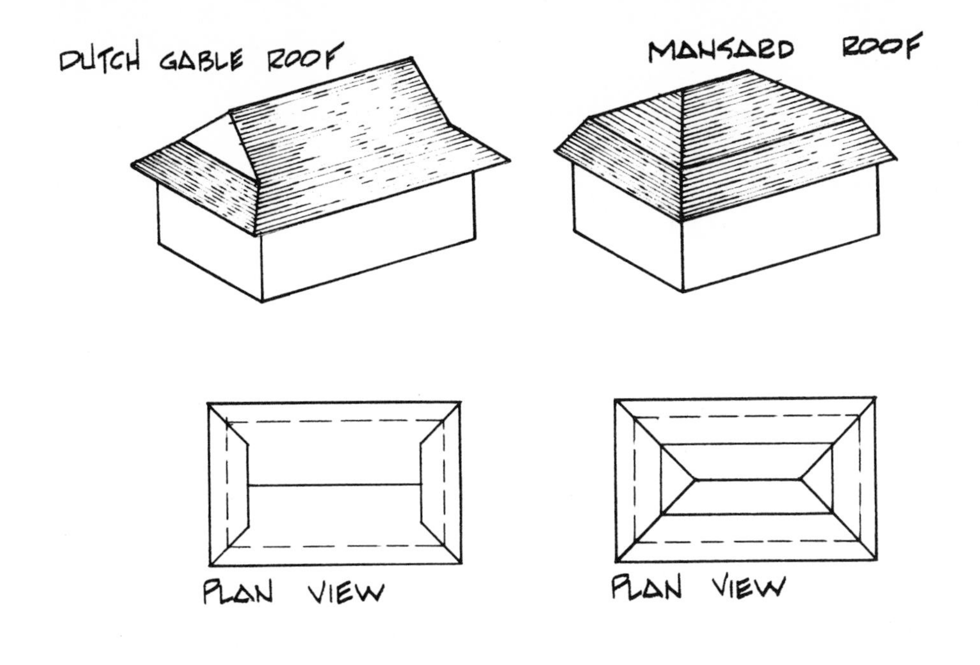
DUTCH GABLE ROOF
MANSARD ROOF
PLAN VIEW
PLAN VIEW

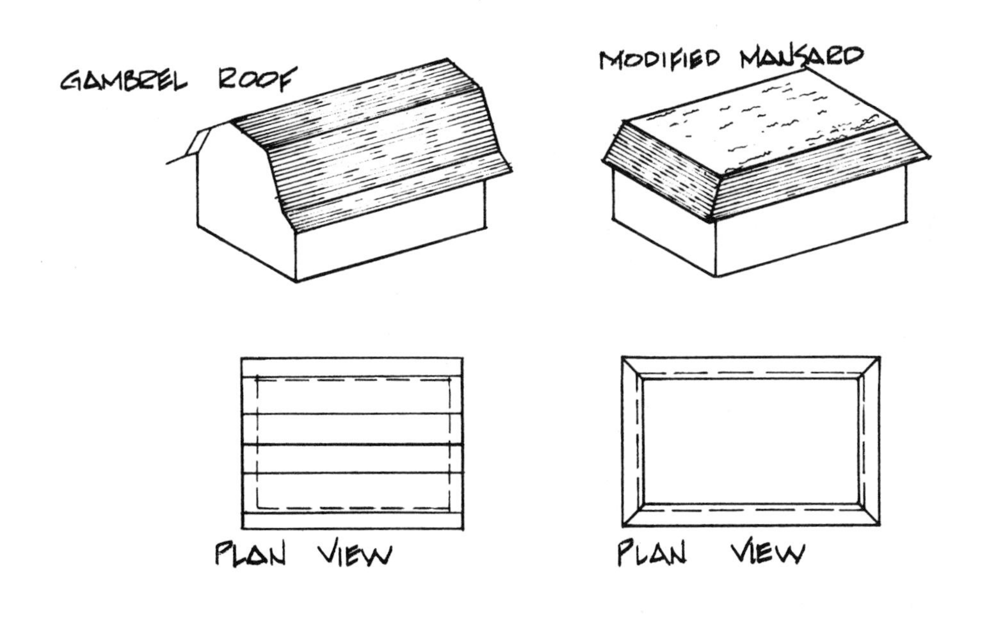
GAMBREL ROOF
MODIFIED MANSARD
PLAN VIEW
PLAN VIEW

40

Roof Members

The framing members of a roof to be estimated include the following:

1. Common rafters—the material estimated for the common rafters also includes material for the various types of jack rafters.
2. Ridge board
3. Hip and valley rafters.

The following illustrations show the members of the four most commonly built roofs.

1. GABLE ROOF

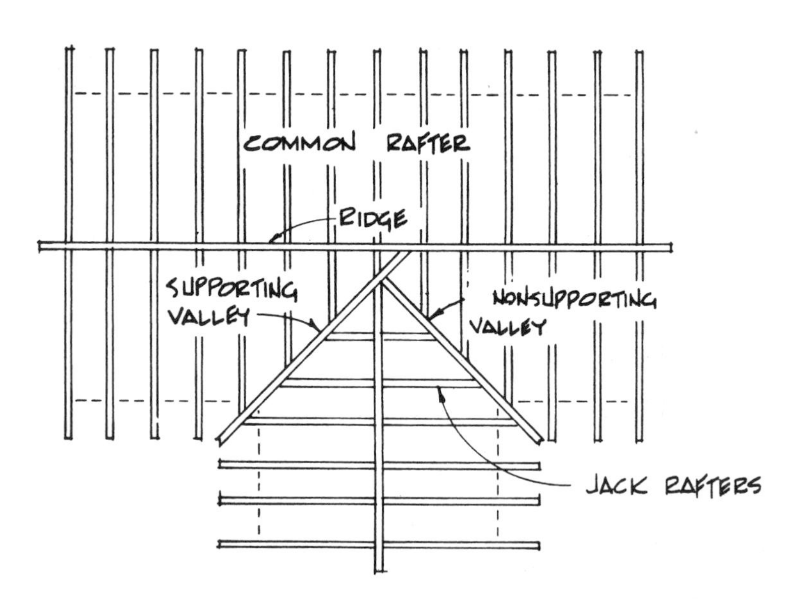

2. HIP ROOF

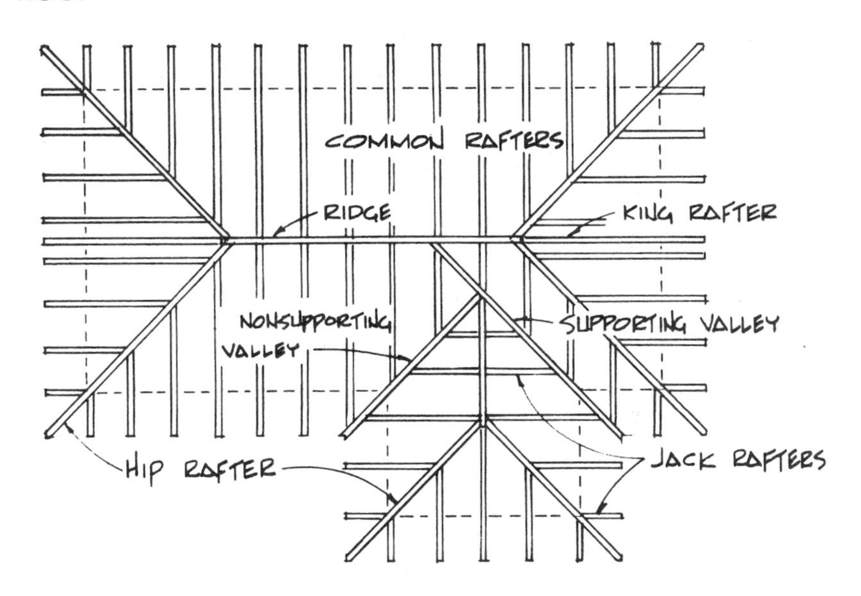

3. DUTCH GABLE ROOF

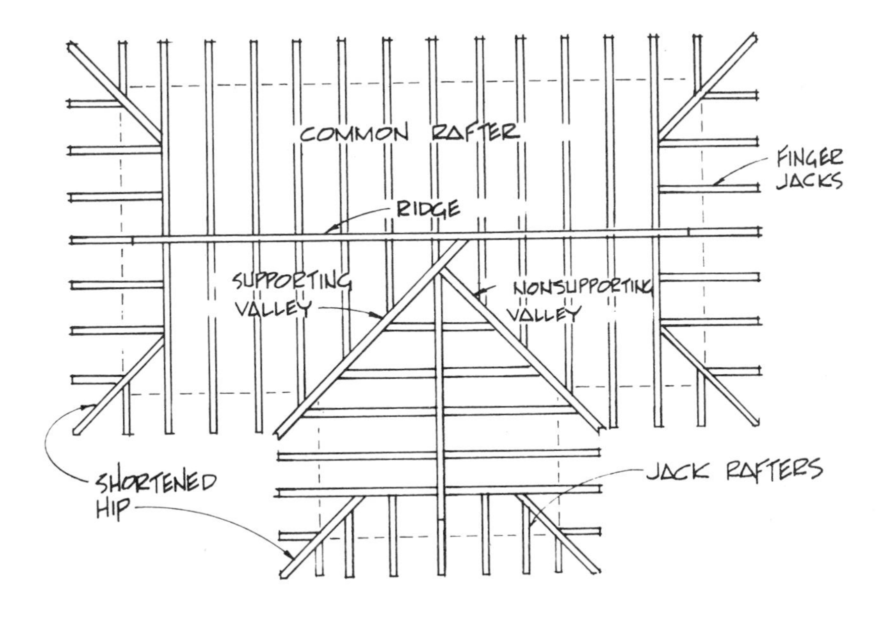

4. FLAT OR SHED ROOF

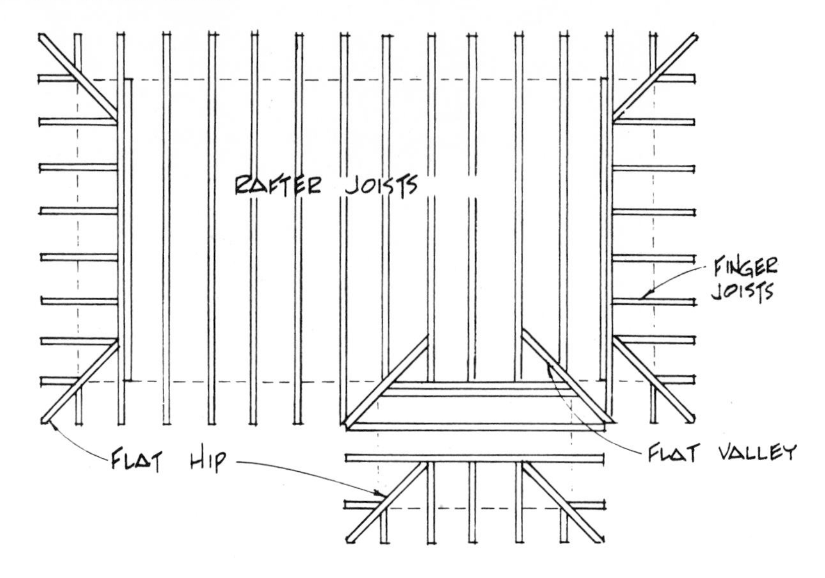

Note: There are various other methods of framing flat or shed roofs.

41

Roof Terminology

The estimator must be familiar with the following terms:

1. Building span—the width of the building
2. Building run—one-half the building span
3. Total rafter span—the span of the building plus the rafter tail overhang
4. Total rafter run—one-half the total rafter span
5. Unit run—the building run based on the unit of one foot (1′-0″)
6. Unit rise—the amount of rise (height) in the rafter for every one-foot unit of run.

ILLUSTRATION OF TERMS 1, 2, 3, AND 4

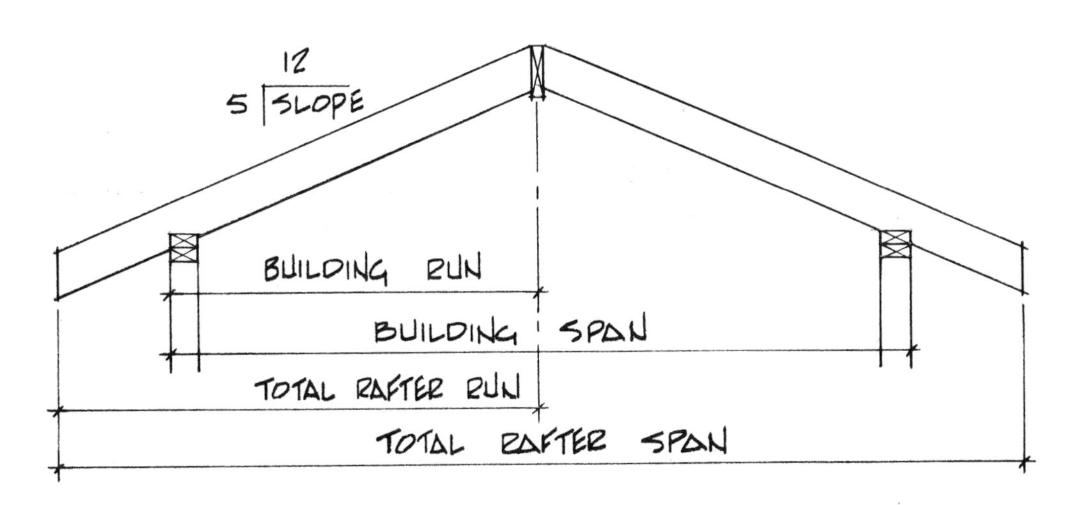

ILLUSTRATION OF TERMS 5 AND 6

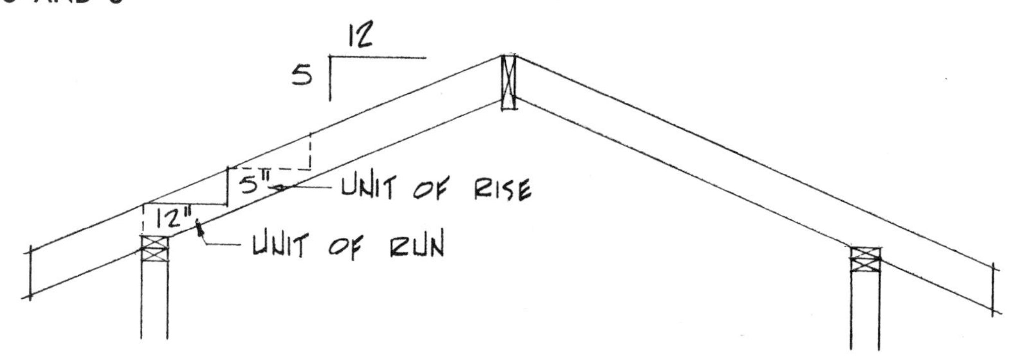

THE PITCH, SLOPE, OR CUT OF A ROOF

The *pitch, slope,* or *cut* of a roof, as far as the estimator is concerned, are synonymous terms. These three terms are used in different localities of the United States to determine the angle of the roof. They are based on a unit run of one foot as illustrated in the diagram above. In this unit the term *slope* takes the place of *cut* and *pitch.*

The estimator can usually find the slope of a roof on the elevation views as illustrated below.

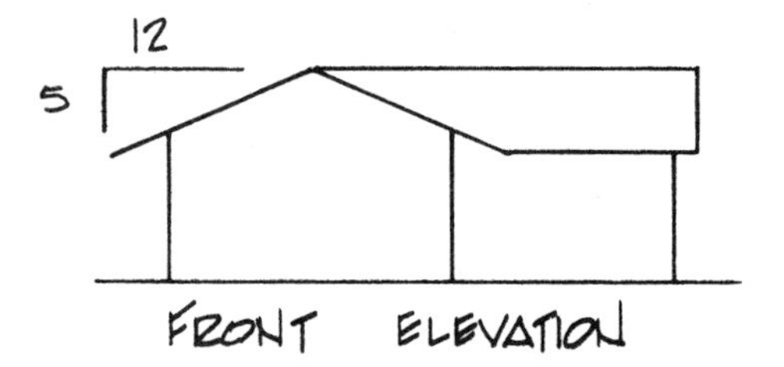

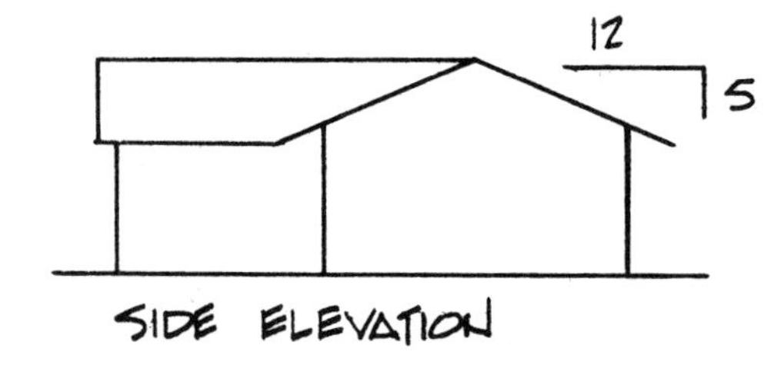

42
Roof Framing Methods

The estimator should be familiar with two methods of framing intersecting roofs, such as a gable on gable or a hip on hip. The following diagrams illustrate these methods.

1. CONVENTIONAL FRAMING

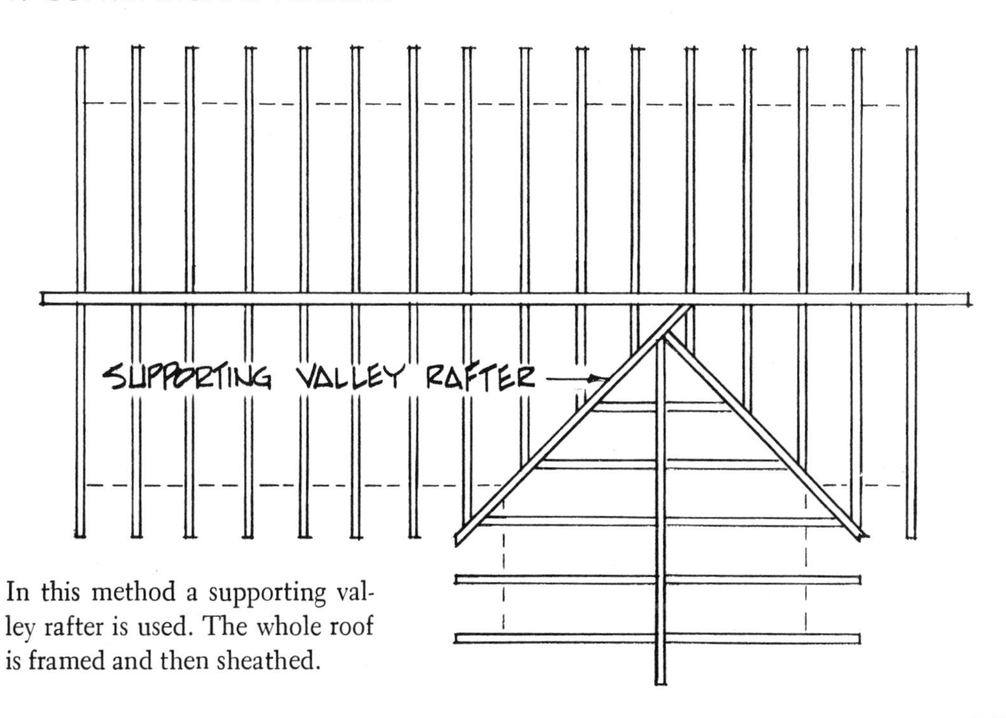

In this method a supporting valley rafter is used. The whole roof is framed and then sheathed.

2. BLIND VALLEY OR CALIFORNIA FRAMED

In the second method, the main roof is framed and sheathed, and then the intersecting roof is framed over the main roof.

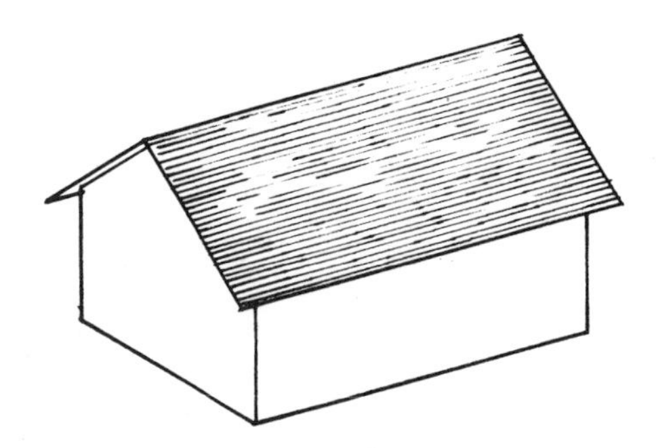

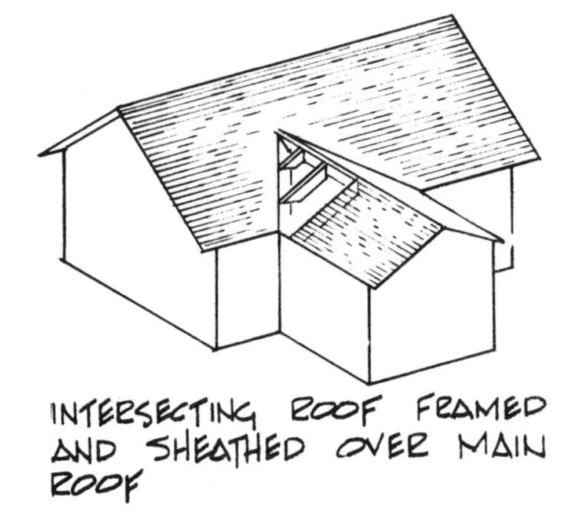

This method of framing is very adaptable when adding on intersecting room additions because the existing roof does not have to be dismantled and rebuilt. Its only drawback is that in certain areas of the country the Blind Valley method of framing intersecting roofs may not provide enough roof support to meet local code requirements.

43

Methods of Estimating Common Rafter Material

1. ***Scaling off the Building Plan.*** This is usually the simplest and fastest method to calculate rafter lengths, but it must be kept in mind that some building prints may not have been reproduced to scale. Also, the architect may have made some dimension changes on the plans but may not have changed the scaled drawing. The architect sometimes puts the initials *N.T.S.* (not to scale) next to the dimension he has changed. In any case, the estimator must be careful when scaling the plans.

The rafter size (width and thickness) is usually given on a section view of the building plans, but the rafter length must be calculated by the estimator. The length of the rafters may be scaled with an architect's scale or a ruler from the elevation views or from a section view as illustrated below.

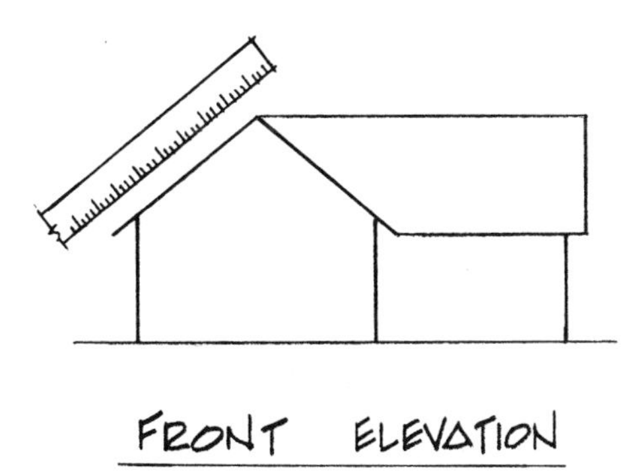

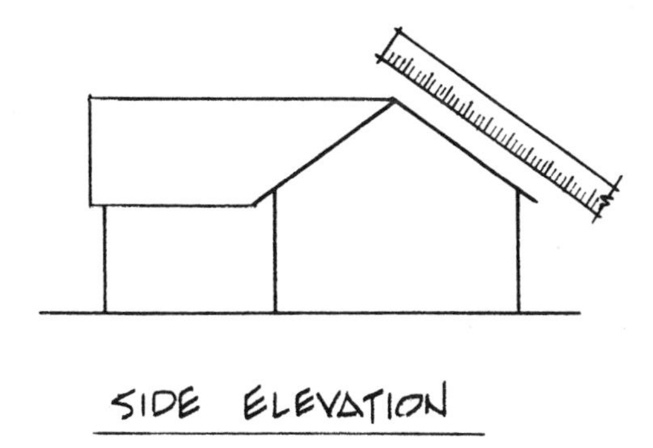

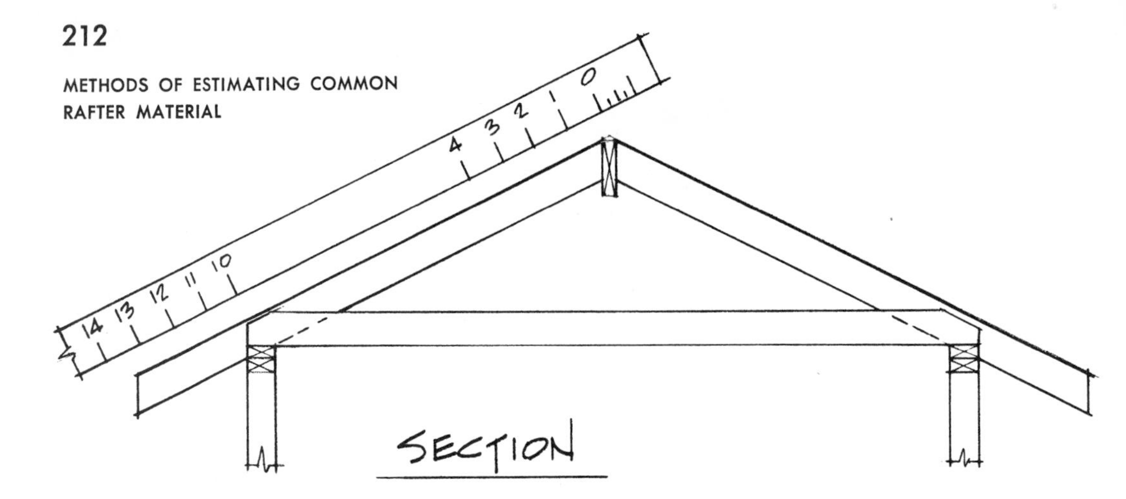

2. *Calculating Rafter Length with the Use of Tables.* There is a mathematical relationship between the length of *run* of a building and the length of rafter material. The difference in length depends on the slope of the roof as illustrated below.

COMMON RAFTER LENGTHS

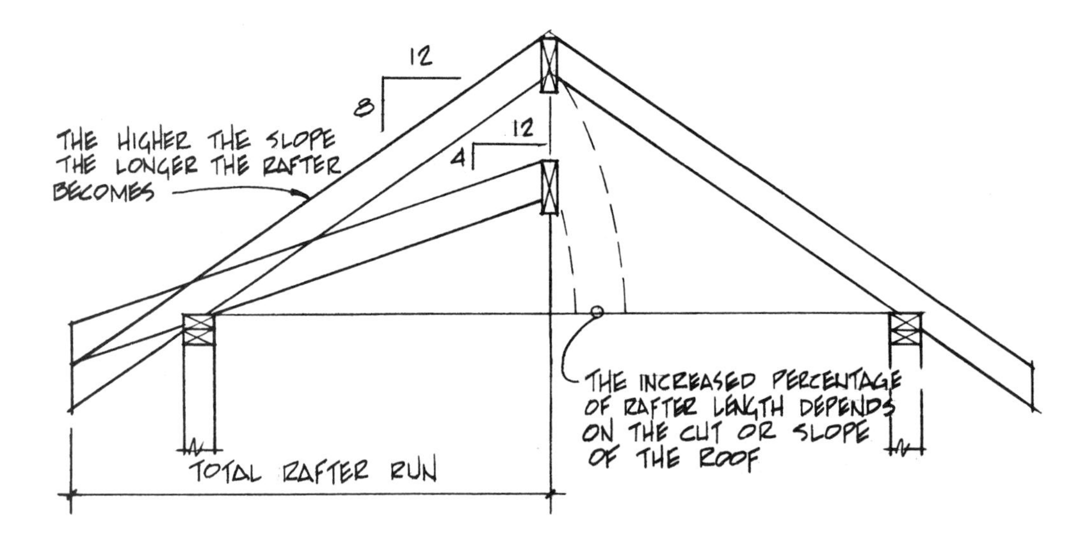

In Table 43-1, the increased percentage of rafter length is given as the rafter constant.

EXAMPLE: The constant for a 4 / 12 slope from the table is 1.05 or 5 percent longer than the total rafter run.

TABLE 43-1 COMMON RAFTER LENGTHS

Roof Slope	*Rafter Constant*	*Roof Slope*	*Rafter Constant*
2-1/2 / 12	1.02	8-1/2 / 12	1.23
3 / 12	1.03	9 / 12	1.25
3-1/2 / 12	1.04	9-1/2 / 12	1.28
4 / 12	1.05	10 / 12	1.30
4-1/2 / 12	1.07	12 / 12	1.41
5 / 12	1.08	14 / 12	1.54
5-1/2 / 12	1.10	16 / 12	1.67
6 / 12	1.12	18 / 12	1.80
6-1/2 / 12	1.14	20 / 12	1.94
7 / 12	1.16	22 / 12	2.09
7-1/2 / 12	1.18	24 / 12	2.24
8 / 12	1.20		

PROCEDURE FOR ESTIMATING COMMON RAFTER LENGTH

1. Determine the roof slope.
2. Select the constant from the common rafter table that corresponds to the roof slope.
3. Calculate the total rafter run.
4. Multiply the common rafter constant by the total rafter run. The result will be the common rafter length.

Example Problem 1:

Estimate the length of the common rafter material in the following illustration.

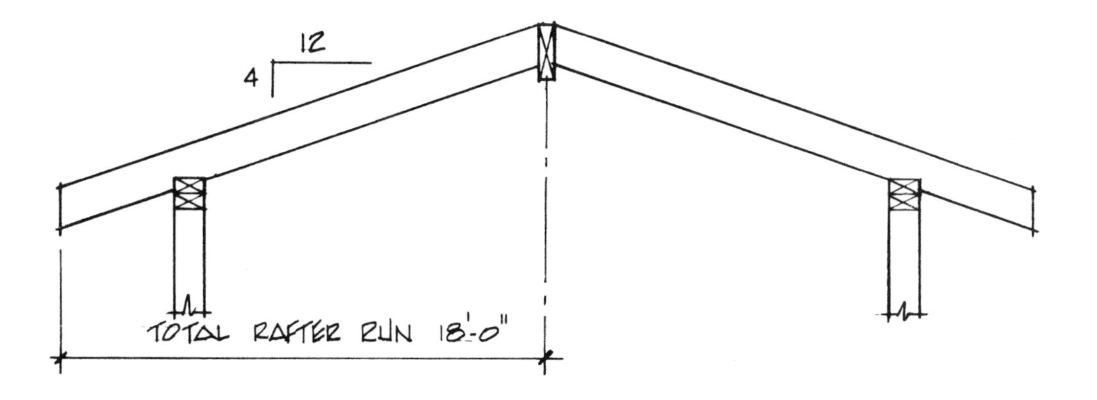

PROCEDURE FOR ESTIMATING COMMON RAFTER LENGTH

1. Determine the roof slope: 4 / 12.
2. Select the constant from the common rafter table that corresponds to the roof slope: 4 / 12 slope = 1.05 constant.
3. Calculate the length of the total rafter run: 18′-0″.
4. Multiply the constant 1.05 by the total rafter run (18′-0″).

```
  1.05
×   18
------
   840
  105
------
 18.90 lin. ft., or a 20′-0″ length of material is needed.
```

Example Problem 2:

Estimate the length of the common rafter material in the following illustration.

PROCEDURE FOR ESTIMATING COMMON RAFTER LENGTH

1. Determine the roof slope: 6 / 12.
2. Select the constant from the common rafter table that corresponds to the roof slope: 6 / 12 slope = 1.12 constant.
3. Calculate the length of the total rafter run.

   Note: In this problem there are two lengths of common rafters, rafter 1 with a total rafter run of 21′-0″ and rafter 2 with a total rafter run of 10′-0″.

   Rafter 1 = 21′-0″ Rafter 2 = 10′-0″
4. Multiply the constant (1.12) by each of the total rafter runs (21′-0″) (10′-0″).

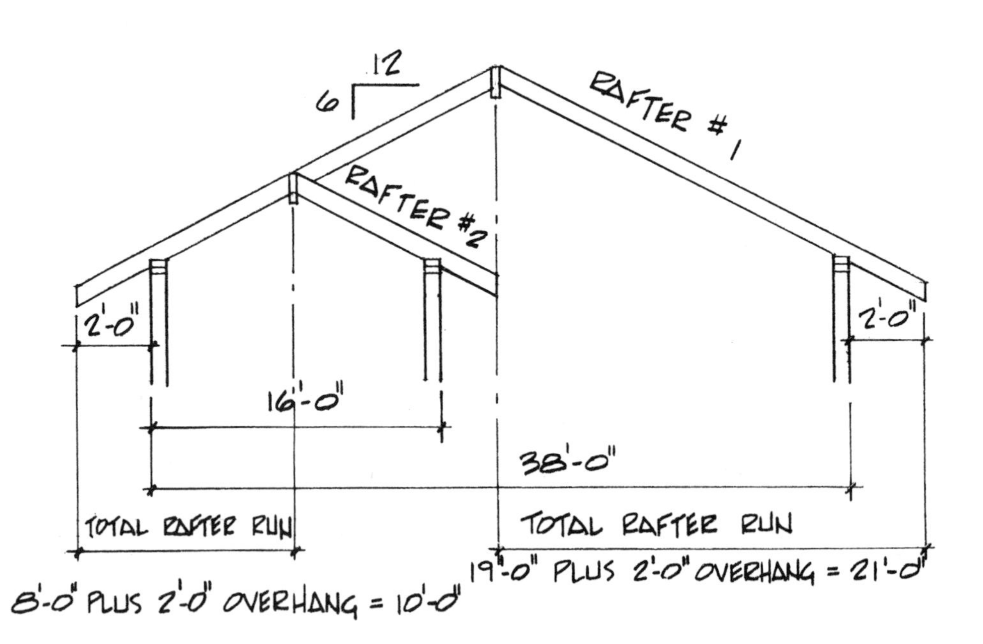

Rafter 1

```
  1.12
×   21
   112
  224
 23.52 lin. ft., or a 24′-0″ length of material is needed.
```

Rafter 2

```
  1.12
×   10
 11.20 lin. ft., or a 12′-0″ length of material is needed.
```

Example Problem 3:

Estimate the length of the common rafter material in the following illustration.

PROCEDURE FOR ESTIMATING COMMON RAFTER LENGTH

1. Determine the roof slope.

   Note: In this problem the front and back of the roof have different roof slopes. The front is a 4 / 12 slope and the back is a 5 / 12 slope.

2. Select the constant from the common rafter table that corresponds to the roof slope.

   4 / 12 slope = 1.05 constant
   5 / 12 slope = 1.08 constant

3. Calculate the length of the total rafter run.

   Note: In this problem the front and back of the roof have different

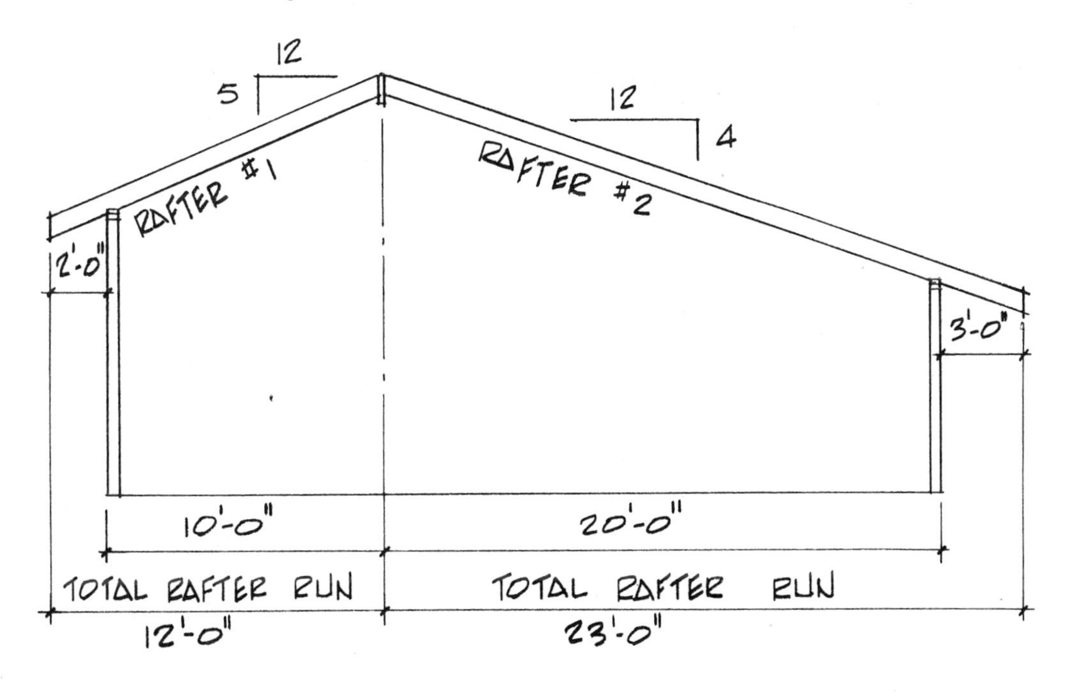

common rafter lengths. Rafter 1 has a total rafter run of 12′-0″ and Rafter 2 has a total rafter run of 23′-0″.

Rafter 1 = 12′-0″ Rafter 2 = 23′-0″

4. Multiply the respective rafter constants (1.08) and (1.05) by the total rafter runs (12′-0″) (23′-0″).

Rafter 1

$$\begin{array}{r} 1.08 \\ \times\ 12 \\ \hline 216 \\ 108 \\ \hline 12.96 \end{array}$$

12.96 lin. ft., or a 14′-0″ length of material is needed.

Rafter 2

$$\begin{array}{r} 1.05 \\ \times\ 23 \\ \hline 315 \\ 210 \\ \hline 24.15 \end{array}$$

24.15 lin. ft., or a 26′-0″ length of material is needed.

44

Estimating the Number of Common Rafters

The number of common rafters needed will depend on the spacing of the rafters. Most rafters will be spaced 16″ O.C. or 24″ O.C. The center-to-center spacing constant for 16″ O.C. is 3/4 (or .75) and for 24″ O.C. is 1/2 (or .50).

PROCEDURE FOR ESTIMATING THE NUMBER OF COMMON RAFTERS

1. Determine the center-to-center spacing constant for the spacing of the rafters.
2. Calculate from the building plans the lineal feet of wall on which the rafters are nailed.
3. Multiply the center-to-center spacing constant by the lineal feet of wall. The result will be the number of rafters for one side of the roof. Double the amount for both sides.

GABLE ROOF

Example Problem 1:

Estimate the number of common rafters for the building illustrated below with the rafters spaced 16″ on center.

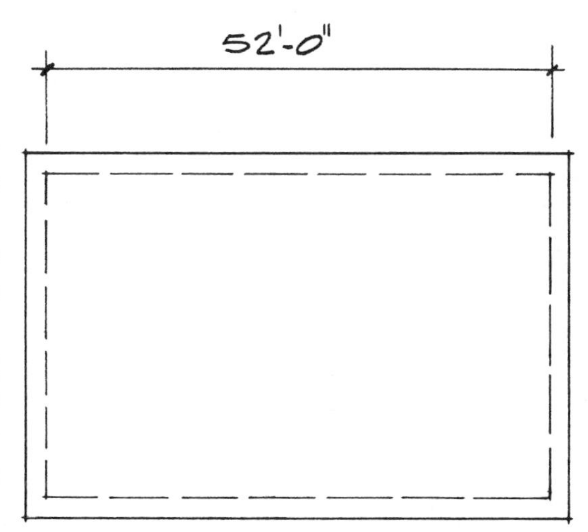

PROCEDURE:

1. Determine the center-to-center spacing constant for rafters 16″ on center: 16″ O.C. = .75.
2. Multiply the spacing constant (.75) by the length of wall on which the rafters are nailed (52 lin. ft.):

```
  52
 .75
 260
364
39.00
```

39.00 plus 1 equals 40 rafters for one side of the roof and 80 rafters for both sides.

Example Problem 2:

Estimate the number of common rafters for the building illustrated below with the rafters spaced 16″ on center.

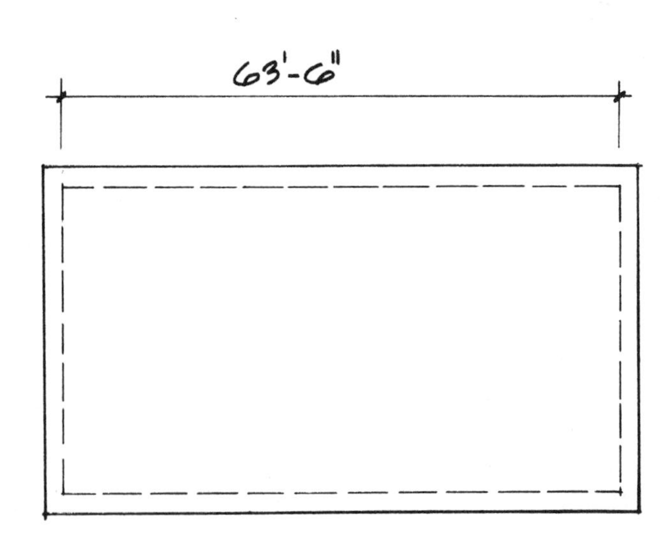

PROCEDURE FOR ESTIMATING THE NUMBER OF COMMON RAFTERS

1. Determine the center-to-center spacing constant for rafters 16″ on center: 16″ O.C. = .75.
2. Multiply the spacing constant (.75) by the length of wall on which the rafters are nailed (63′-6″ or decimal equivalent of 63.5):

```
  63.5
× .75
  3175
 4445
47.625
```

47.625 or 48 plus 1 equals 49 rafters for one side of the roof or 98 rafters for both sides.

Example Problem 3:

Estimate the number of common rafters for the building illustrated below with the rafters spaced 24″ on center.

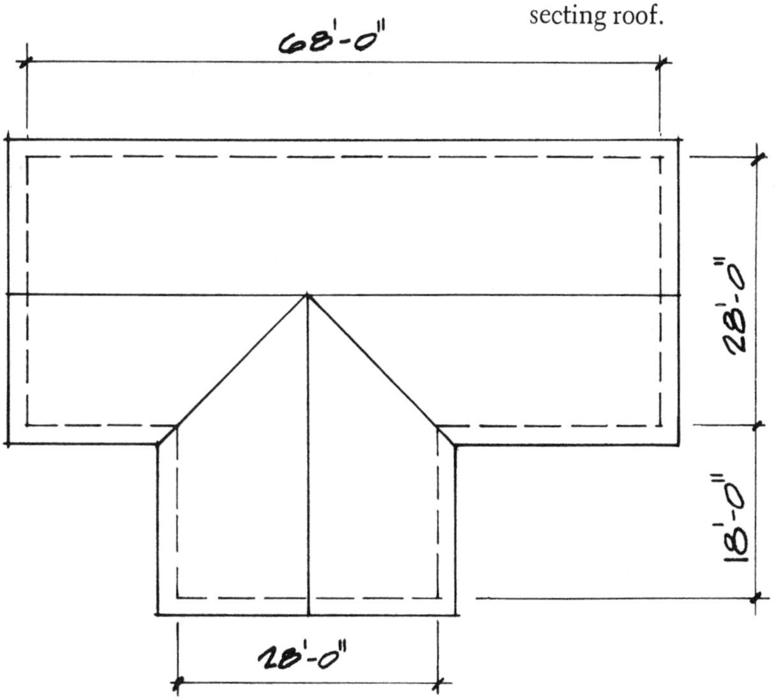

This problem has two parts, first the rafters to be estimated for the main roof, and second the rafters to be estimated for the intersecting roof.

68'-0"

28'-0"

18'-0"

28'-0"

PROCEDURE FOR ESTIMATING THE NUMBER OF RAFTERS FOR THE MAIN ROOF OF THE BUILDING

1. Determine the center-to-center spacing constant for rafters spaced 24″ on center: 24″ O.C. = .50.
2. Multiply the spacing constant (.50) by the length of wall on which the common rafters are nailed (68 lin. ft.):

$$\begin{array}{r} 68 \\ \times\ .50 \\ \hline 34.00 \end{array}$$

34.00 plus 1 equals 35 rafters for one side of the building or 70 rafters for both sides.

The common rafter material will take care of all the rafters (commons and jacks) covering the main portion of the building as illustrated below.

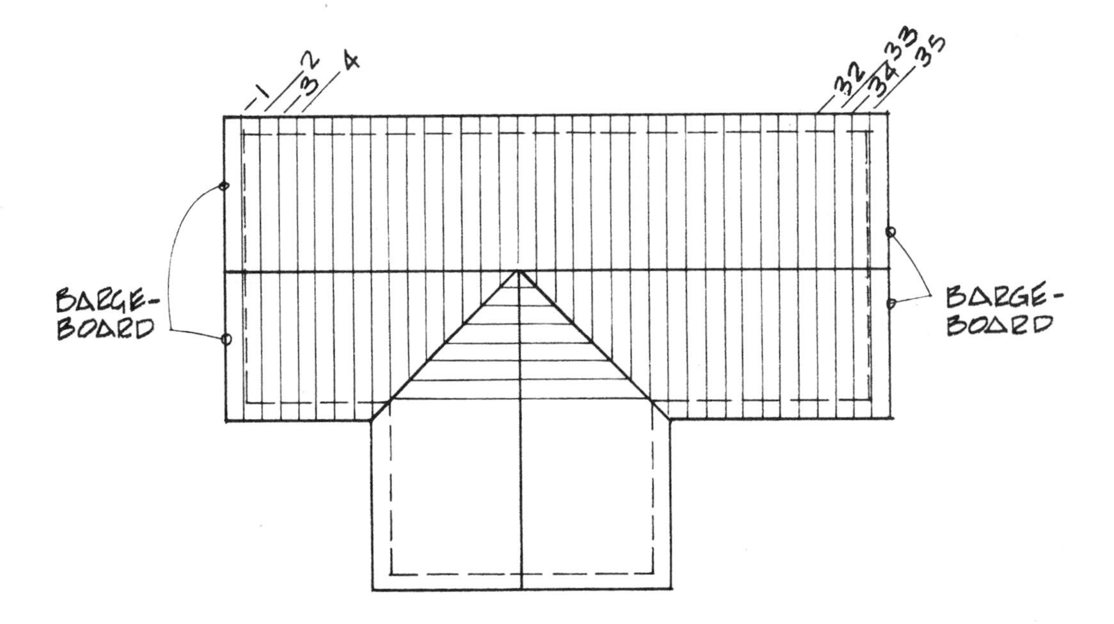

In the second part of the problem, the rafters needed to cover the intersecting building extension are estimated.

PROCEDURE FOR ESTIMATING THE NUMBER OF RAFTERS FOR THE INTERSECTING GABLE ROOF

1. The center-to-center spacing constant has been established: 24″ centers = .50.
2. Multiply the spacing constant (.50) by the length of wall on which the common rafters are nailed. In this problem the building extension is 18′-0″.

$$\begin{array}{r} 18 \\ \times\ .50 \\ \hline 9.00 \end{array}$$

9.00 plus 1 equals 10 rafters for one side of the building or 20 rafters for both sides. Using this procedure for estimating the rafters on the building extension provides a few extra available lengths of rafter material for any unforeseeable problem that may arise on the job.

Answer:

70 rafters needed for the main roof
+ 20 rafters needed for the intersecting roof

90 rafters total

Note: In this problem, the main roof span and the intersecting roof span are both 28′-0″. This means that the rafters for both sections are of the same length. In most cases, the spans are probably not the same and there would be two different lengths of common rafter material, one length for the main roof and one for the intersecting roof.

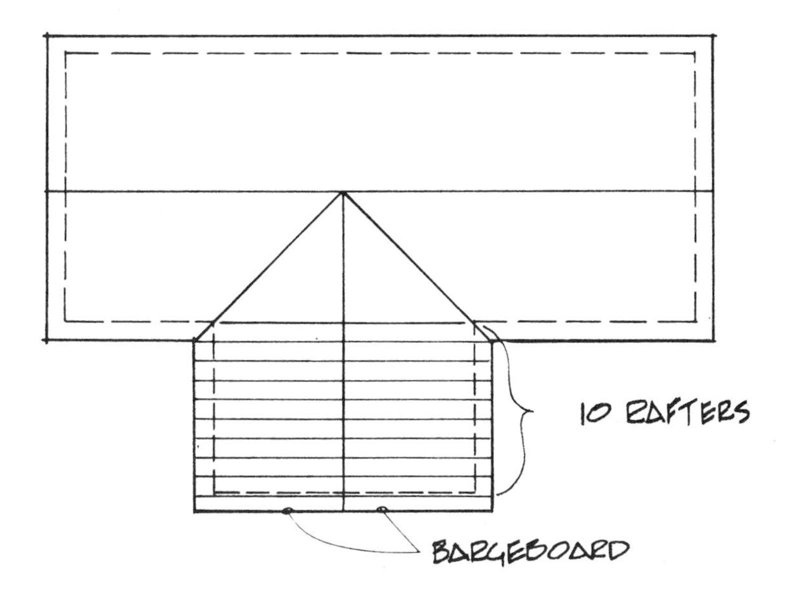

Because the span of the main and intersecting roofs are the same, the estimator can simplify this problem by adding the length of the intersecting extension (18′-0″) to the length of the main roof (68′-0″). At this point two extra rafters must be added, one for the main roof section and one for the intersecting roof section. As the rafters are spaced 2′-0″ O.C., these two rafters may be included by adding an extra 4 feet to the total above.

$$68 + 18 = 86 + 4 \text{ ft. extra} = 90 \text{ lin. ft.}$$

This may then be multiplied by the center-to-center spacing constant for 24″ centers (.50) for the answer:

$90 \times .50 = 45$ rafters for one side of the roof

or

90 rafters for both sides.

45

Estimating Common and Jack Rafter Material for a Hip Roof

The hip roof, at a given slope, has the same roof area as a gable roof. This means that the rafter material needed for a hip roof will be the same as that needed for a gable roof. The various lengths of material needed for jack rafters will be cut out of the lengths of common rafter material. Usually the longest jack rafter and the shortest jack rafter may be cut from one length of common rafter material as illustrated below.

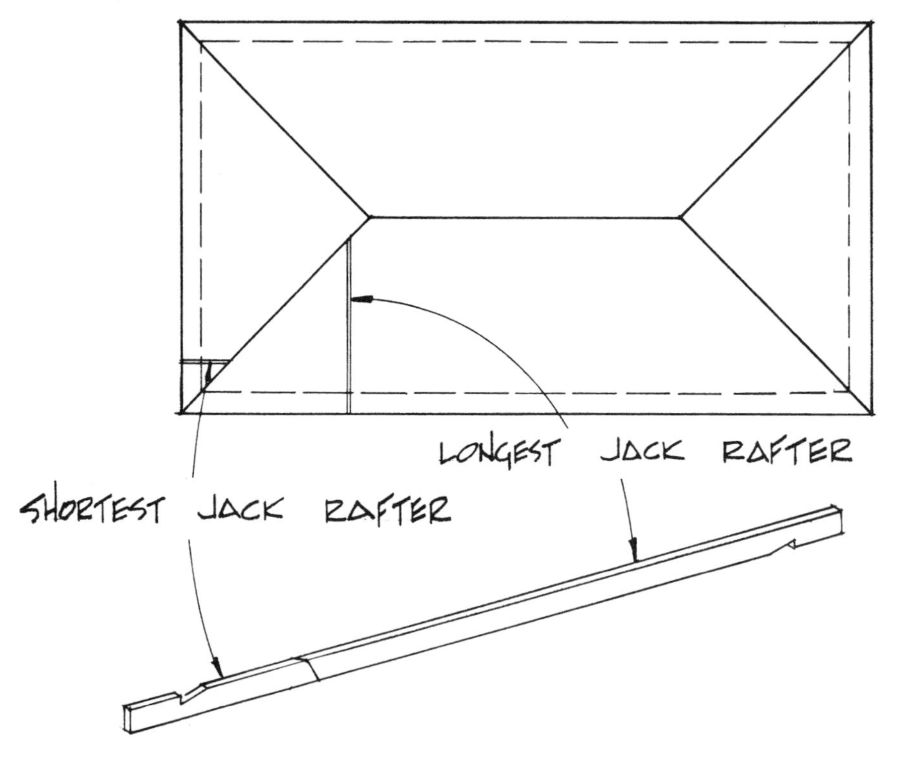

Example Problem:

Estimate the number and length of rafter material needed for the hip roof illustrated below.

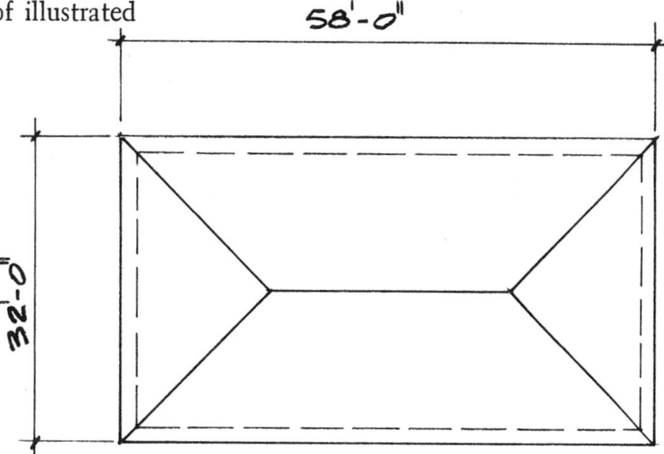

INFORMATION NEEDED:
RAFTER SLOPE 5:12
RAFTER SIZE 2×6
RAFTER SPACING 16" O.C.

PROCEDURE FOR ESTIMATING THE NUMBER OF COMMON RAFTERS NEEDED FOR A HIP ROOF

1. Determine the center-to-center spacing constant for rafters spaced 16″ on center: 16″ O.C. = .75.
2. Multiply the spacing constant (.75) by the length of the building on which the common and jack rafters are nailed (58′-0″) for the answer:

$$\begin{array}{r} 58 \\ \times\ .75 \\ \hline 290 \\ 406\ \ \\ \hline 43.50 \end{array}$$

43.50 or 44 plus 1 equals 45 rafters for one side of the roof or 90 rafters for both sides.

PROCEDURE FOR ESTIMATING COMMON RAFTER LENGTH

1. Select the constant from the common rafter table that corresponds to the roof slope:

   5 / 12 slope = 1.08 constant

2. Calculate the length of the total rafter run: 16′-0″.
3. Multiply the roof constant (1.08) by the total rafter run 16′-0″:

$$\begin{array}{r} 1.08 \\ \times\ \ 16 \\ \hline 648 \\ 108\ \ \\ \hline 17.28 \end{array}$$

17.28 lin. ft. or an 18′-0″ length of material is needed.

4. **The answer:**

   90 pieces 2 × 6 × 18′

46

Estimating Hip and Valley Rafter Lengths

Hip rafters, valley rafters, and ridge boards are always 2 inches wider than the rafter width. For example, if the rafters are 2 × 6 material, the hips, valleys, and ridges will be 2 × 8. Hip and valley rafter lengths can be found with the help of the data in Table 46-1. It is used in the same manner as the table for finding common rafter lengths.

PROCEDURE FOR ESTIMATING HIP OR VALLEY RAFTER LENGTHS

1. Determine the roof slope.
2. Determine the hip and valley constant for the roof slope.
3. Calculate the total rafter run.
4. Multiply the hip and valley constant by the total rafter run. The result will be the hip or valley length.

TABLE 46-1 HIP AND VALLEY RAFTER LENGTHS

Roof Slope	*Constant*	*Roof Slope*	*Constant*
2-1/2 / 12	1.43	8 / 12	1.56
3 / 12	1.44	8-1/2 / 12	1.58
3-1/2 / 12	1.45	9 / 12	1.60
4 / 12	1.46	9-1/2 / 12	1.62
4-1/2 / 12	1.47	10 / 12	1.64
5 / 12	1.48	12 / 12	1.74
5-1/2 / 12	1.49	14 / 12	1.84
6 / 12	1.50	16 / 12	1.94
6-1/2 / 12	1.52	18 / 12	2.06
7 / 12	1.54	20 / 12	2.18
7-1/2 / 12	1.55	22 / 12	2.32
		24 / 12	2.46

Example Problem 1:

Estimate the length of the hip rafters in the following illustration.

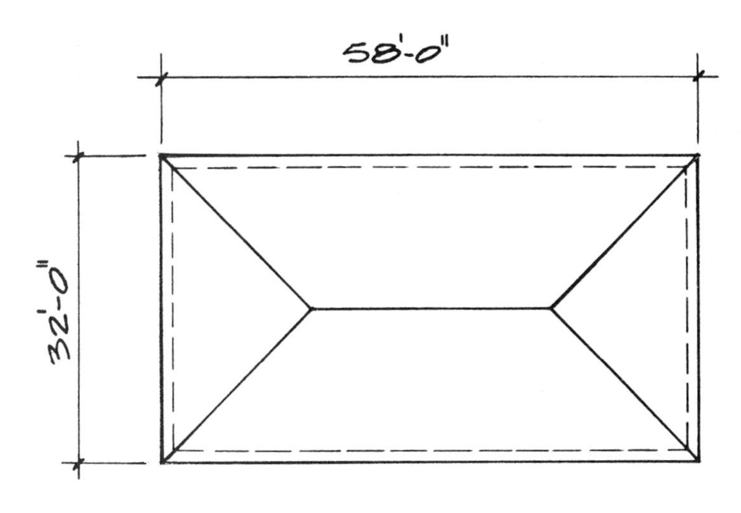

Note: Roof slope 5 / 12
Hip size 2 × 10

PROCEDURE:

1. Determine the roof slope: 5 / 12 slope.
2. Select the constant from the hip and valley table that corresponds to the roof slope: 5 / 12 slope = 1.48 constant.
3. Calculate the total rafter run: 16'-0".
4. Multiply the hip and valley constant (1.48) by the total rafter run (16'-0"):

```
   1.48
 ×   16
 ------
    888
   148
 ------
  23.68
```

23.68 lin. ft. or a 24'-0" piece of material is needed.

Answer: 4 pieces 2 × 10 × 24'

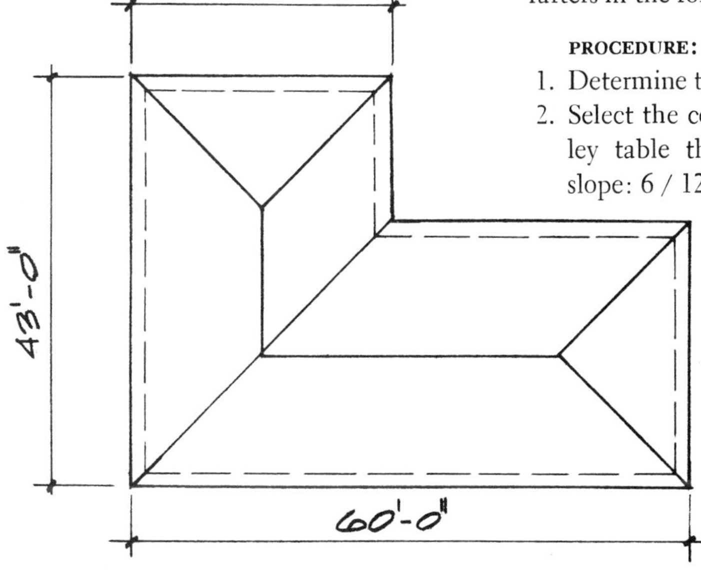

Example Problem 2:

Estimate the length of the hip and valley rafters in the following illustration.

PROCEDURE:

1. Determine the roof slope: 6 / 12 slope.
2. Select the constant from the hip and valley table that corresponds to the roof slope: 6 / 12 slope = 1.50 constant.

3. Calculate the total rafter run: 14′-0″.
4. Multiply the hip and valley constant (1.50) by the total rafter run (14′-0″):

$$\begin{array}{r} 1.50 \\ \times \quad 14 \\ \hline 600 \\ 150 \\ \hline 21.00 \end{array}$$

21.00 lin. ft. or a 22′-0″ piece of material is needed.

Answer: Hips 5 pieces 2 × 8 × 22′
Valleys 1 piece 2 × 8 × 22′

Example Problem 3:

Estimate the number, size, and length of the hips and valleys in the following illustration.

Note: Roof slope 4 / 12
Common rafter size 2 × 6

PROCEDURE:

There are two total rafter spans to consider when estimating this problem: 20′-0″ and 30′-0″. The total rafter runs for these would be 10′-0″ and 15′-0″.

1. Determine the roof slope: 4 / 12 slope.
2. Select the constant from the hip and valley table that corresponds to the rafter slope: 4 / 12 slope = 1.46 constant.
3. Calculate the total rafter run (in this case there are two, 10′-0″ and 15′-0″).
4. Multiply the hip and valley constant (1.46) by the total rafter runs:

$$\begin{array}{r} 1.46 \\ \times \quad 10 \\ \hline 14.60 \end{array}$$

14.60 lin. ft. or a 16′-0″ piece of material is needed.

$$\begin{array}{r} 1.46 \\ \times \quad 15 \\ \hline 730 \\ 146 \\ \hline 21.90 \end{array}$$

21.90 lin. ft. or a 22′-0″ piece of material is needed.

5. Check the roof plan for the number of hips and valleys of each size.

Answer: 4 hips 2 × 8 × 22′
2 hips 2 × 8 × 16′
2 valleys 2 × 8 × 16′

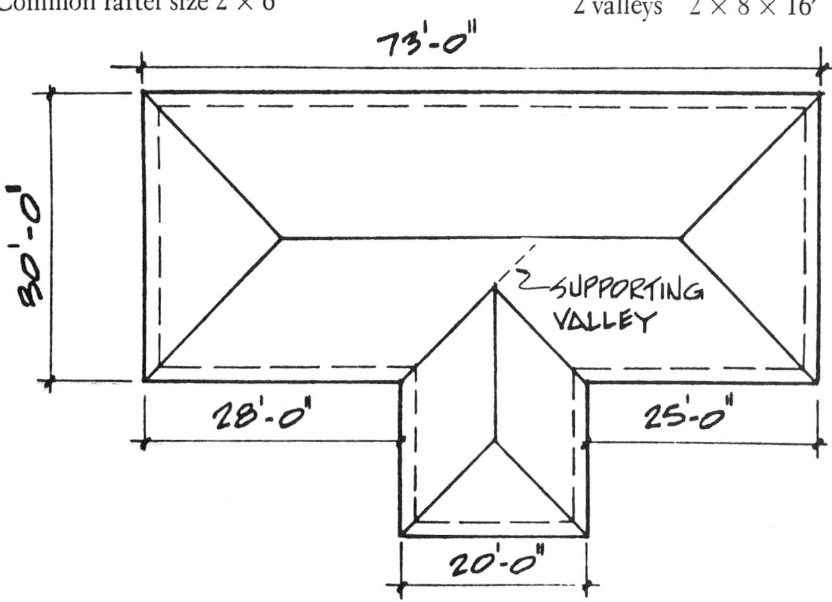

47

Estimating Ridge Board Material

The ridge is the highest framing member of the roof. Ridges are usually made of material that is one inch or two inches thick and two inches wider than the rafter material. In other words, if the rafter size is 2 × 6, the ridge board, depending on which thickness is being used, will be 1 × 8 or 2 × 8 material. The easiest method to estimate ridge lengths is to draw the roof members directly on the scaled floor plan. The ridge lengths can then be scaled for gable roofs, hip roofs, and intersecting gable and hip roofs. If scaling the plans is not possible, the following methods may be used for estimating the various lengths of ridge material.

PROCEDURE FOR ESTIMATING RIDGE BOARD MATERIAL FOR A STRAIGHT GABLE ROOF

The length of the ridge board for a straight gable roof is the length of the building. If the ridge board is to support the bargeboard on the gable overhang, as illustrated below, the length of the ridge will be the length of the building plus the gable overhang.

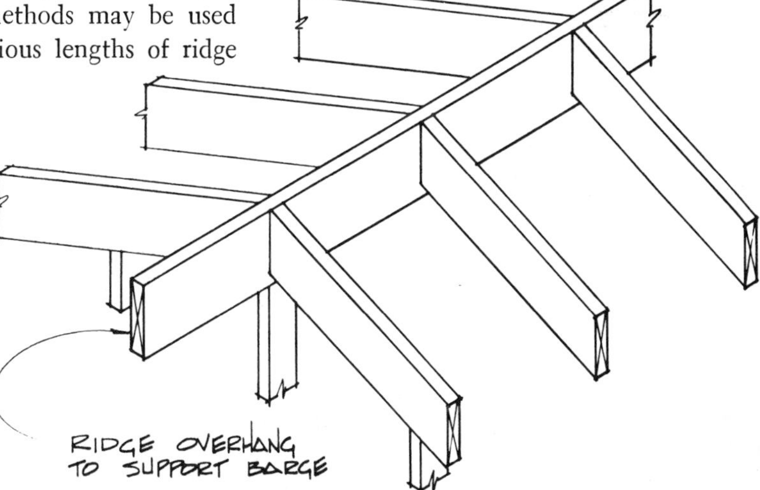

Example Problem:

Estimate the lineal feet of ridge board material needed in the following illustration.

The length of the building (58′-0″ plus the 2′-0″ overhang on each gable) equals the length of the ridge (62′-0″).

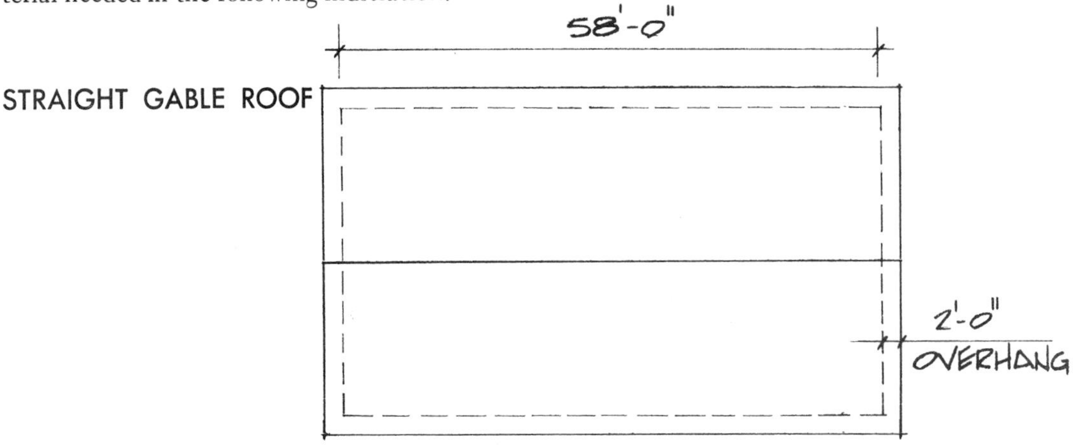

PROCEDURE FOR ESTIMATING RIDGE BOARD MATERIAL FOR A INTERSECTING GABLE ROOF

The length of the ridge board for an intersecting gable is found by adding the building extension length to the total rafter run of the extension as illustrated below.

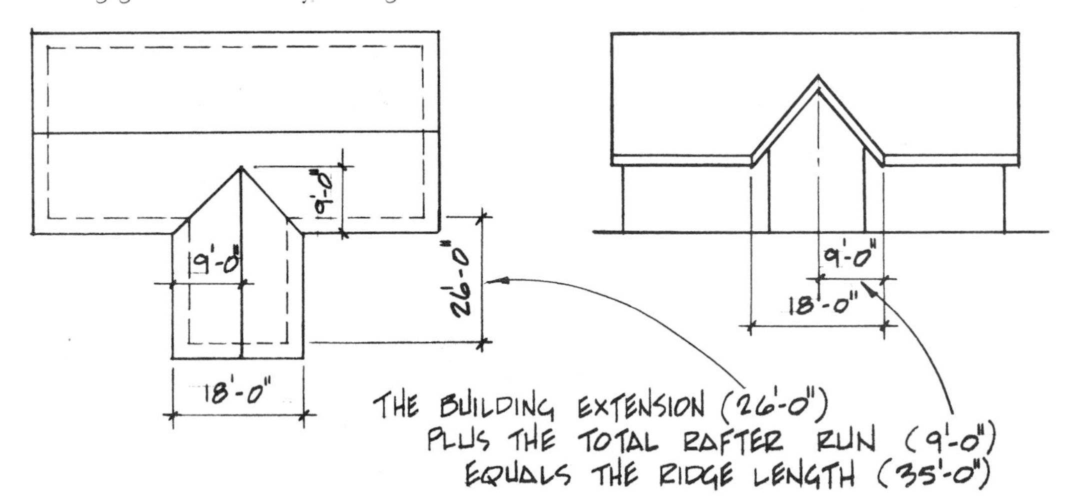

PROCEDURE FOR ESTIMATING RIDGE BOARD MATERIAL FOR A HIP ROOF

The ridge length for a full hip roof is found by subtracting the width or span of the building from the length of the building as illustrated below.

EXAMPLE:

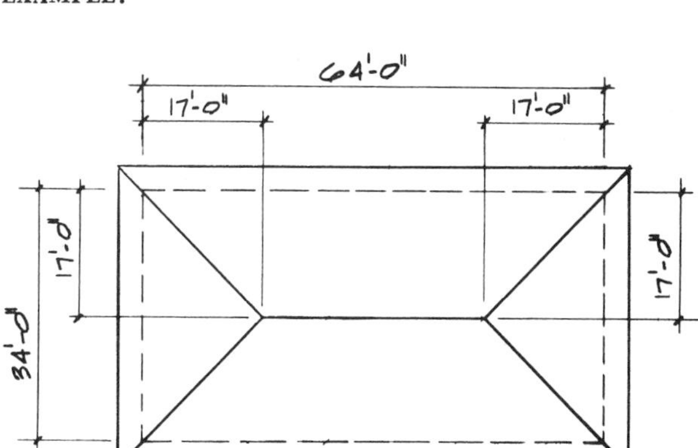

The width or span of the building (34'-0") is subtracted from the length of the building (64'-0"):

$$\begin{array}{r} 64 \\ -\ 34 \\ \hline 30 \end{array}$$

30 lin. ft. of ridge board.

PROCEDURE FOR ESTIMATING RIDGE BOARD MATERIAL FOR AN INTERSECTING HIP ROOF

The length of ridge board for an intersecting hip roof is the same as the length of the intersecting building extension as illustrated below.

EXAMPLE:

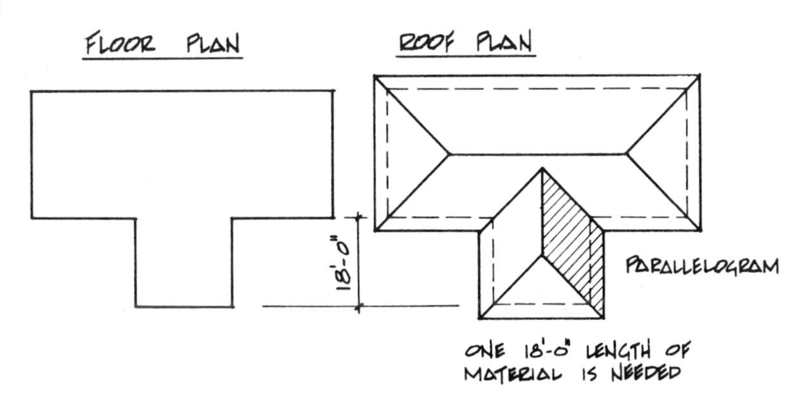

Estimate the ridge lengths in the following roof illustrations.

PROBLEM 1:

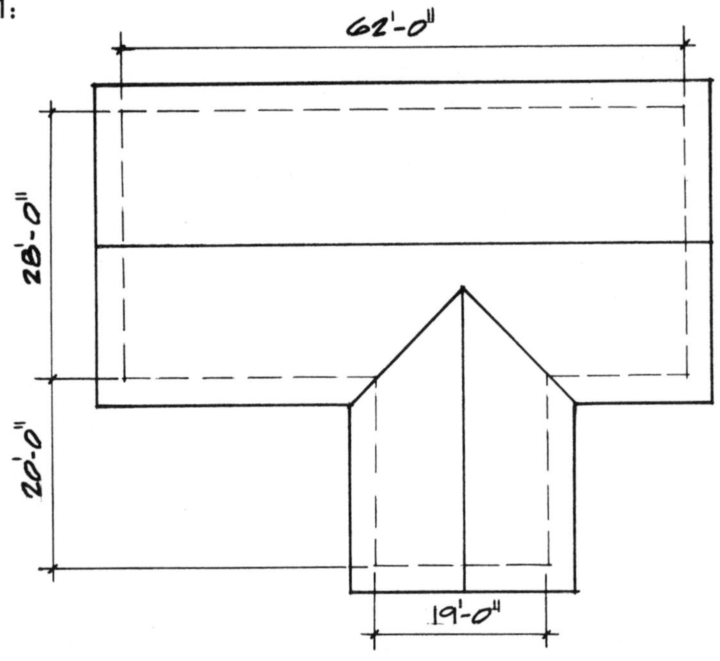

PROBLEM 2:

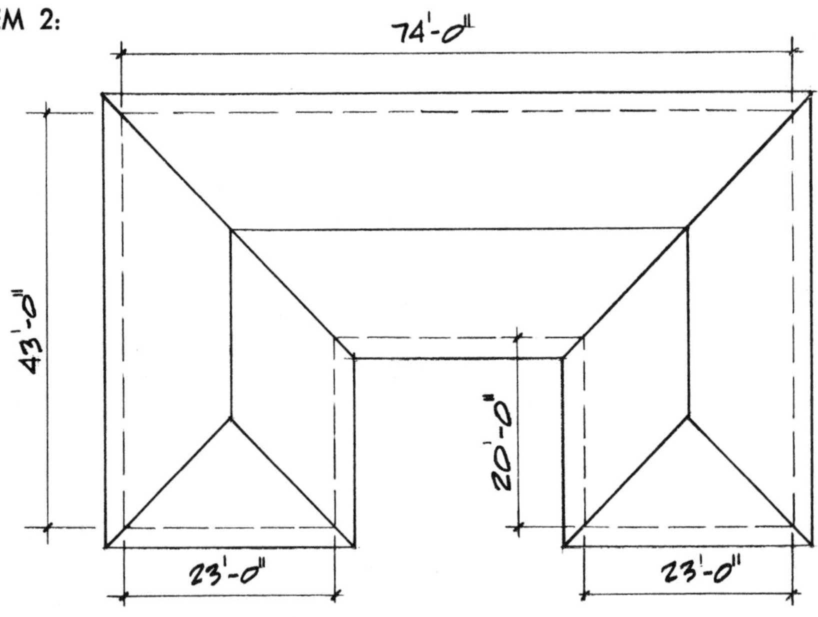

PROBLEM 3:

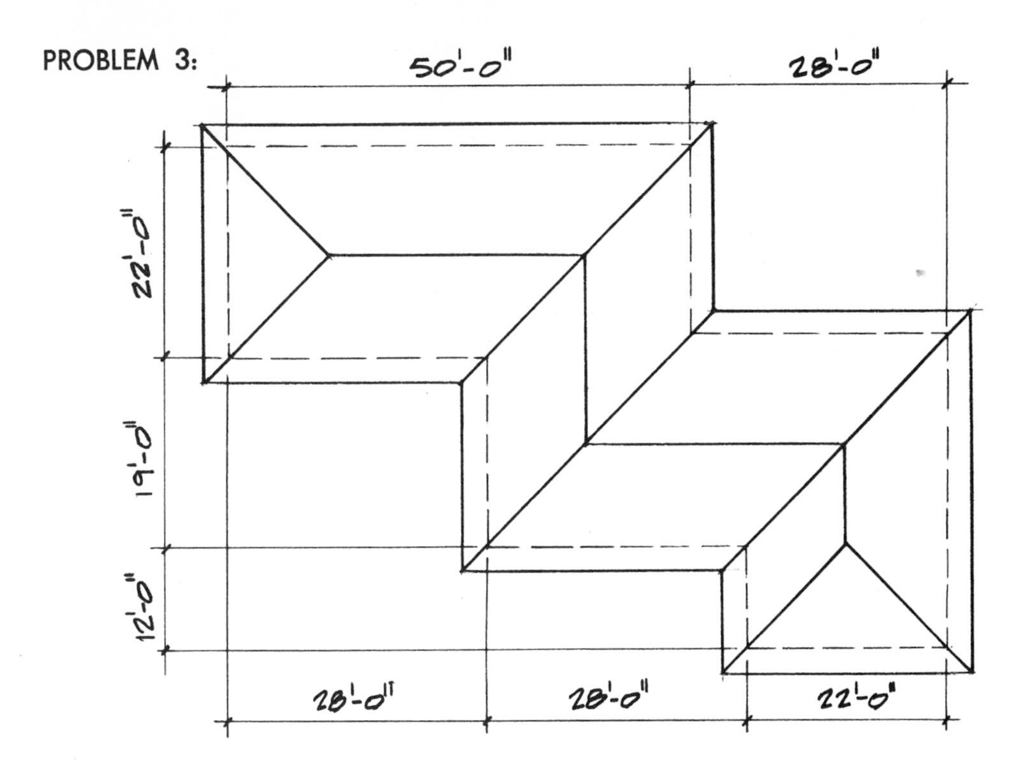

PROBLEM 4:

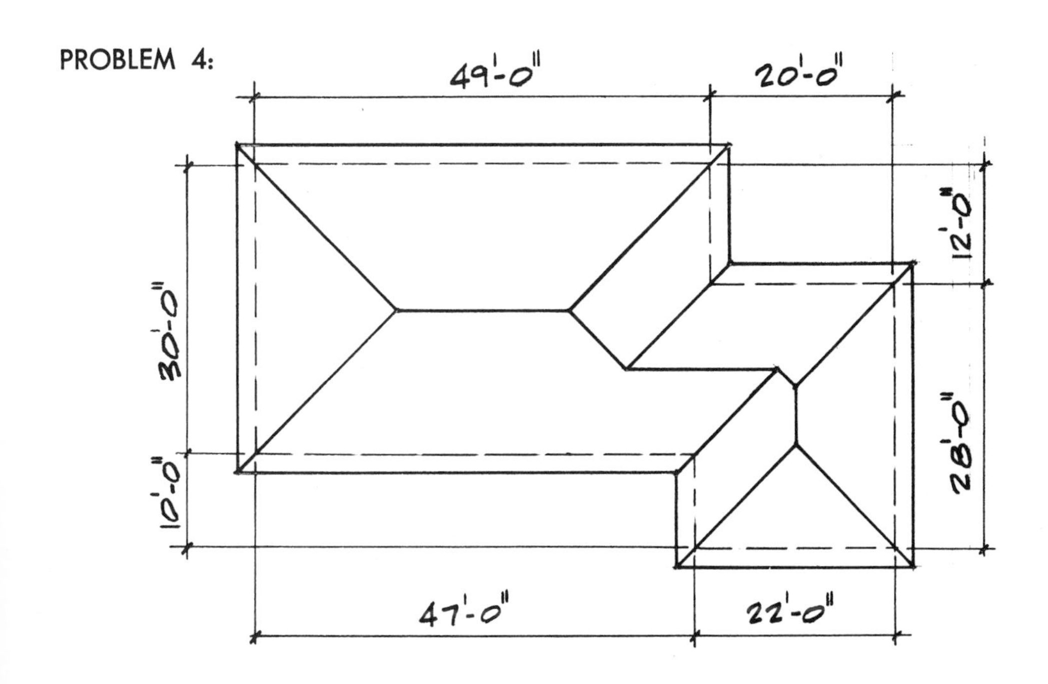

48

Estimating Fascia and Bargeboard Material

Fascia board is the horizontal trim member along the edge of the roof. Gable roofs have a combination of fascia and bargeboard, whereas hip roofs have only fascia board material to estimate.

Fascia board is usually one or two inches thick, of various widths, and rough or smooth surfaced depending on the aesthetic look the architect is trying to achieve. Usually a good grade of dry material such as spruce, hemlock, pine, redwood, cedar, or select Douglas fir is ordered for fascia and bargeboard. The length of fascia or bargeboard material depends on the type of wood available in the area. Some types only come in 16- and 18-foot lengths. Other types are available in 24- or 26-foot lengths.

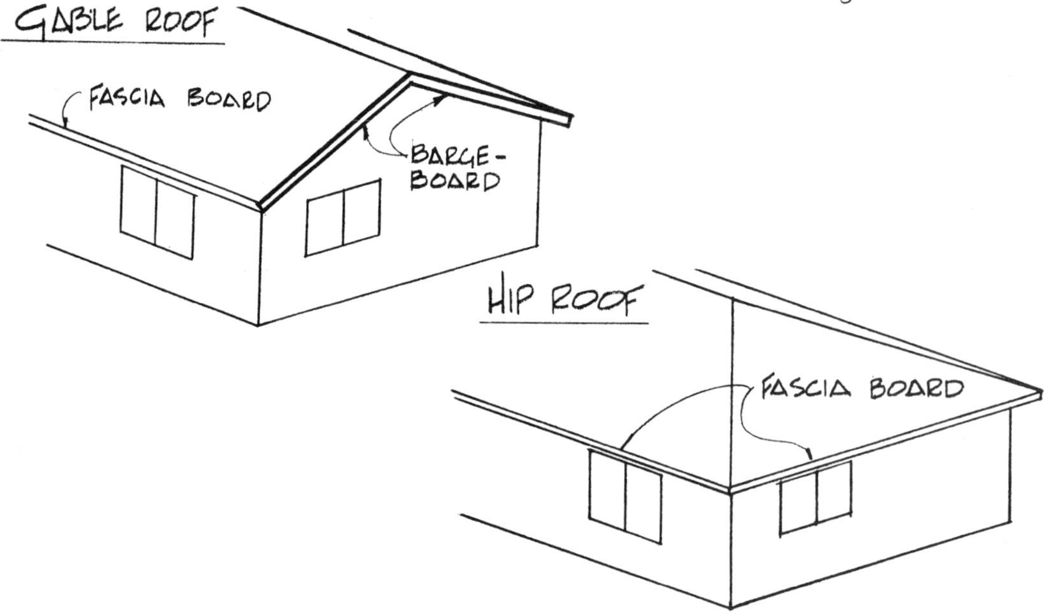

The fascia board is usually spliced on the center of a rafter tail as illustrated below.

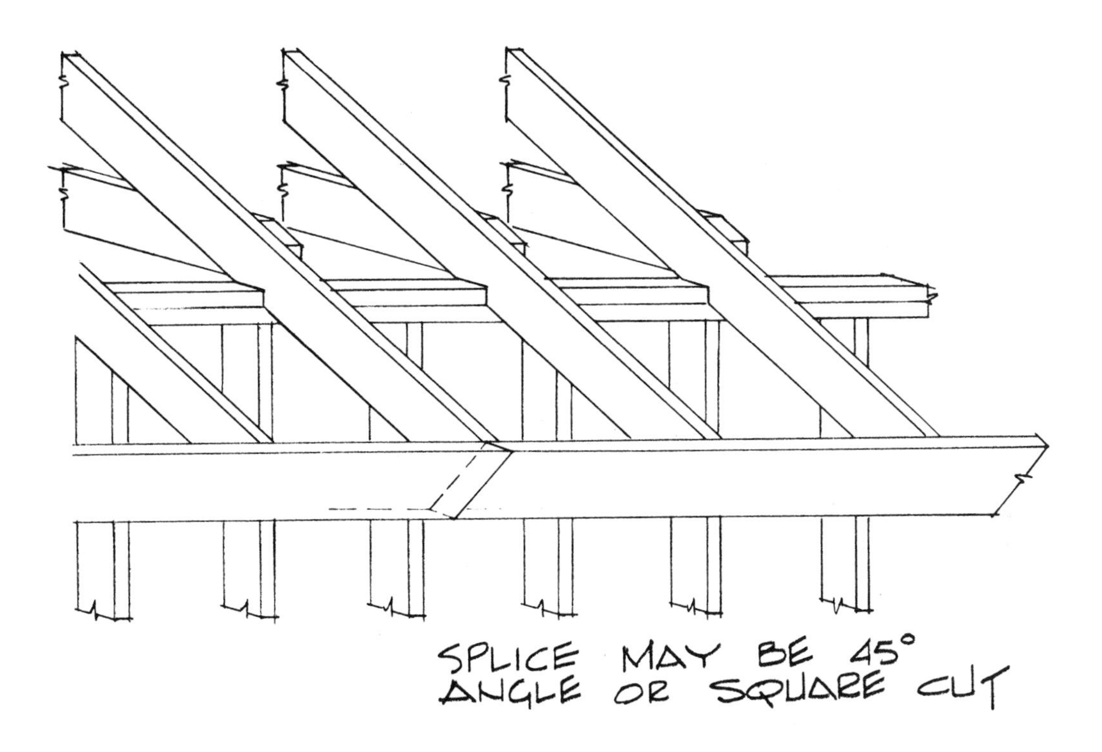

To estimate the necessary lengths of material when dealing with long runs of spliced fascia board, the estimator must know exactly where the rafter tails fall. But the actual layout position of the rafters will be determined by the carpenter on the job and could vary from the estimator's theorized position, thus shifting the layout one way or the other. Because of this, a few extra feet should be added to each length of fascia material. The extra length will also allow the carpenter enough material to cut off the cracked or checked ends on each piece of fascia board for a neater appearance.

Whether the rafter layout is 16″ O.C. or 24″ O.C., the two extra feet of material will allow the layout to shift one way or the other and still have enough material to finish the job.

Note: Sometimes to reduce waste, fascia material can be ordered after rafters are in position or when actual layout is known.

EXAMPLE OF CHANGING LAYOUT:

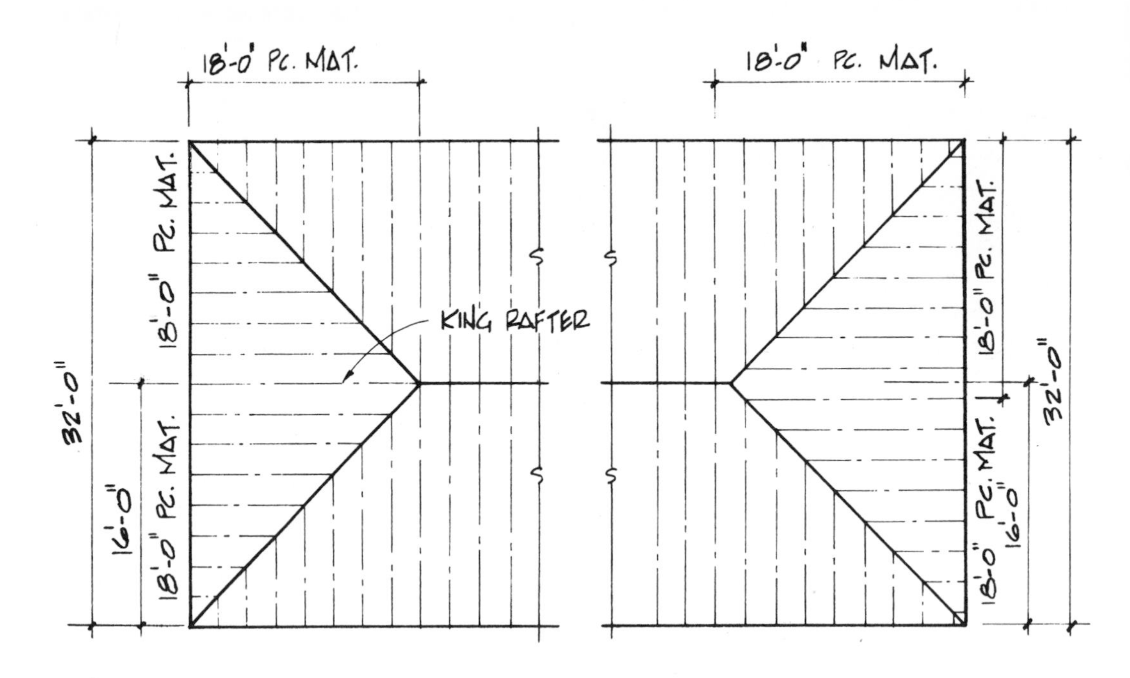

BARGEBOARD

The bargeboard is the trim board used on the ends of a gable roof.

The length of material for bargeboard is the same as the common rafter length. The thickness and width of barge material may vary or may be the same as the common rafter. When estimating barge material, the estimator should check the floor plan, roof plan, or elevation views to determine the number, type, and size of barge members. He can scale the elevation views for the length of barge pieces or check the length of the common rafters already estimated to compile a list of barge material. Some buildings have more than one building span and will require barge material of various lengths.

EXAMPLE 1:

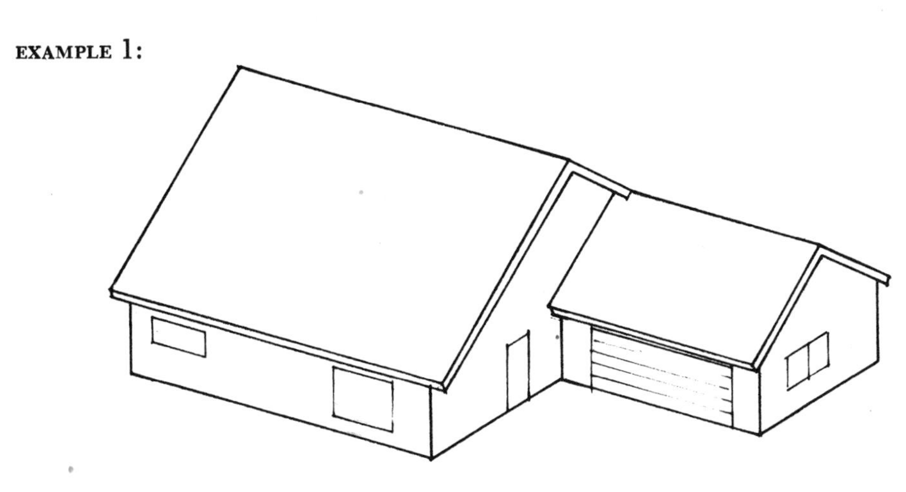

EXAMPLE 2:

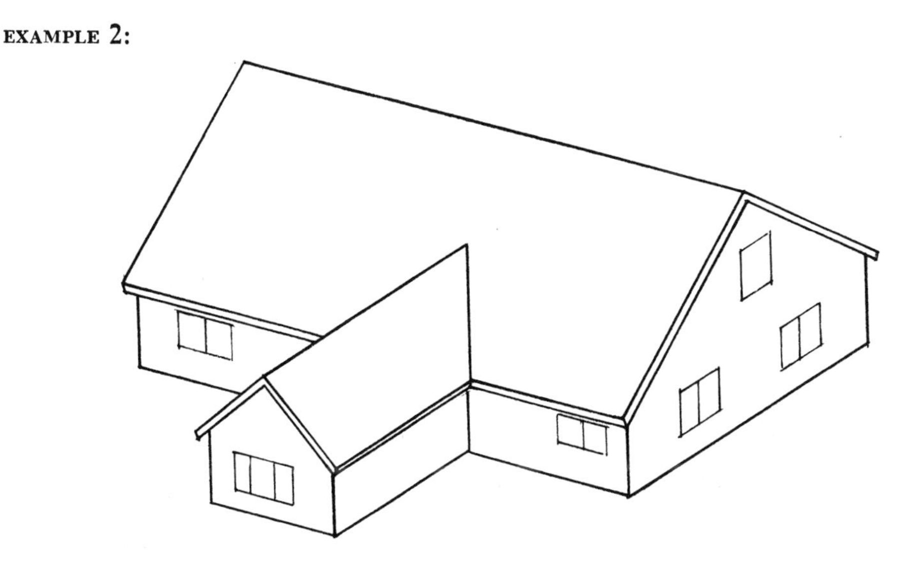

EXAMPLE 3:

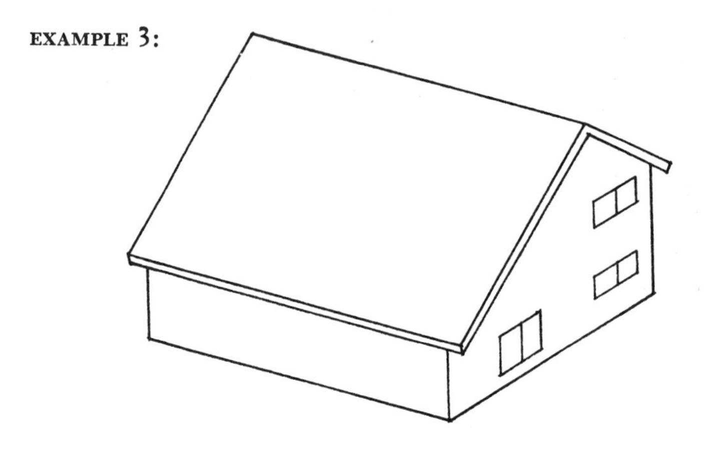

Note: The barge is usually not spliced unless a single length of material is not available.

List the fascia and barge material needed for the following examples:

EXAMPLE 1:

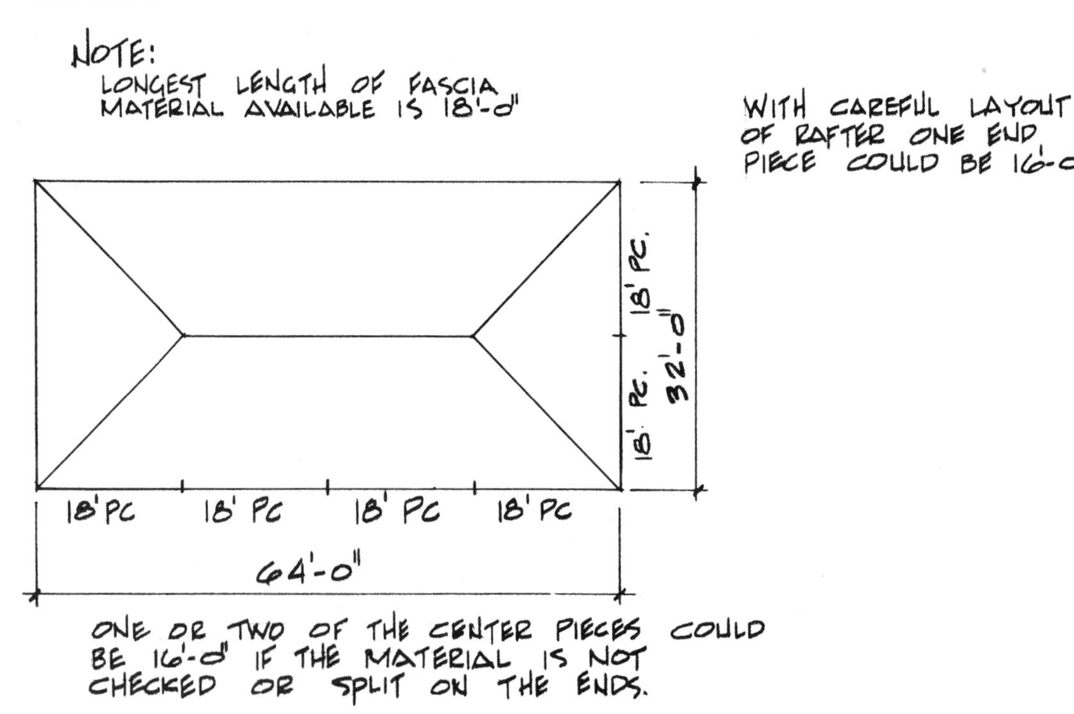

EXAMPLE 2:

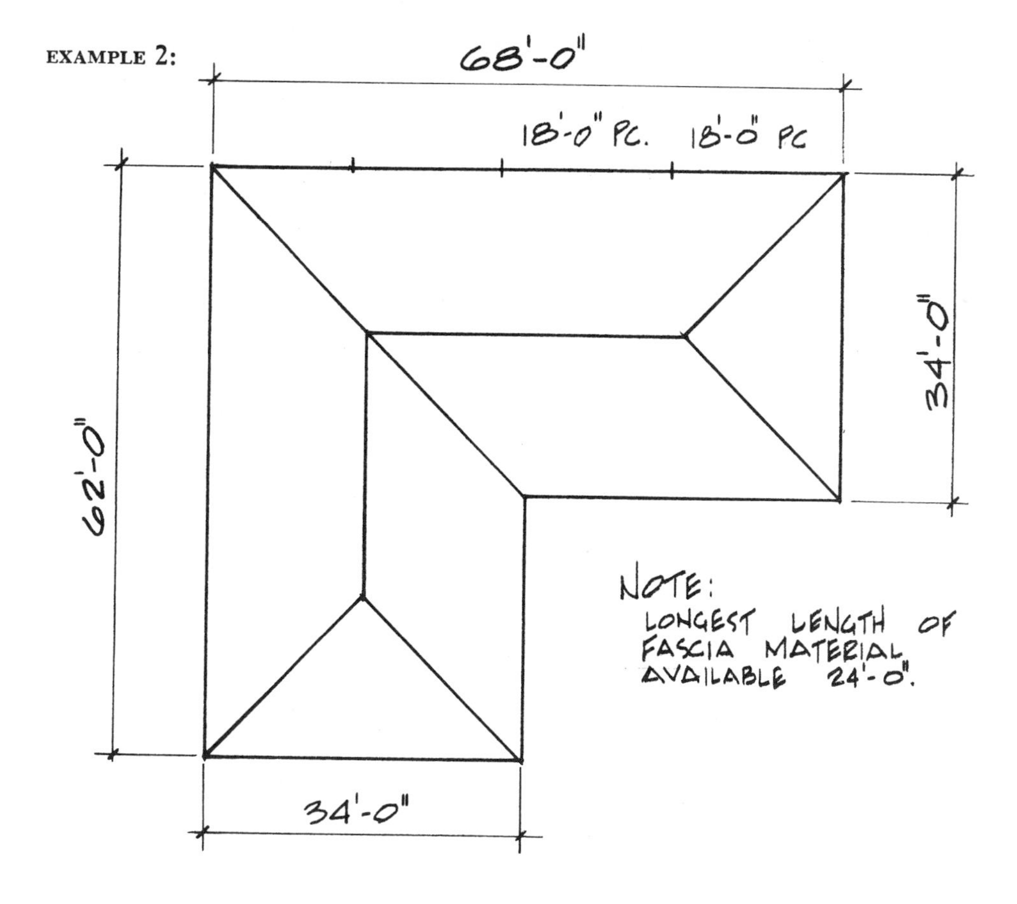

EXAMPLE 3:

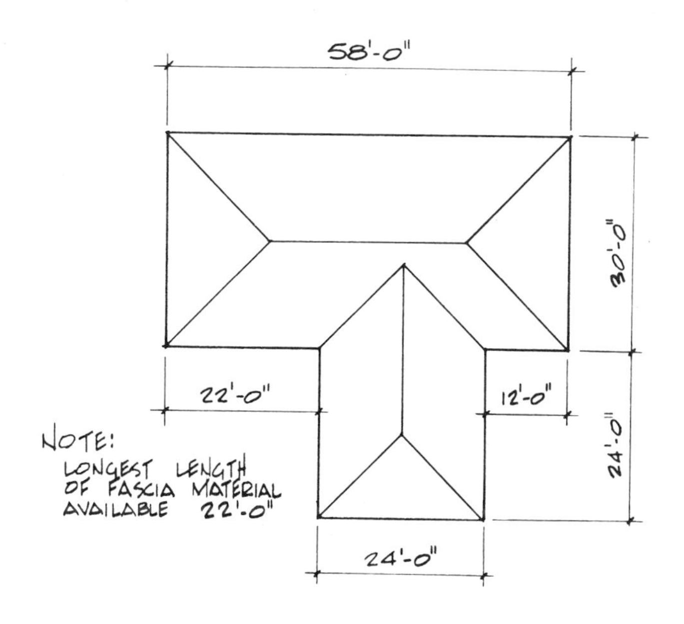

EXAMPLE 4:

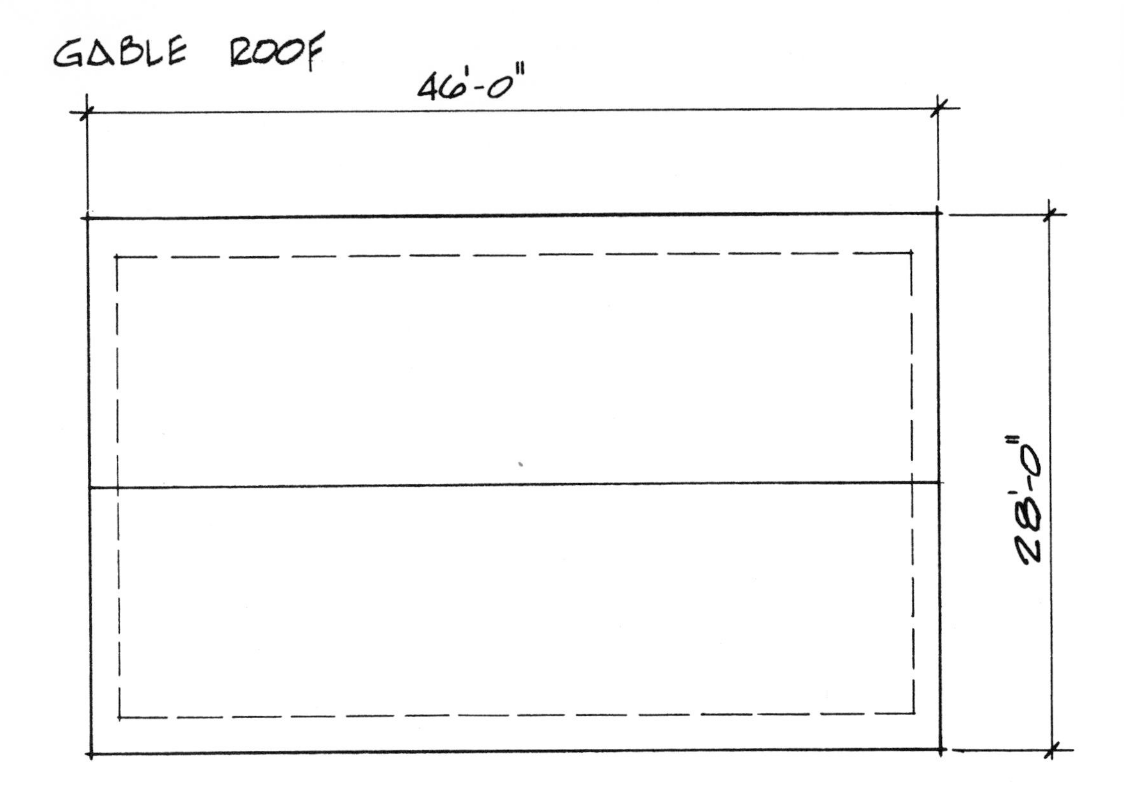

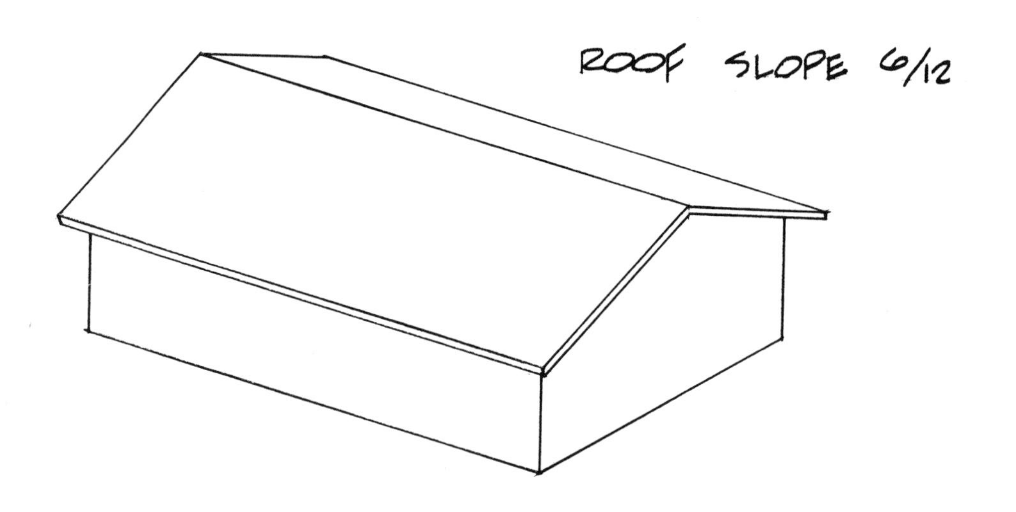

EXAMPLE 5: In the following gable roof examples, the bargeboard lengths will be given. They are found by checking the common rafter lengths or scaling an elevation view.

GABLE ROOF

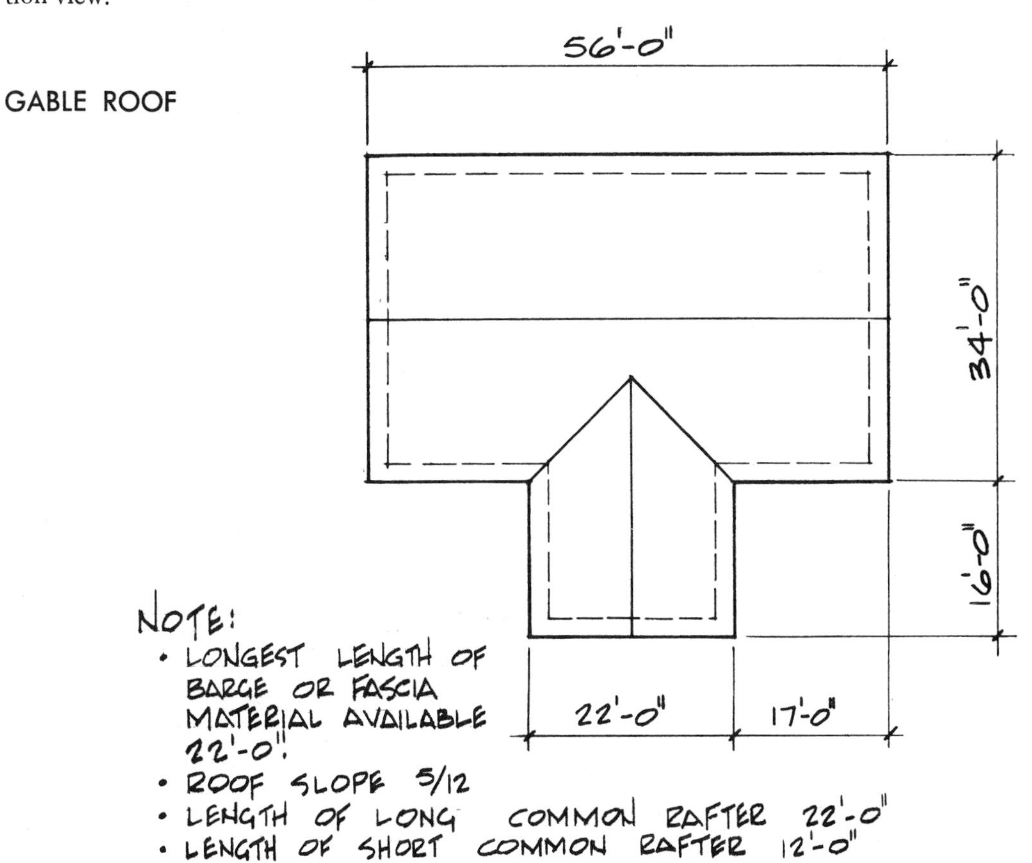

49

Estimating Gable Lookouts

Gable lookouts are used on a closed cornice or rake to support the bargeboard and provide nailing backing for board or plywood soffit material. The gable lookouts for a closed cornice are the same size as the rafter material and, depending on local code requirements, may be spaced 16, 24, or 32 inches on center. The length of the lookouts will vary depending on the rafter layout. Because of this, the estimator must approximate their length when estimating lookout material.

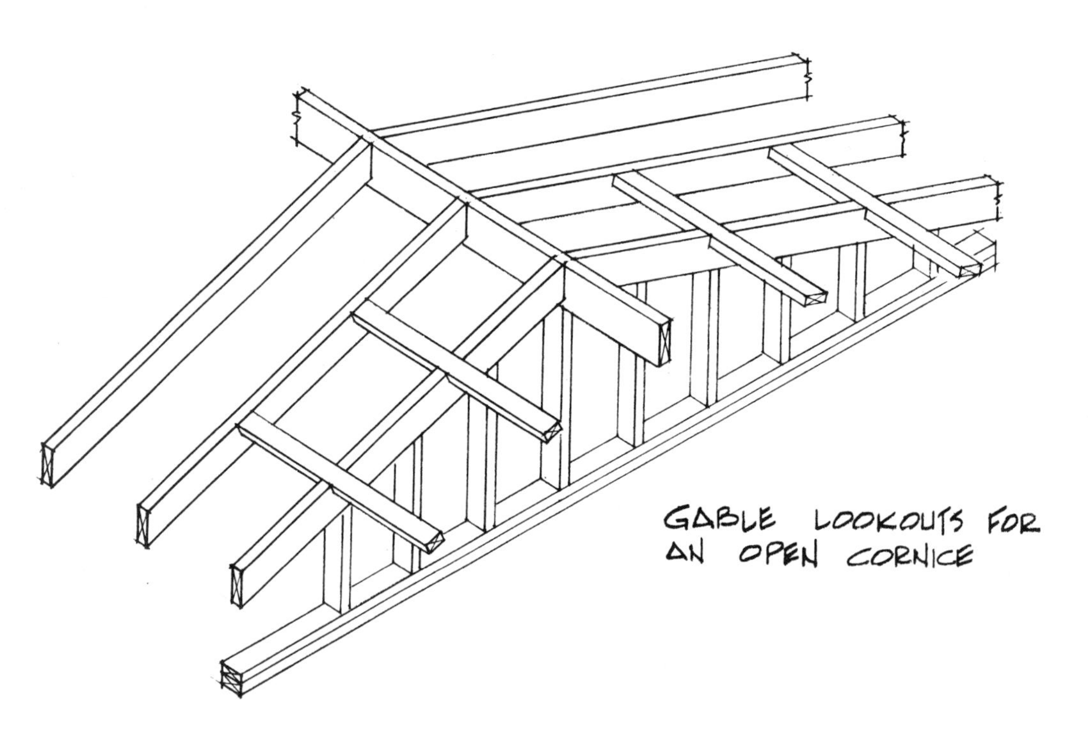

Gable lookouts are used on an open cornice to support the bargeboard. They are usually made from 2 × 4 inch material and are spaced from 24 to 36 inches on center.

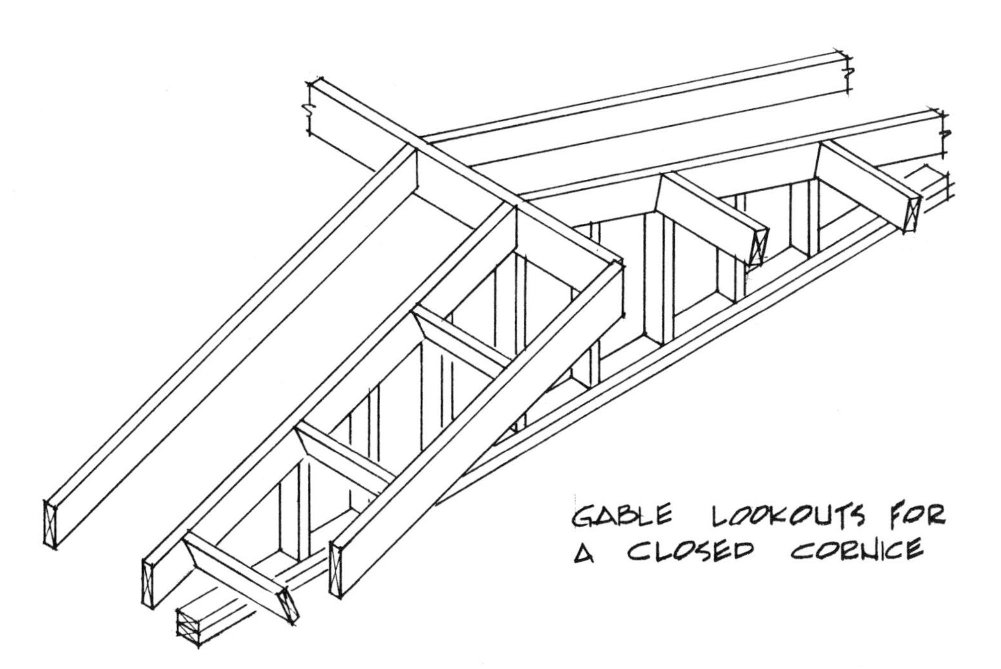

PROCEDURE FOR ESTIMATING GABLE LOOKOUTS

1. To estimate the material needed for gable lookouts, you must determine the center-to-center spacing of the lookouts. This spacing is usually not regulated by local code requirements.
2. The common rafter length, excluding the rafter tail overhang, is then divided by the spacing figure. The result will be the number of lookouts needed for one side of the gable.
3. Multiply the number of lookouts for one gable rafter by the number of gable rafters of the same length. The result will be the total number of gable lookouts needed.
4. Determine the length of the lookouts by scaling from the building plans or by visually approximating and multiplying this length by the number of lookouts for the total lineal footage of material needed.

50 Problems—Estimating Roof Framing Material

PROBLEM 1:

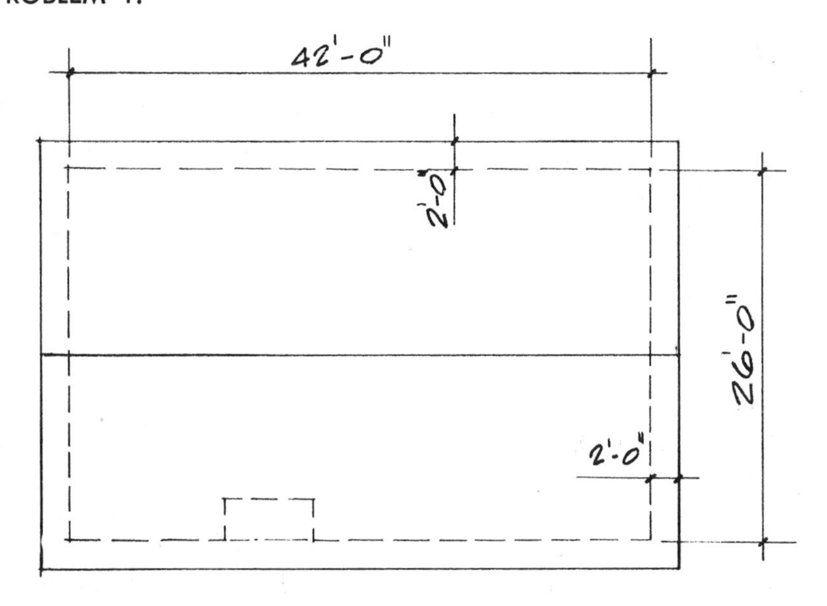

Information needed:

Roof slope 6 / 12
Rafter size 2 × 6
Rafter spacing 2'-0" O.C.
Fascia and bargeboard 2 × 8

ESTIMATE THE FOLLOWING:

1. Number, size, and length of common rafters ______
2. Lineal feet and size of ridge board material ______
3. Lineal feet and size of fascia board ______
4. Number of pieces, size, and length of bargeboard ______

PROBLEM 2:

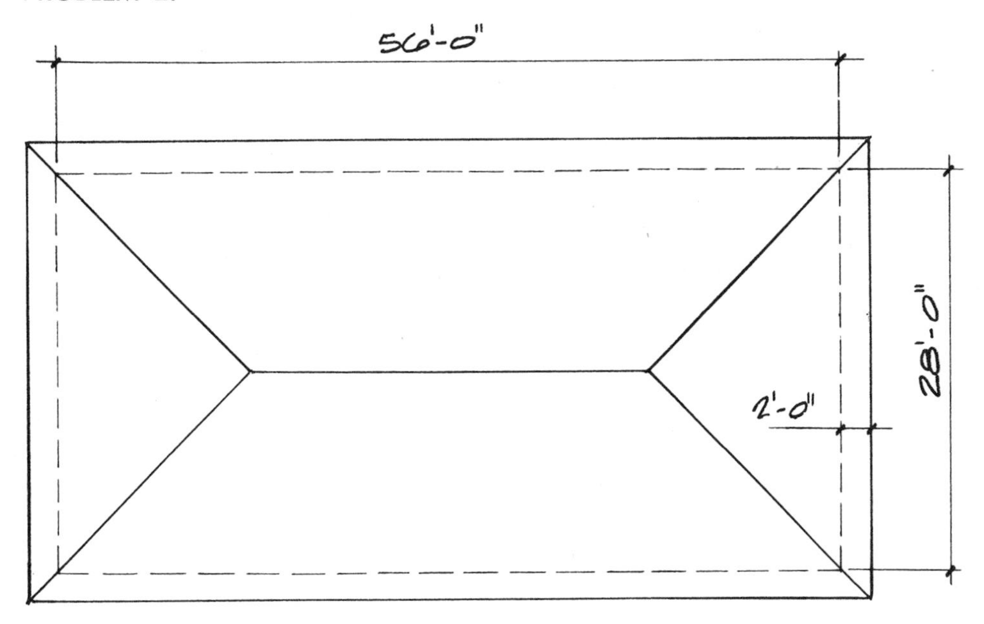

Information needed:

Roof slope 5 / 12
Rafter size 2 × 6
Rafter spacing 16" O.C.
Fascia board 2 × 6

ESTIMATE THE FOLLOWING:

1. Number, size, and length of common rafters ______
2. Number, size, and length of hip rafters ______
3. Lineal feet of ridge board ______
4. Lineal feet and size of fascia board ______

PROBLEM 3:

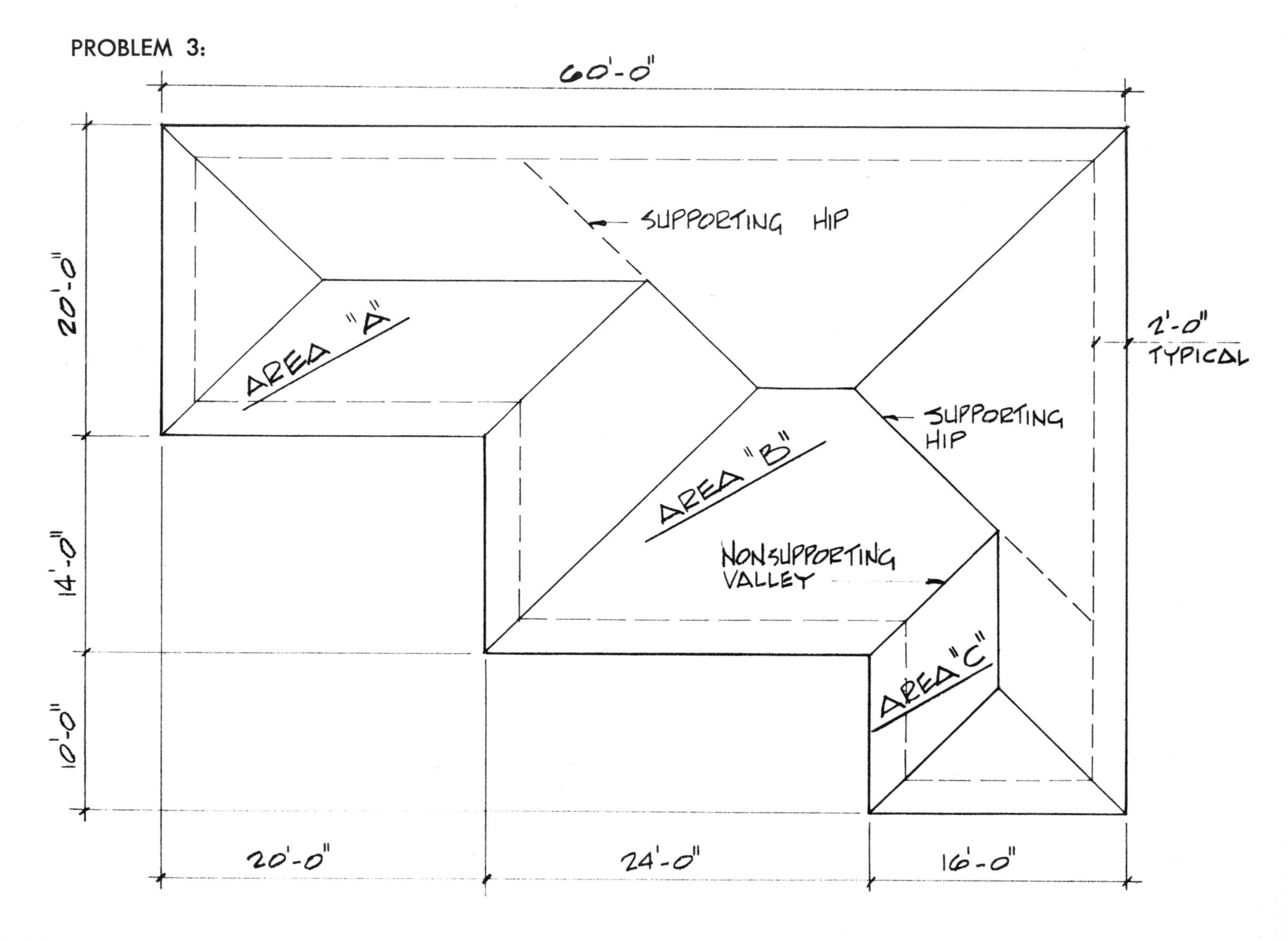

60'-0"
SUPPORTING HIP
AREA "A"
2'-0" TYPICAL
SUPPORTING HIP
AREA "B"
NONSUPPORTING VALLEY
AREA "C"
20'-0"
14'-0"
10'-0"
20'-0"
24'-0"
16'-0"

Information needed:

Roof slope 5-1/2 / 12
Rafter size 2 × 8
Rafter spacing 16″ O.C.
Fascia board 2 × 10

ESTIMATE THE FOLLOWING:

1. Number, size, and length of common rafters for area A ______________
   area B ______________
   area C ______________
2. List the size, length, and number of hip and valley rafters
   area A hips ______________
   area B hips ______________
   area C hips ______________
   area A valley ______________
   area C valley ______________
3. Size and length of 3 ridge board members area A ______________
   area B ______________
   area C ______________
4. Lineal feet and size of fascia board ______________

PROBLEM 4:

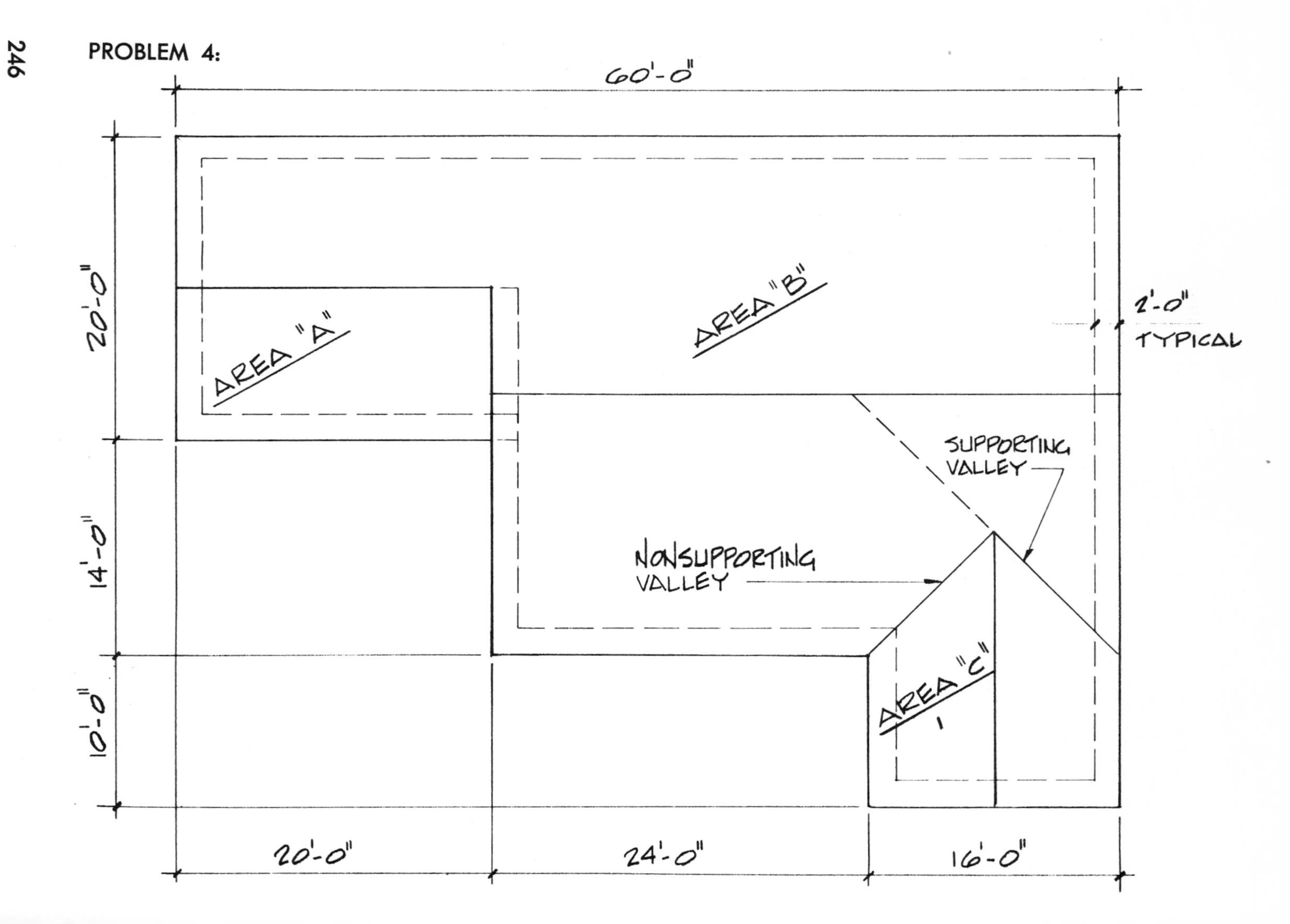
60'-0"
2'-0"
TYPICAL
AREA "A"
AREA "B"
AREA "C"
SUPPORTING VALLEY
NONSUPPORTING VALLEY
20'-0"
14'-0"
10'-0"
20'-0"
24'-0"
16'-0"

Information needed:

Roof slope 4 / 12
Rafter size 2 × 6
Rafter spacing 16″ O.C.
Fascia and bargeboard 2 × 8

ESTIMATE THE FOLLOWING:

1. Number, size, and length of common rafters for area A ________
   area B ________
   area C ________
2. List the size and length of supporting and nonsupporting valley rafters
   supporting ________
   nonsupporting ________
3. Lineal feet of ridge board needed for area A ________
   area B ________
   area C ________
4. Lineal feet and size of fascia material ________
5. Number, size, and length of barge members for area A ________
   area B ________
   (long length) ________
   (short length) ________
   area C ________

51

Types of Roof Sheathing Material

Roof sheathing is an underlayment of wood board, plywood, or fiberboard material that is nailed to the rafters in preparation for some sort of roofing material. Following are the various types of roof sheathing used in today's home industry.

1. Common sheathing boards usually 1″ × 4″, 1″ × 6″, or 1″ × 8″.

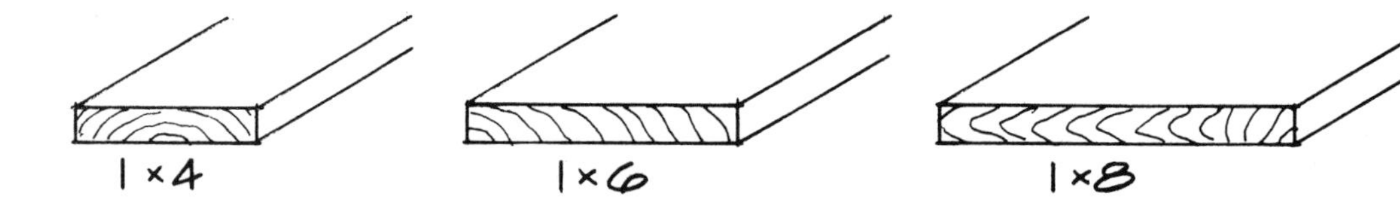

2. Shiplap boards usually 1″ × 6″ or 1″ × 8″ V-groove. Boards may be purchased surfaced smooth or resawn for rustic look.

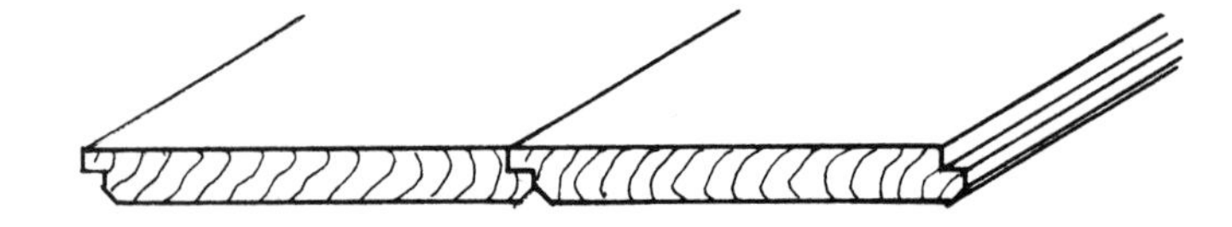

3. Tongue-and-groove boards usually 1″ × 6″ or 1″ × 8″ V-groove. Boards may be purchased surfaced smooth or resawn for rustic look.

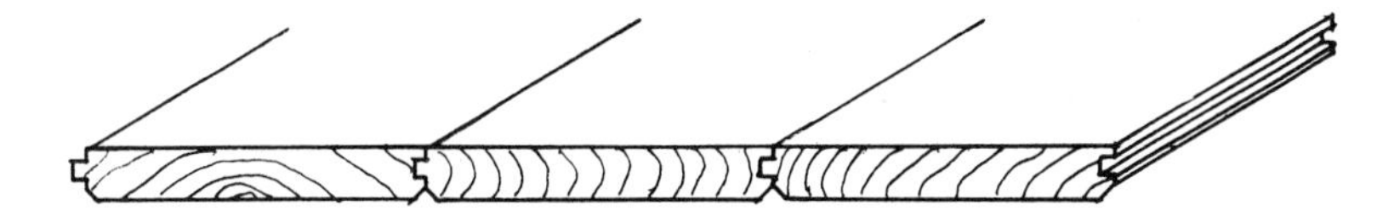

4. Select decking tongue-and-groove boards (2″ × 6″ or 2″ ×8″ are usually used where underside will be exposed, as in a vaulted or cathedral ceiling). Boards may be purchased surfaced smooth or resawn for rustic look.

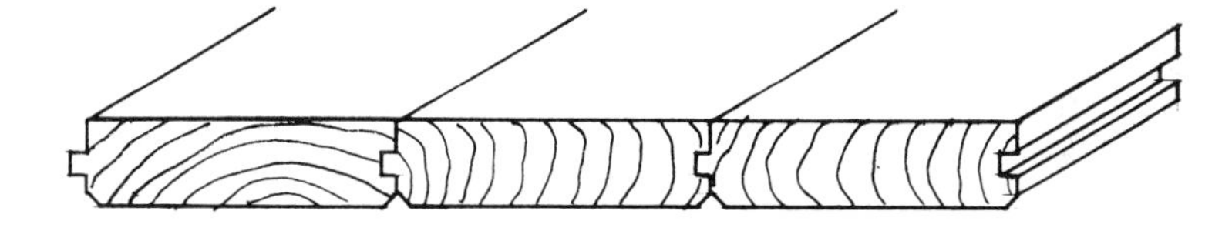

5. Plywood roof sheathing usually 1/2″ × 4′ × 8′. When using plywood, correct rafter spacing and layout is critical for least amount of waste.

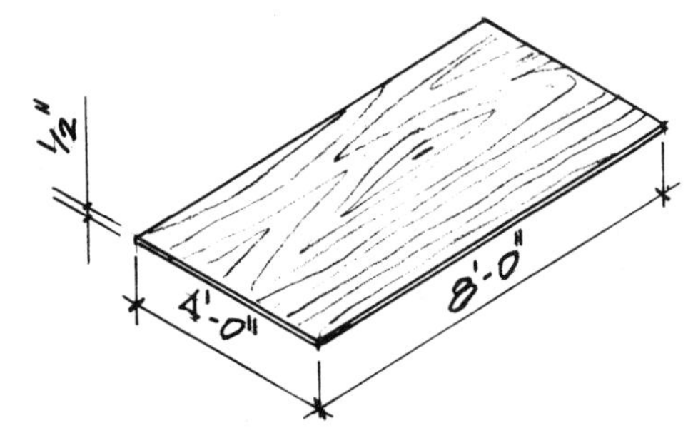

6. Fiberboard roof sheathing is a thermo-insulating material that may be purchased in 2″ × 2′ × 8′ tongue-and-groove panels or in 3″ × 2′ × 8′ tongue-and-groove panels. It is usually used on flat-pitched or moderately curved roofs. It comes with a white face for exposure and may be covered with a built-up or a shingle roof.

Note: Maximum framing centers for 2″ material 32″ O.C.
Maximum framing centers for 3″ material 48″ O.C.

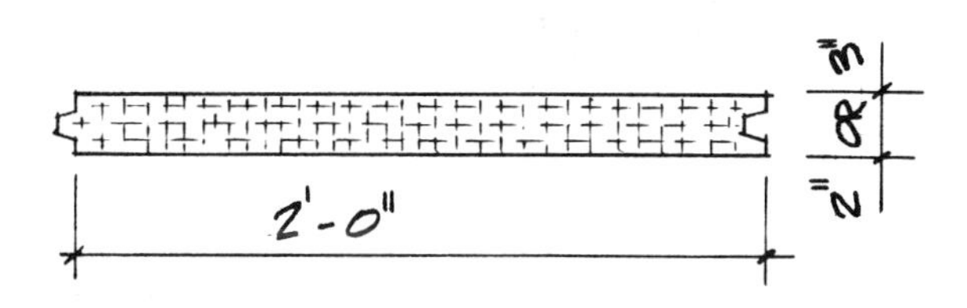

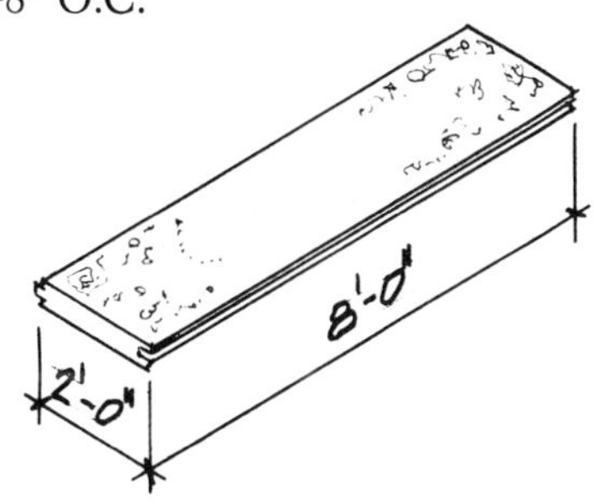

52

Eave Details

To estimate roof sheathing, an estimator has to understand the different types of overhang or eave details. Below are various eave details and the materials used for their construction. These eave details may be used as is, or combined to give the aesthetic look desired for a particular style of construction.

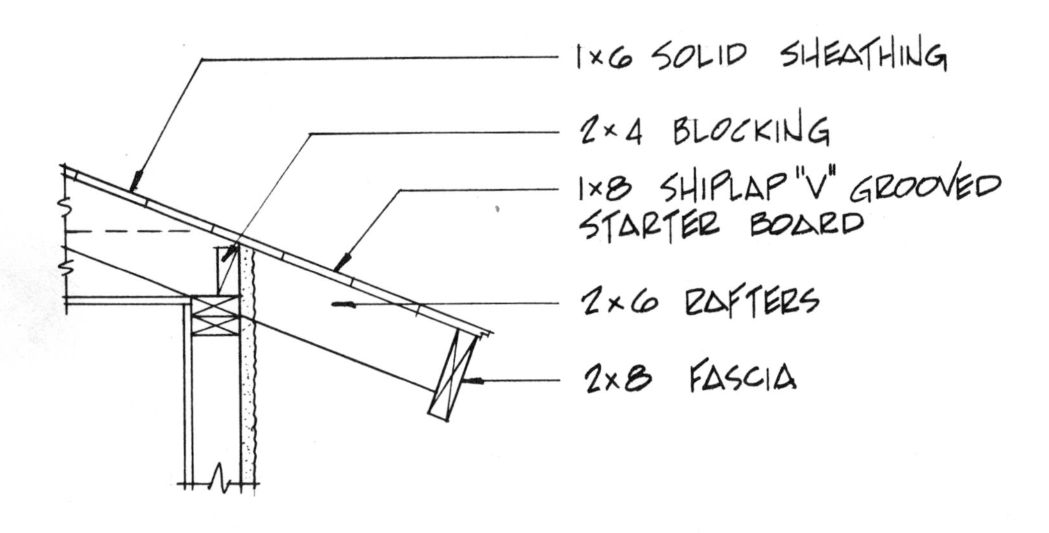

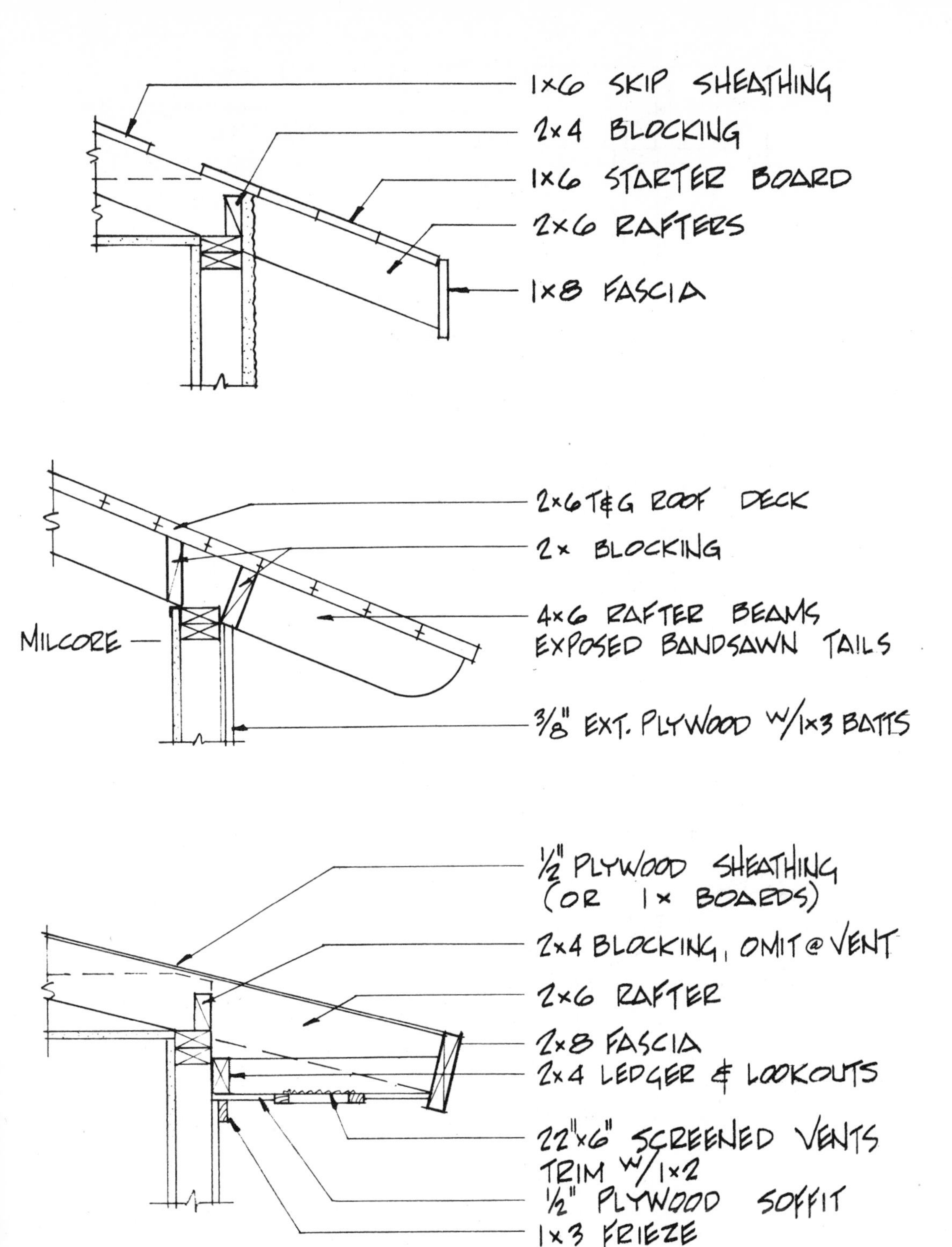
1x6 SKIP SHEATHING
2x4 BLOCKING
1x6 STARTER BOARD
2x6 RAFTERS
1x8 FASCIA
2x6 T&G ROOF DECK
2x BLOCKING
4x6 RAFTER BEAMS
EXPOSED BANDSAWN TAILS
MILCORE
3/8" EXT. PLYWOOD W/1x3 BATTS
1/2" PLYWOOD SHEATHING
(OR 1x BOARDS)
2x4 BLOCKING, OMIT @ VENT
2x6 RAFTER
2x8 FASCIA
2x4 LEDGER & LOOKOUTS
22"x6" SCREENED VENTS
TRIM W/1x2
1/2" PLYWOOD SOFFIT
1x3 FRIEZE

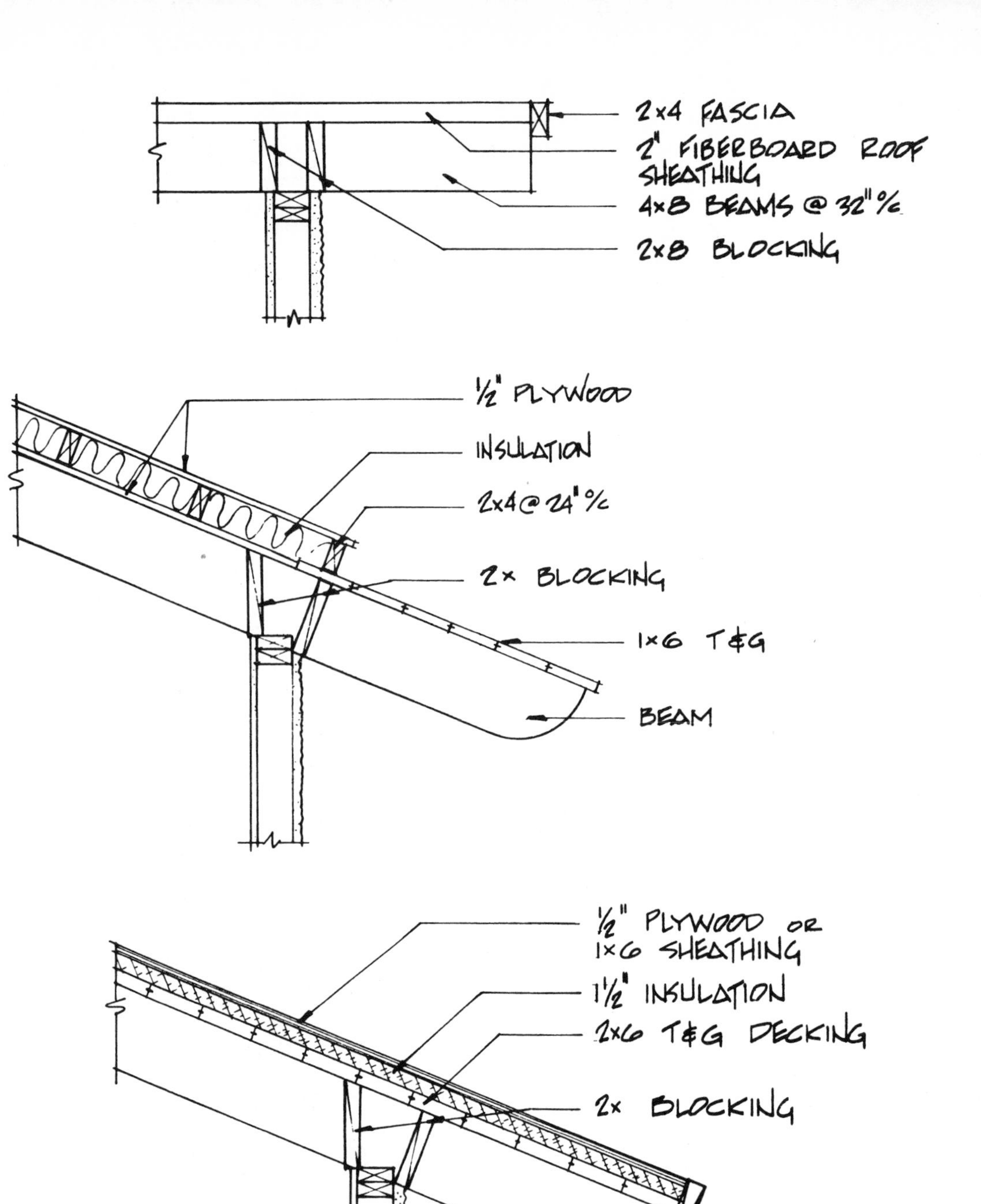
2×4 FASCIA
2" FIBERBOARD ROOF SHEATHING
4×8 BEAMS @ 32" O/C
2×8 BLOCKING
1/2" PLYWOOD
INSULATION
2×4 @ 24" O/C
2× BLOCKING
1×6 T&G
BEAM
1/2" PLYWOOD OR 1×6 SHEATHING
1 1/2" INSULATION
2×6 T&G DECKING
2× BLOCKING
FASCIA
RAFTER BEAM

53

Estimating Starter Board, Solid, and Skip Sheathing

Some roofs will be completely covered with one type of roof sheathing material. Other roofs may have a different type of roof sheathing on the building overhang. If this is the case, the roof sheathing material for the building overhang will have to be calculated separately.

EXAMPLE:

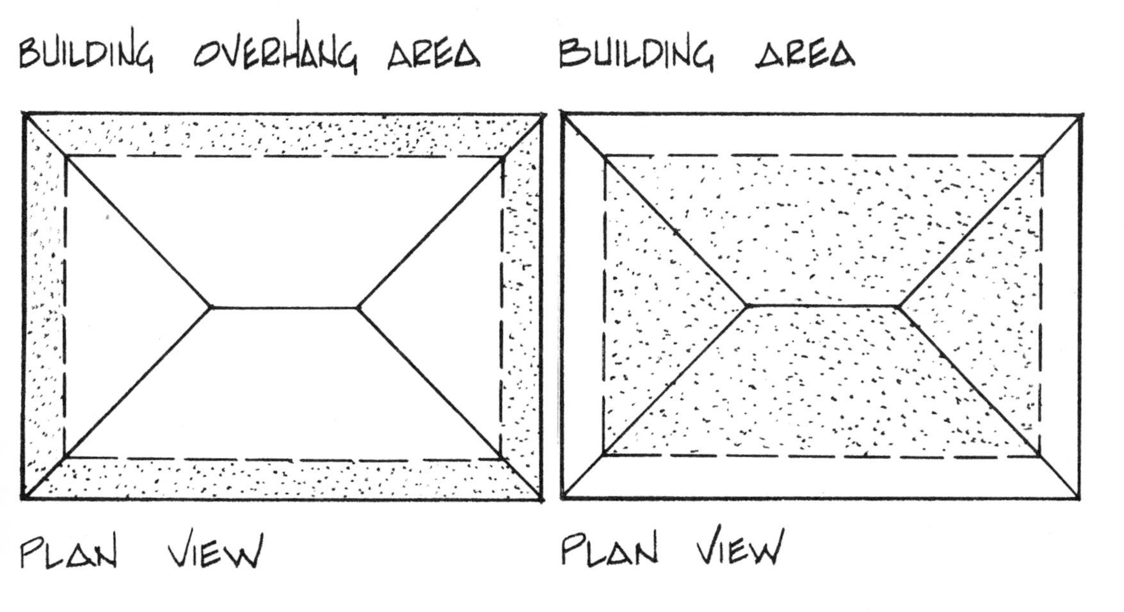

ESTIMATING STARTER BOARD, SOLID, AND SKIP SHEATHING

There are also roofs that have the sheathing boards spaced apart. This is called *skip sheathing* and is usually used when the roofing material is wood shingle, wood shake shingle, and certain types of tile shingles.

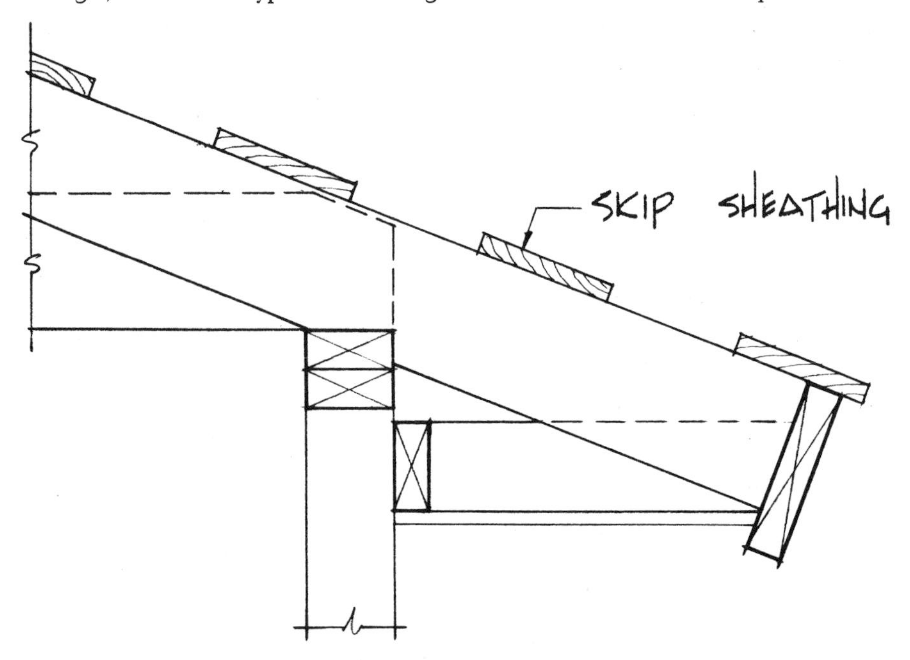

Two methods are used to estimate the amount of roof sheathing needed to cover a roof: the first method, scaling the building plan, can be used easily on relatively simple-shaped roofs. The second method uses a table that helps calculate the square foot area of the roof, and from this the amount of roof sheathing may be estimated.

METHOD 1

PROCEDURE:

1. From a section or elevation view, scale the rafter length on which the sheathing is nailed.
2. Change the rafter length to inches.
3. Divide the total inches calculated above by the width of the sheathing material used. The result will be the number of boards needed to cover the roof from the eave to the ridge.
   If using 1 × 6 sheathing divide by 5-1/2″ or its decimal equivalent 5.5.
   If using 1 × 8 sheathing divide by 7-1/4″ or its decimal equivalent 7.25.
4. Double the number of boards for both sides of the roof.
5. Multiply the number of boards by the length of the roof.
6. Add 5% for cutting waste.

Example Problem:

Estimate the lineal feet of roof sheathing needed to cover the whole roof in the following illustration.

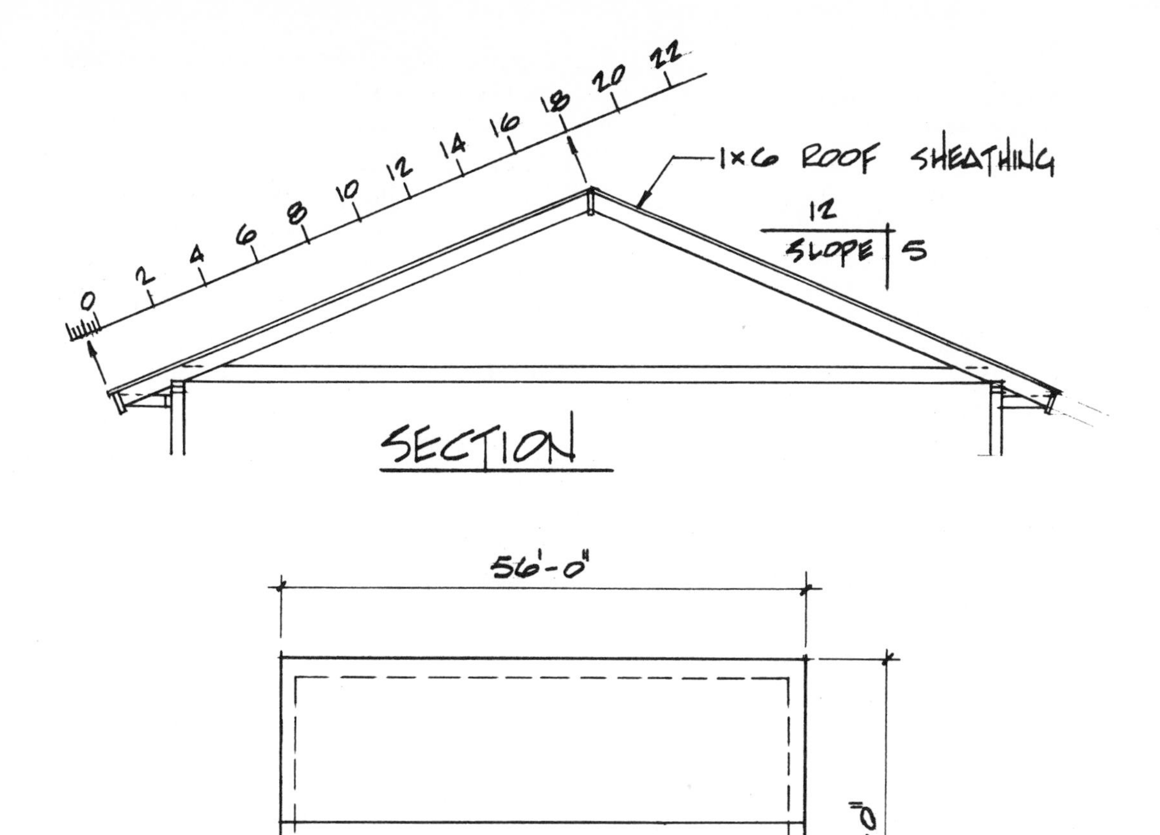

PROCEDURE FOR ESTIMATING ROOF SHEATHING

1. Scale the rafter length from the section view as illustrated: 18′-6″ (decimal equivalent 18.5).
2. Change the rafter length to inches.

$$\begin{array}{r} 18'\,6'' \\ \times\ 12' \\ \hline 36 \\ 18 \\ \hline 216 + 6'' = 222 \text{ inches} \end{array}$$

3. Divide 222 inches by 5.5 (the width of a 1 × 6 sheathing board) to get the number of boards: 222 ÷ 5.5 = 40.3 or 41 boards.
4. Double the number of boards for both sides of the roof: 41 × 2 = 82 boards.
5. Multiply the number of boards (82) by the length of the roof (56′-0″): 82 × 56 = 4,592 lin. ft. of sheathing.
6. Add 5% for cutting waste.

```
    4,592
  × 1.05
   22960
   0000
  4592
4,821.60 or 4,822 lin. ft. of 1 × 6
sheathing
```

METHOD 2

ESTIMATING ROOF SHEATHING FROM THE SQUARE FOOT AREA OF THE ROOF

The square foot area of the roof may be found by using the same table that was used to find common rafter lengths illustrated below. The constant from Table 53-1 corresponding to the roof slope, multiplied by the horizontal square foot area of the roof will give the square foot area of the sloped roof. This method is quite accurate and will give the roof area no matter how complicated the roof design. It must be kept in mind, however, that if the roof has both roof sheathing and starter board, the estimating problem, as in the first method, would have two parts. First, the amount of starter board would have to be estimated, and then the amount of roof sheathing.

TABLE 53-1 Estimating Square Foot Roof Areas

Roof Slope	*Constant*	*Roof Slope*	*Constant*
2-1/2 / 12	1.02	8 / 12	1.20
3 / 12	1.03	8-1/2 / 12	1.23
3-1/2 / 12	1.04	9 / 12	1.25
4 / 12	1.05	9-1/2 / 12	1.28
4-1/2 / 12	1.07	10 / 12	1.30
5 / 12	1.08	12 / 12	1.41
5-1/2 / 12	1.10	14 / 12	1.54
6 / 12	1.12	16 / 12	1.67
6-1/2 / 12	1.14	18 / 12	1.80
7 / 12	1.16	20 / 12	1.94
7-1/2 / 12	1.18	22 / 12	2.09
		24 / 12	2.24

PROCEDURE FOR ESTIMATING STARTER BOARD

1. Calculate the number of boards needed to cover the eave.
2. Multiply the number of boards by the perimeter of the roof line.
3. Add 5% for cutting waste.

Example Problem:

Estimate the lineal feet of 1 × 6 tongue-and-groove starter board in the following illustration.

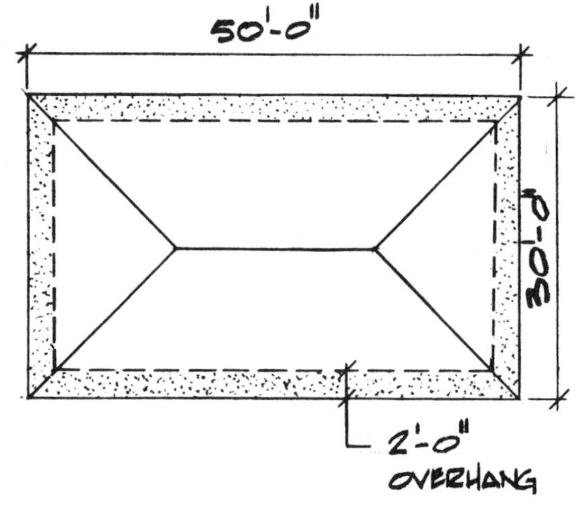

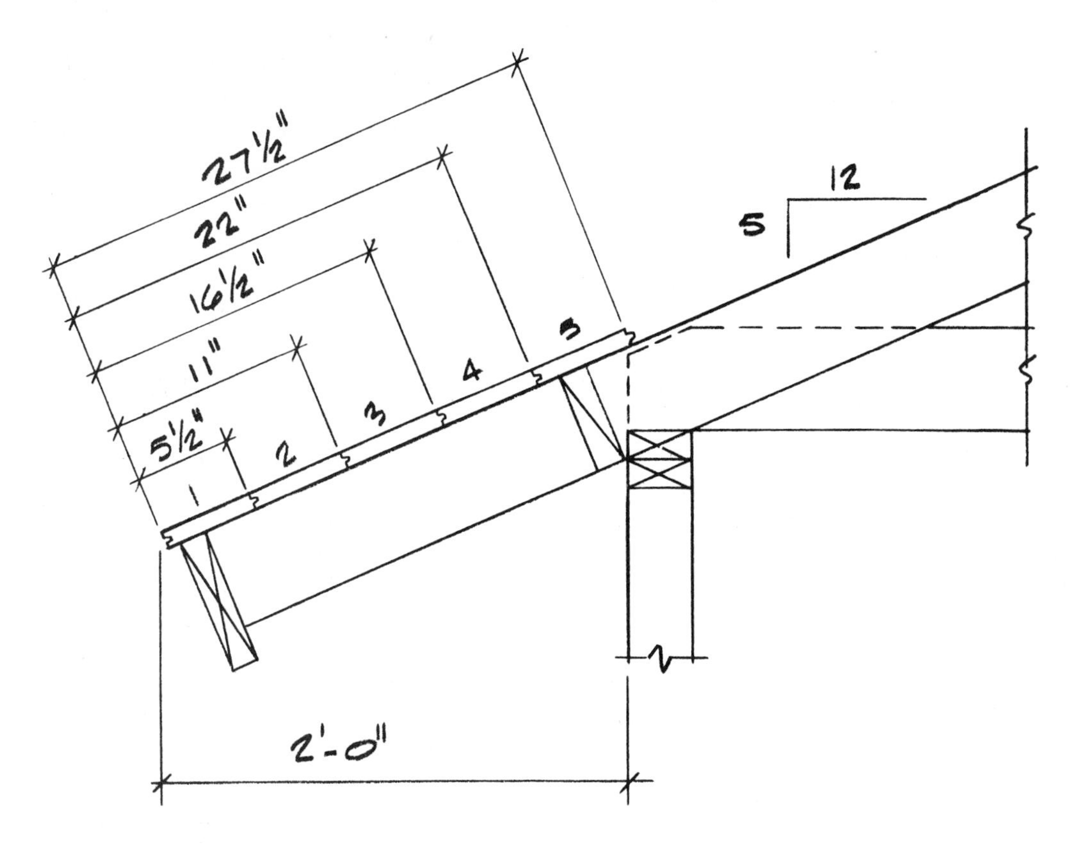

PROCEDURE FOR ESTIMATING STARTER BOARD

1. Calculate the number of boards to cover the eave. The estimator may scale from a section or elevation view the approximate number of inches the starter board is to cover as illustrated above (approximately 28"). This may then be divided by the width of the material used, in this case 5.5 for 1 × 6 starter board. The result is the number of boards to cover the eave: 28 ÷ 5.5 = 5 boards.
   Another method for estimating the number of boards to cover the eave is to find the constant from the roof area table that corresponds to the 5 / 12 roof slope: 5 / 12 slope = 1.08 constant.
   This figure is then multiplied by the overhang 2'-0" or 24" to find the approximate length of eave to be covered by starter boards: 1.08 × 24" = 25.92 inches or approximately 26" to be covered.
   This can then be divided by the 5.5 width of the starter board: 26 ÷ 5.5 = 4.7 or 5 boards.
2. Multiply the number of boards to cover the eave (5) by the perimeter of the roof line (160 lin. ft.): 5 × 160 = 800 lin. ft.
3. Add 5% for cutting waste.

```
    800
 × 1.05
   4000
  8000
840.00 lin. ft. of starter board material
```

Example Problem:

Now estimate the amount of roof sheathing to cover the rest of the roof as illustrated below.

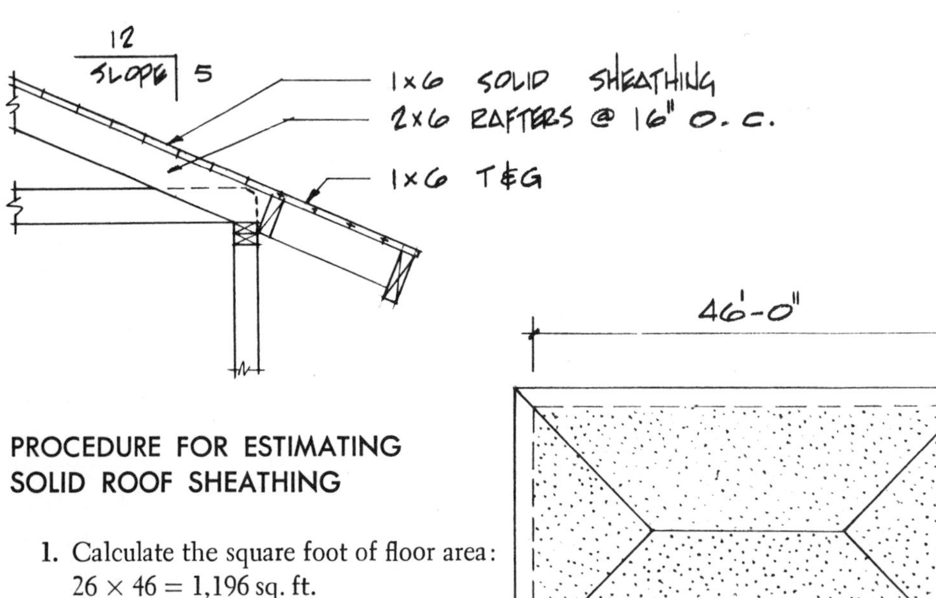

PROCEDURE FOR ESTIMATING SOLID ROOF SHEATHING

1. Calculate the square foot of floor area: 26 × 46 = 1,196 sq. ft.
2. Multiply the floor area (1,196) by the roof constant for a 5 / 12 slope (1.08).

```
   1,196
 ×  1.08
    9568
  11960
1,291.68 or 1,292 sq. ft. of roof area
```

3. Multiply the roof area (1,292) by 2 (the number of lineal feet of 1 × 6 material in a sq. ft.).

```
   1,292
 ×     2
   2,584 lin. ft.
```

Note: If 1 × 8 sheathing is used, multiply the roof area by 1-1/2 (1-1/2 lin. ft. of 1 × 8 material in a sq. ft.).

4. Multiply the 2,584 lin. ft. by the waste factor from the following table (53-2).

TABLE 53-2 Waste Factors for Roof Sheathing

Size	*Waste Factor for Square-Edge Boards*	*Waste Factor for T&G or Shiplap Boards*
1 × 6	1.22	1.35
1 × 8	1.22	1.32

```
     2,584
  ×   1.22
      5168
     5168
    2584
3,152.48 or 3,154 lin. ft. of 1 × 6 material
```

Example Problem:

Estimate the lineal feet of starter board in the following illustration.

For a hip roof, multiply the number of boards to cover the overhang by the perimeter of the roof line, and add 5% for waste.

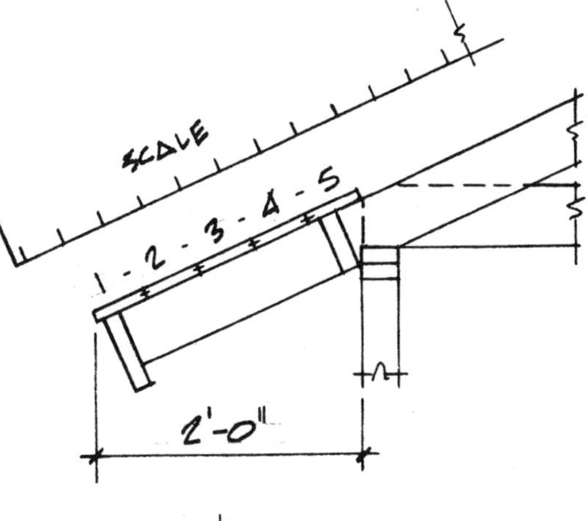

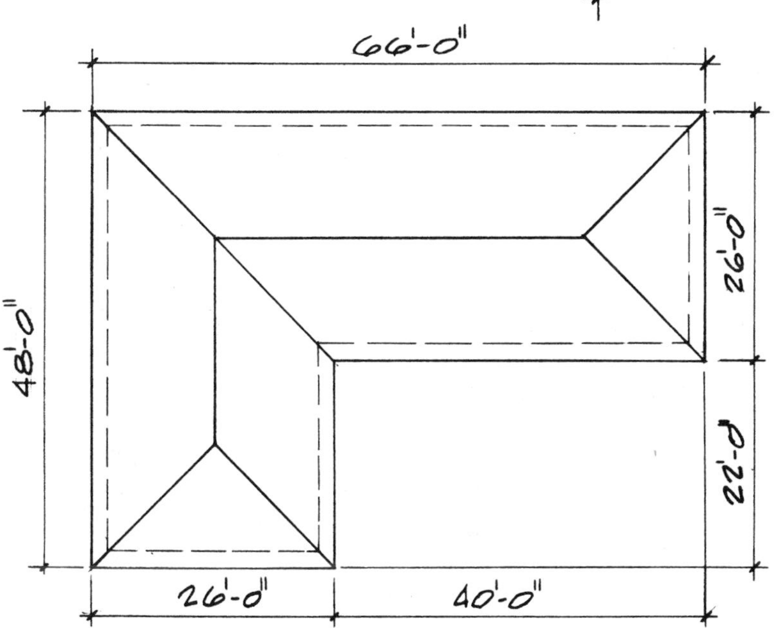

PROCEDURE FOR ESTIMATING STARTER BOARD FOR A HIP ROOF

1. Scale or calculate the number of boards to cover the overhang: 5 boards.
2. Calculate the perimeter of the roof line: 66′ + 26′ + 40′ + 22′ + 26′ + 48′ = 228 lin. ft.
3. Multiply the perimeter (228) by the number of boards to cover the overhang (5): 228 × 5 = 1,140.
4. Add 5% for cutting waste: 1,140 × 1.05 = 1,197 or 1,198 lin. ft. of 1 × 6 material.

Example Problem:

Estimate the lineal feet of starter board in the following illustration.

For a gable roof, multiply the number of boards to cover the overhang by the perimeter of the roof line, taking into account the length of the rafters on the gable ends of the building as illustrated below. Add 5% for waste.

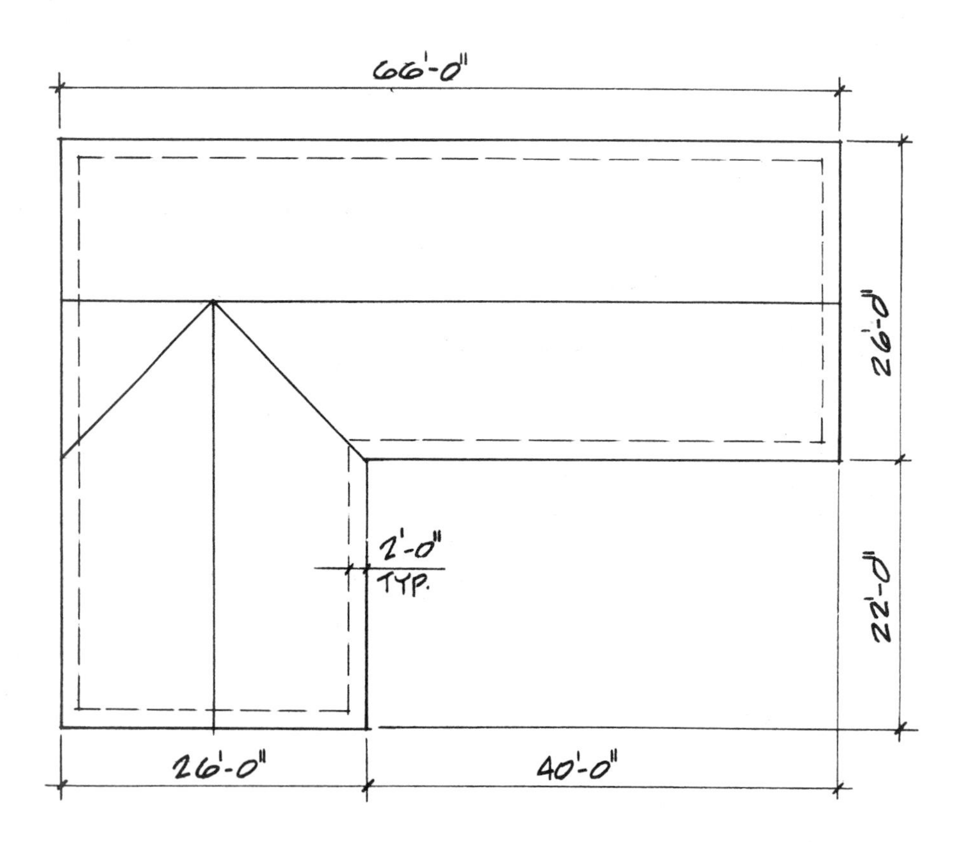

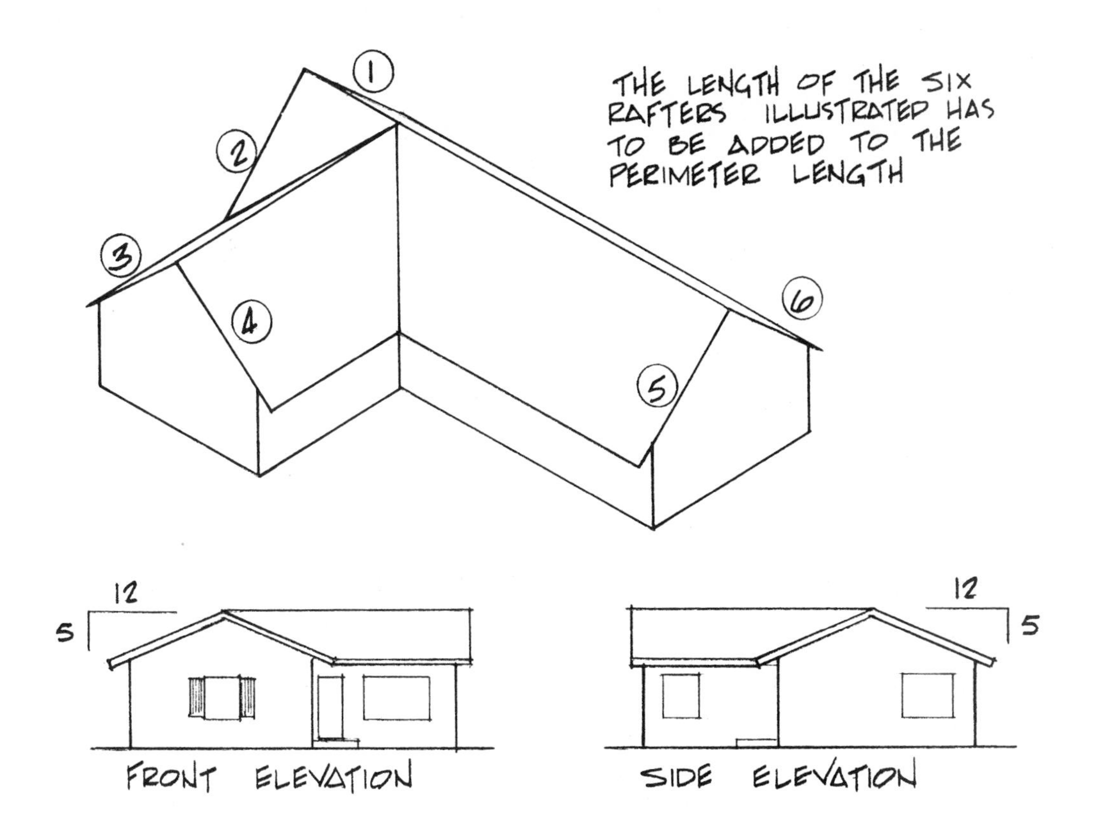

Information needed:

Roof slope 5 / 12
Rafter constant for a 5 / 12 slope (1.08)
Starter board material 1 × 6 tongue-and-groove
Overhang 2′-0″ TYP.

PROCEDURE:

1. Find the rafter length by multiplying the rafter constant 1.08 by the total rafter run of 13′-0″.

$$\begin{array}{r} 1.08 \\ \times \quad 13 \\ \hline 14.04 \text{ lin. ft.} \end{array}$$

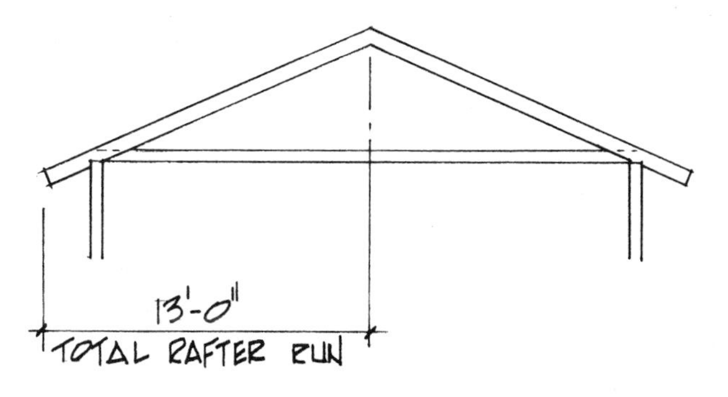

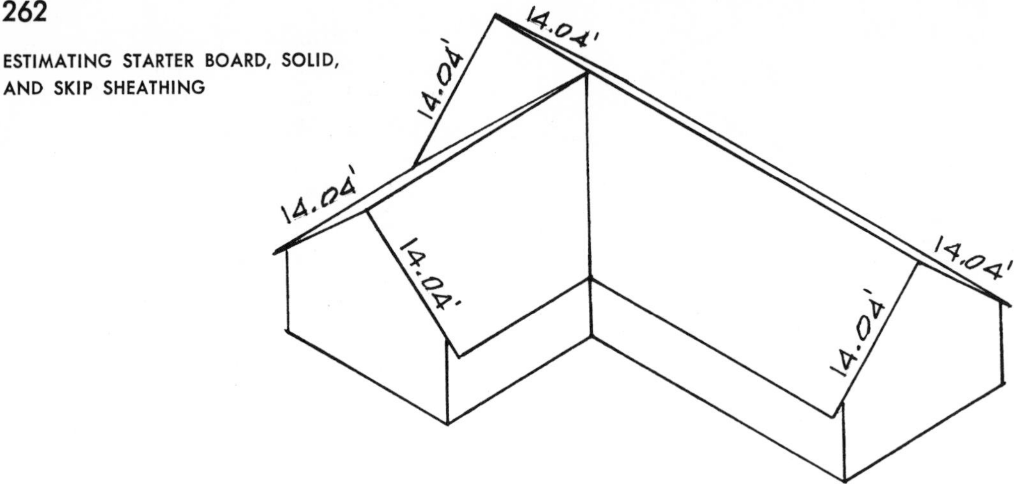

2. Multiply the rafter length 14.04 by the number of gable rafters of the same length (6): $6 \times 14.04 = 84.24$ (for convenience, round to the nearest foot) 85 lin. ft.

3. Add the 85 lineal feet from above to the other perimeter measurements.

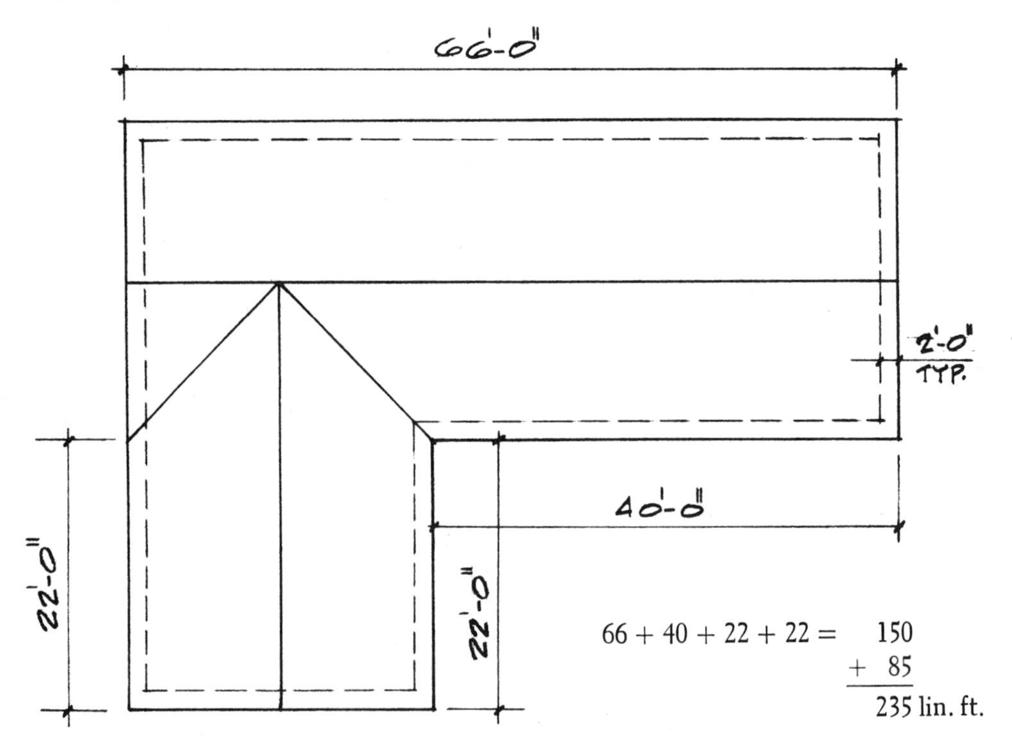

$$66 + 40 + 22 + 22 = 150$$
$$+\ 85$$
$$235 \text{ lin. ft.}$$

4. Multiply the 235 lineal feet from above by the number of 1 × 6 boards it will take to cover a 2′-0″ overhang: 235 × 5 = 1,175 lin. ft.

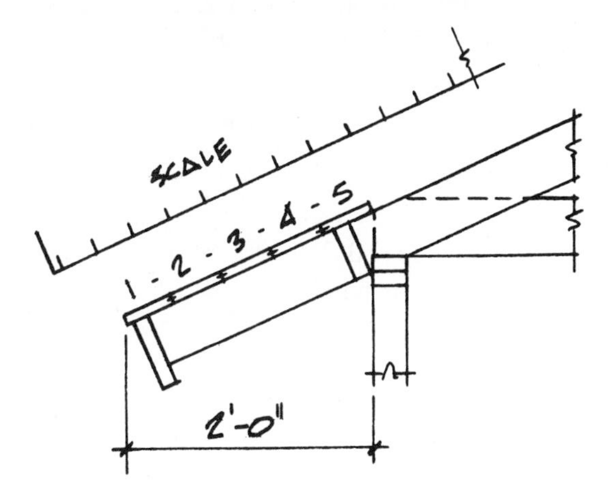

5. Add 5% for waste.

1,175 × 1.05 = 1,233.75 or
1,234 lin. ft. of 1 × 6 T&G material

SKIP SHEATHING

When skip sheathing is used, the spacing of the sheathing boards will depend on the type and size of roofing material used. The usual spacing for wood shingles and wood shakes when using 1 × 6 sheathing material is illustrated below.

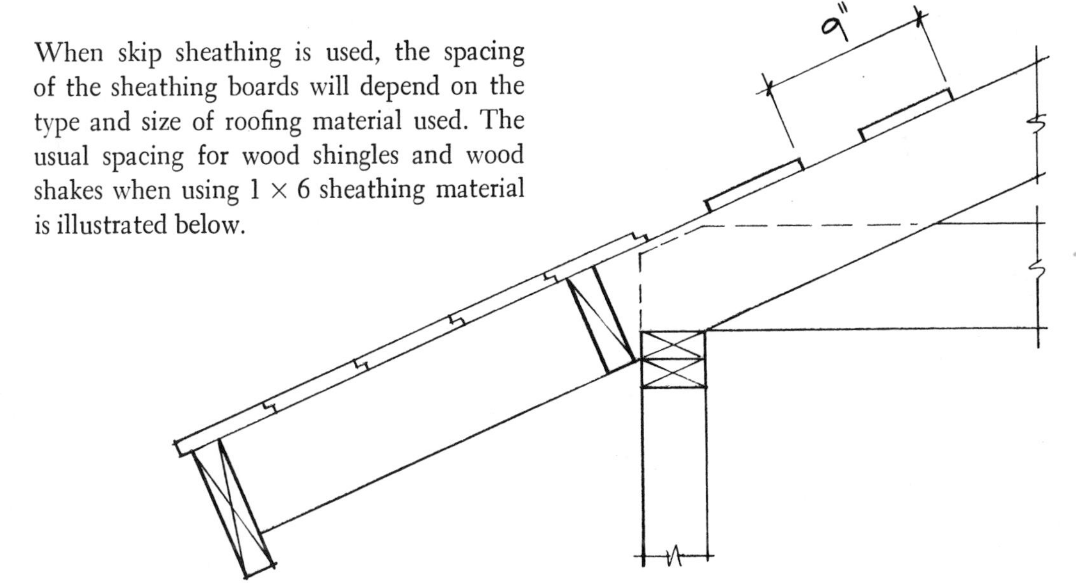

ESTIMATING PROCEDURE FOR SKIP SHEATHING

1. The procedure for estimating skip sheathing is the same as for estimating solid roof sheathing on p. 258.
2. Take 66% of the amount estimated above for the skip sheathing. An allowance has already been taken for the size differential of material and cutting waste.

EXAMPLE:

1. The answer to the previous example problem for estimating solid roof sheathing was 3,154 lin. ft.
2. For the amount of skip sheathing multiply 3,154 lineal feet by 66%.

```
  3,154
×   .66
-------
  18924
 18924
-------
```

2,081.64 or 2,082 lin. ft. of 1 × 6 skip sheathing material.

54

Problems—Roof Sheathing and Starter Board Material

PROBLEM 1:

ESTIMATE THE FOLLOWING:

1. Square foot area the roof covers (flat plane) __________
2. Square foot area of the roof __________
3. Lineal feet of 1 × 8 sheathing to cover roof (including 22 percent waste) __________

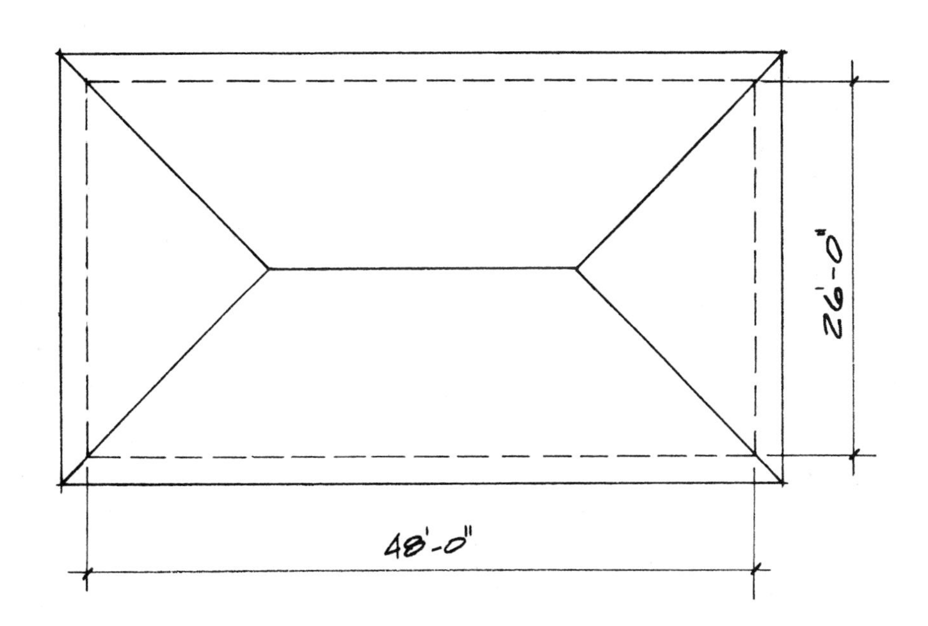

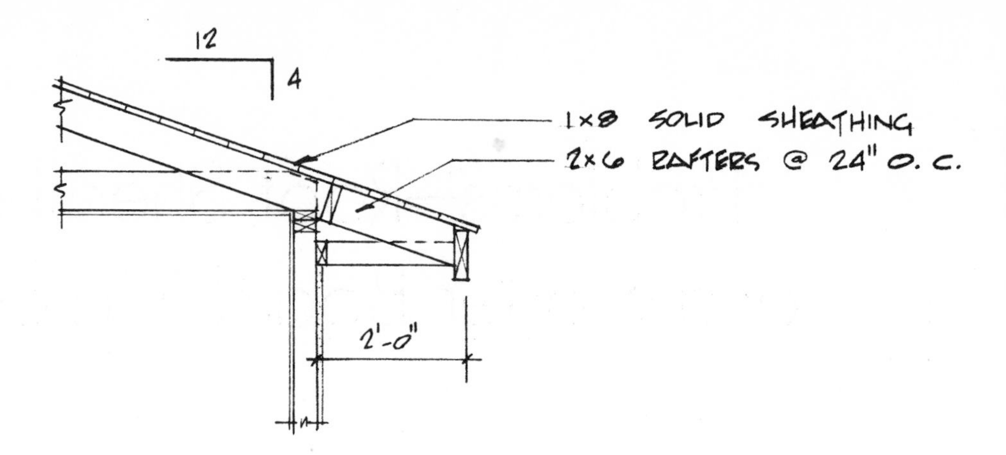

PROBLEM 2:

ESTIMATE THE FOLLOWING:

Starter board

1. Lineal feet of perimeter eave line __________
2. Number of 1 × 6 boards to cover 2′-0″ eave __________
3. Lineal feet of 1 × 6 T&G starter board (including 5% waste) __________

Roof sheathing

4. Square footage of building floor area __________
5. Square footage of roof area above floor area __________
6. Lineal feet of 1 × 6 solid roof sheathing (including 22 percent waste) __________

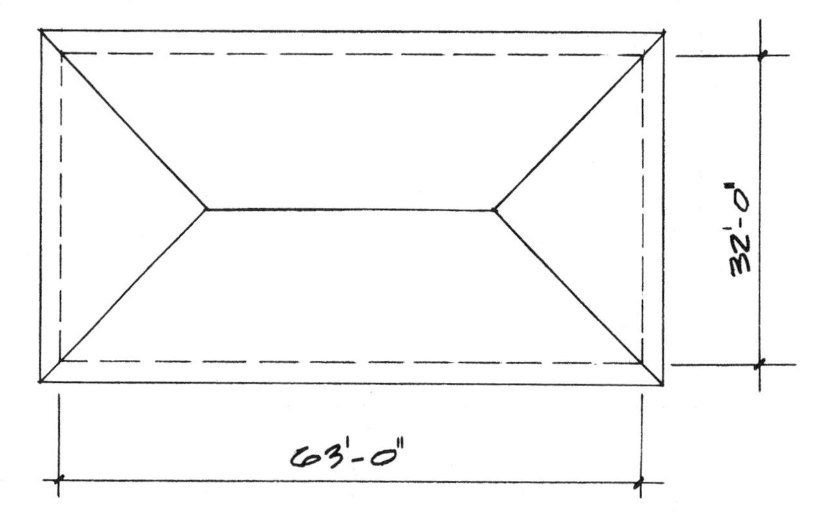

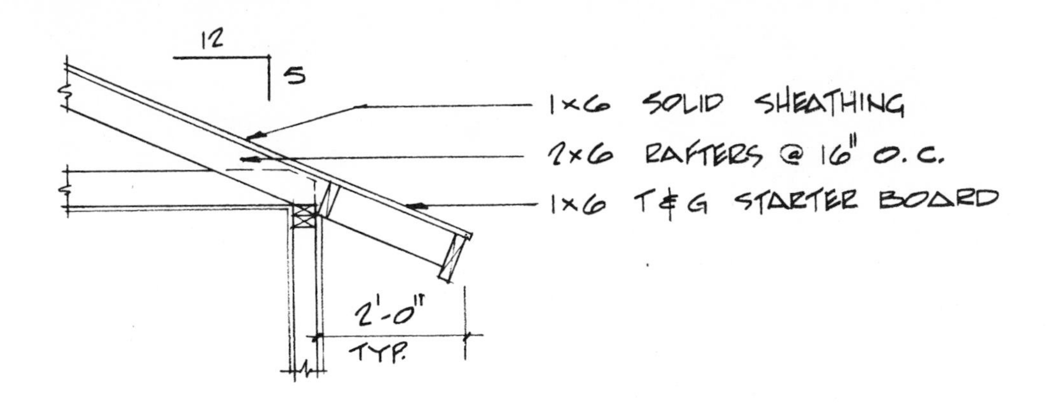

PROBLEM 3:

ESTIMATE THE FOLLOWING:

Starter board

1. Length of common rafter ____________
2. Lineal feet of eave line plus lineal feet of gable rake ____________
3. Number of 1 × 8 shiplap starter boards to cover 2′-6″ eave ____________
4. Lineal feet of 1 × 8 shiplap starter board (including 5 percent waste) ____________

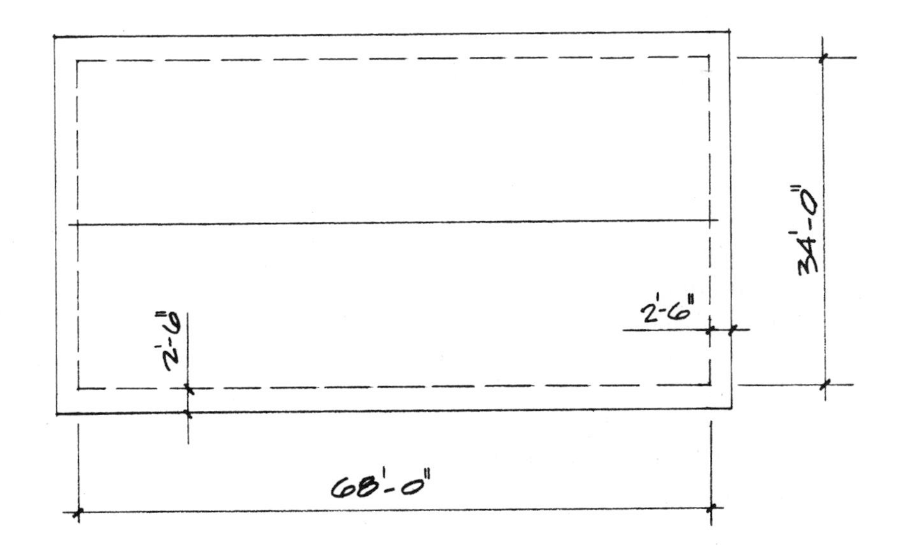

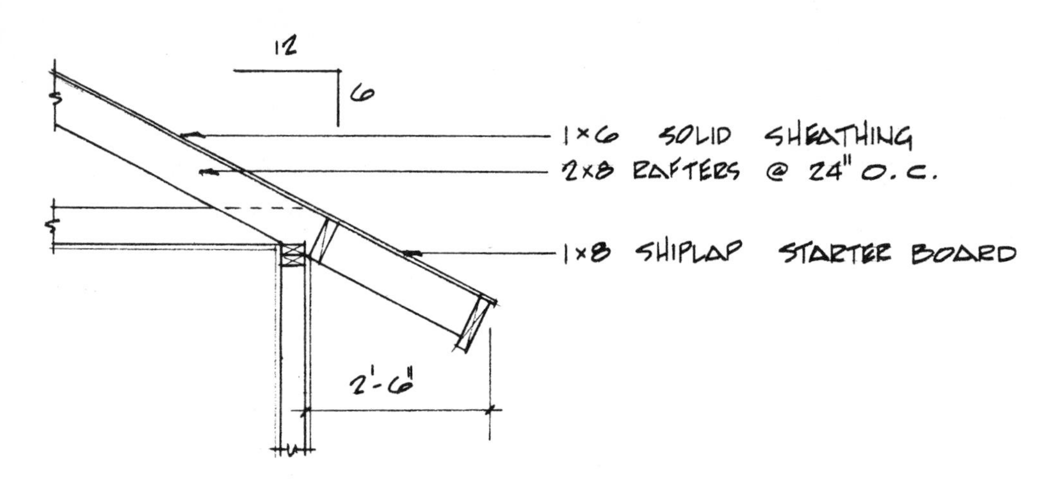

Roof sheathing

5. Square footage of building floor area __________
6. Square footage of roof area above floor area __________
7. Lineal feet of 1 × 6 solid sheathing (including 22 percent waste) __________

PROBLEM 4:

ESTIMATE THE FOLLOWING:

Starter board

1. Lineal feet of perimeter eave line __________
2. Number of 1 × 6 T&G boards to cover a 2′-0″ eave __________
3. Lineal feet of 1 × 6 T&G starter board to cover eave (including 5% waste) __________

Roof sheathing

4. Square footage of building floor area __________
5. Square footage of roof area above floor area __________
6. Lineal feet of 1 × 6 solid sheathing (including 22% waste) __________
7. 66% of solid sheathing above equals skip sheathing needed __________

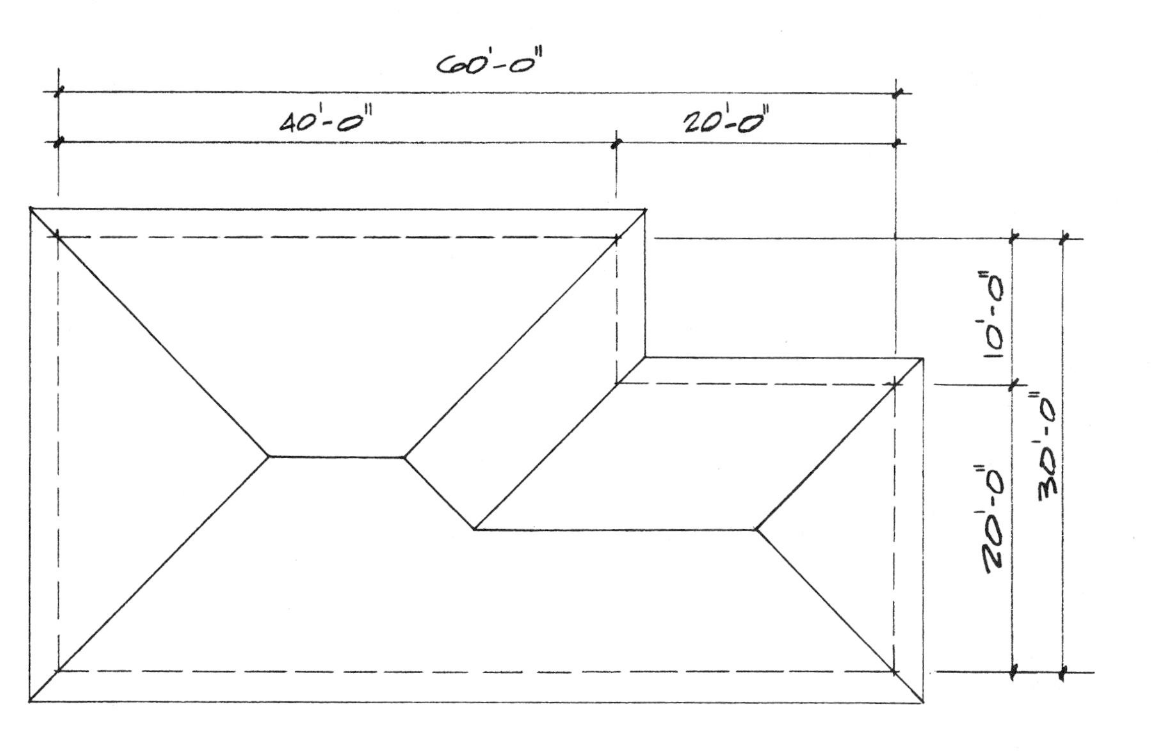
60'-0"
40'-0"
20'-0"
10'-0"
20'-0"
30'-0"

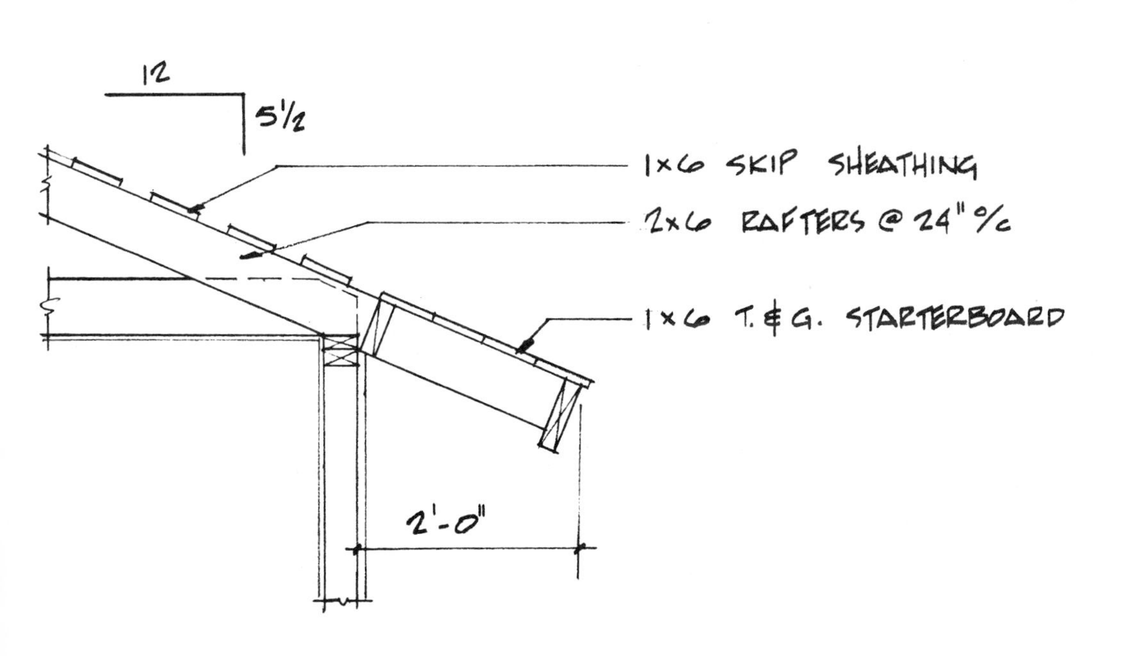
12
5½
1×6 SKIP SHEATHING
2×6 RAFTERS @ 24" o/c
1×6 T. & G. STARTERBOARD
2'-0"

PROBLEM 5:

ESTIMATE THE FOLLOWING:

Starter board

1. Lineal feet of perimeter eave line ______
2. Number of 1 × 6 shiplap boards to cover 2′-6″ eave ______
3. Lineal feet of 1 × 6 shiplap to cover eave (including 5% waste) ______
4. Lineal feet of gable rake 1′-0″ wide ______
5. Number of 1 × 6 shiplap boards to cover 1′-0″ gable rake ______
6. Lineal feet of 1 × 6 shiplap to cover gable rakes (including 5% waste) ______
7. Total lineal feet of 1 × 6 shiplap starter board ______

Roof sheathing

8. Square footage of building floor area ______
9. Square footage of roof above building ______
10. Lineal feet of 1 × 6 solid sheathing (including 22 percent waste) ______
11. 66% of solid sheathing above equals skip sheathing needed ______

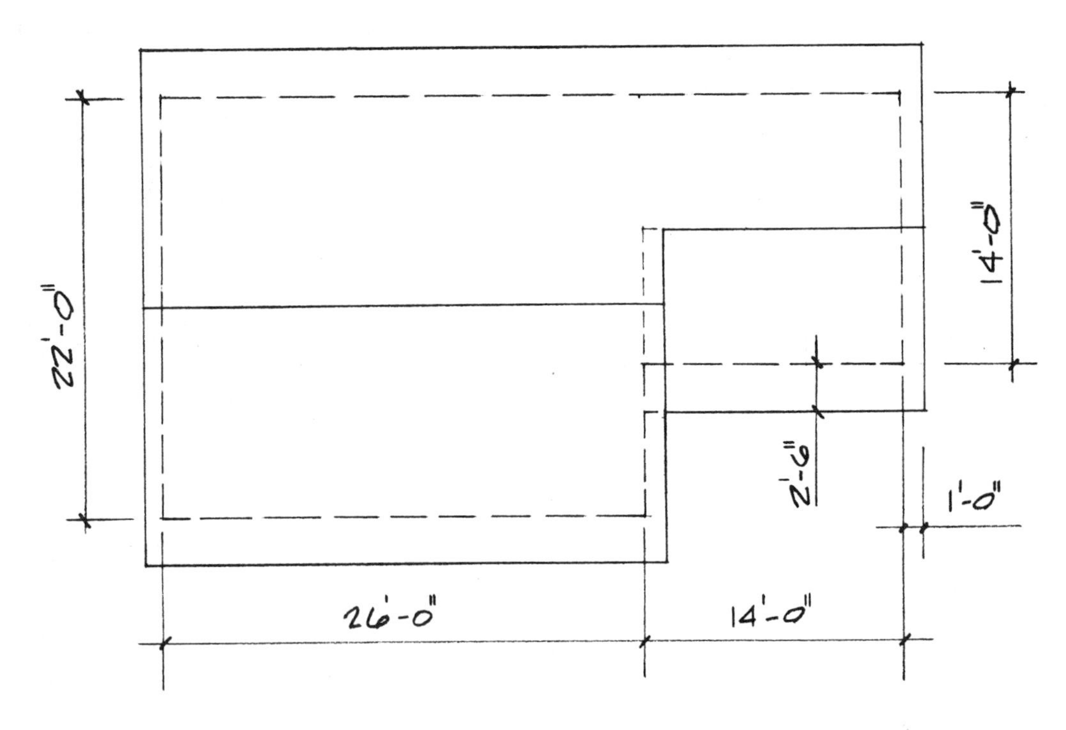

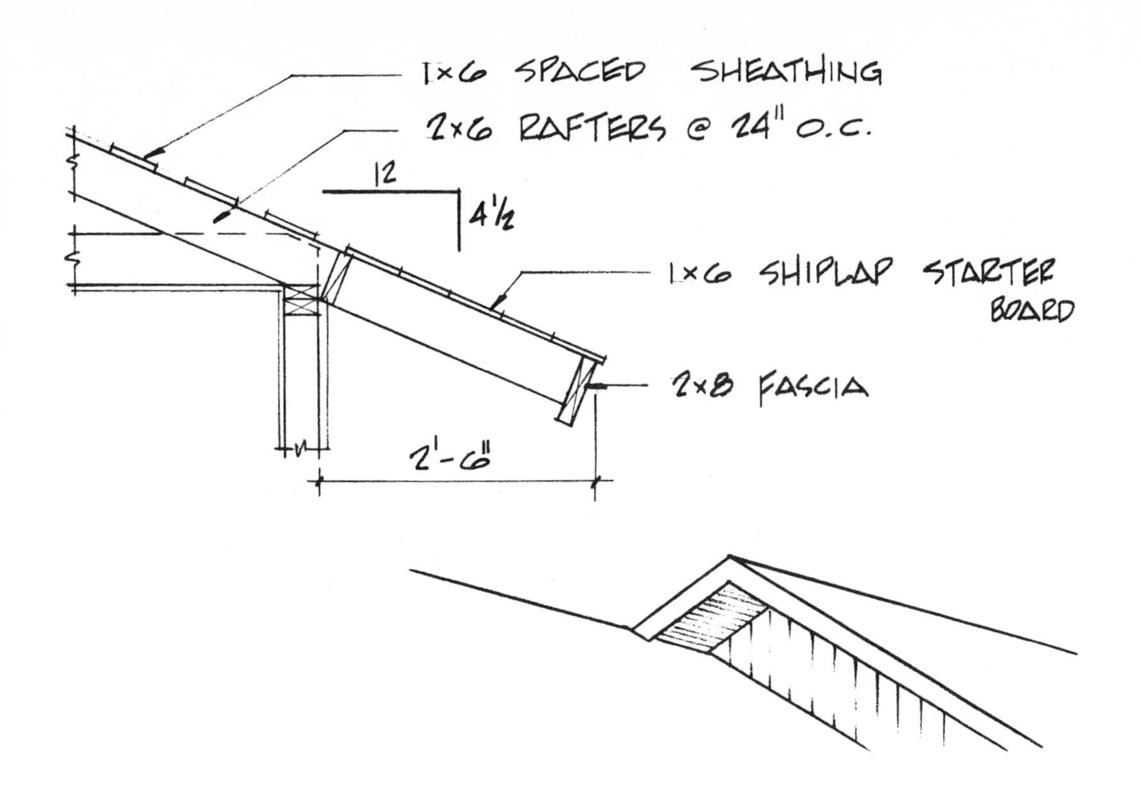
1×6 SPACED SHEATHING
2×6 RAFTERS @ 24" O.C.
12
4½
1×6 SHIPLAP STARTER BOARD
2×8 FASCIA
2'-6"

VII
ROOFING MATERIALS

55

Types of Roofing Materials

Various types of roofing materials are available today for residential and commercial construction that provide a long-lasting, waterproof protection against the elements. Some of these materials are required by local code to be fire retardant. From the list below we will concern ourselves with the three most commonly used materials for residential construction: composition shingles, wood shingles, and built-up roofing.

ROOFING MATERIALS

Some of the following are rated more fire retardant than others:

1. Composition shingles
   a. Asphalt shingle (most commonly used)
   b. Fiberglass shingle (fire retardant)
   c. Asphalt mineral paper (commonly called roll roofing).
2. Wood shingles
   a. Wood shingle—may be treated to be fire retardant
   b. Wood shake shingle—may be treated to be fire retardant.
3. Built-up roofing (felt paper and hot asphalt combination)
   With an asbestos cap sheet (fire retardant)
4. Asbestos shingle (fire proof)
5. Concrete tile and cement fiber shingle (fire proof)
6. Clay shingles (fire proof)
7. Slate shingles (fire proof)
8. Aluminum and various metal shingles (fire proof)

All the roofing materials above are figured in what is called a *square* in the building trades. A square consists of enough roofing material to cover a 10′-0″ × 10′-0″ area, or 100 square feet of roof as illustrated below.

The estimator must have the following general information to be familiar with the most widely used residential roofing materials.

Composition Shingles

Composition shingles are flexible shingles made of asphalt or fiberglass and surfaced with a granule layer. They are purchased by the square and are specified by the weight of the shingles per square.

Weight specifications.

3 Bundles per sq.	(235 lbs. per square) (approximately 90% of asphalt shingles used are 235 lbs.) (255 lbs. per square)
4 Bundles per sq.	(290 lbs. per square)

Note: Shingles that differ in weight also differ in number of bundles per square. Therefore, shingles should always be ordered by the square and *not* by the bundle, for the sake of accurate coverage. Individual shingle size is 12″ × 36″, usually with 3 tabs to a shingle.

EXAMPLE:

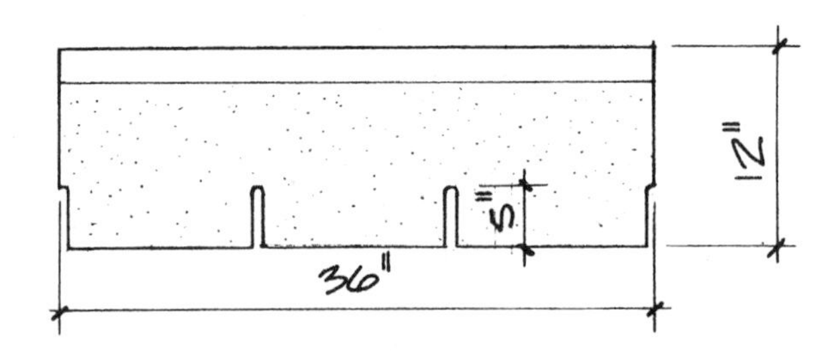

When composition shingles are used, the roof is solid sheathed and the shingles are applied with a 5″ exposure and a 2″ head lap. An underlayer of 15-lb. felt covers the roof before the shingles are nailed.

Hip and ridge shingles are available in 9″ × 12″ size. There are 3 bundles of hip and ridge shingles per square, equaling 150 lineal feet with a 5″ exposure.

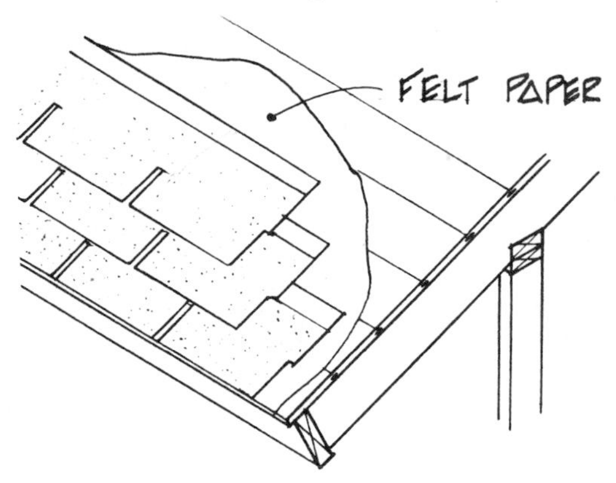

Composition Roofing

Composition roofing is usually used on flat or low-pitched roofs (roofs less than 4 / 12 slope). It consists of a minimum of three layers of 15-lb. felt paper; the first layer of felt paper is nailed or stapled to the roof sheathing and hot asphalt is mopped on between the remaining layers. The top layer is flood coated with hot asphalt and covered with approximately 300 lbs. of small rock per square. Large rock may also be

spread for a decorative effect (using two 80-lb. bags per square).

Wood Shingles

Wood shingles are graded #1 (blue label) and #2 (red label). The #1 shingle has vertical grain, no knots, and a tendency to lie flat after weathering. The #2 shingle has slash or flat grain, knots in the upper portion of the shingle, and a tendency to cup or warp when exposed to weather. Wood shingles come 16″, 18″, and 24″ long; and there are four bundles to a square (100 sq. ft.). One bundle of hip and ridge shingles equals 15 lineal feet.

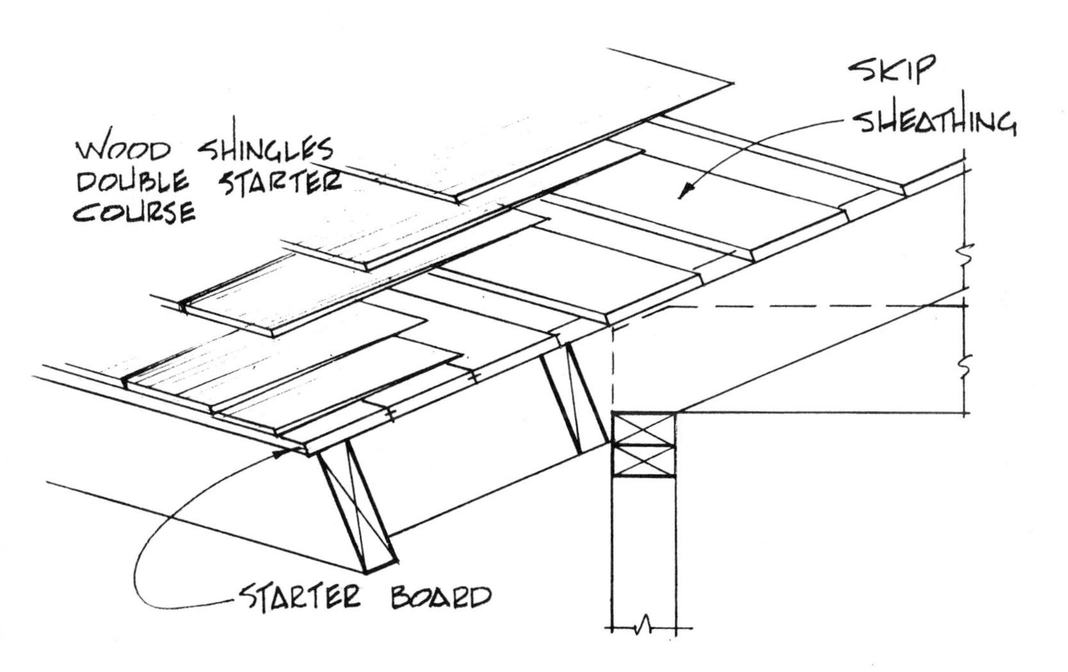

Note: Wood shingles do not need felt paper underlayment for slopes over 4 / 12.

The shingle exposure is regulated by the slope of the roof:

16″ shingles usually have a 5″ exposure
18″ shingles usually have a 5-1/2″ exposure
24″ shingles usually have a 7-1/2″ exposure

Wood Shake Shingles

Shake shingles are graded medium or heavy.

Medium Shake Shingles are 24″ long and have a thickness at the butt of 1/2″ to 3/4″. Medium shakes come either 4 or 5 bundles to a square.

Heavy Shake Shingles are 24″ long and have a shingle-butt thickness of 3/4″ to 1-1/4″. Heavy shakes *always* come 5 bundles to a square.

Shake shingles have a layer of 30-lb. 18″ wide felt paper between each row of shingles. Felt paper may be purchased:

1. 1 roll of 30-lb. felt paper 18″ wide covers approximately 1 square of roof.
2. 1 roll of 30-lb. felt paper 36″ wide covers approximately 2 squares of roof.

Note: 36″ wide felt paper breaks into two 18″ wide rolls for use.

One bundle of hip and ridge shakes equals 15 lineal feet. In some areas a more economical shorty-shake shingle is available for the starter course. Shorty shingles cover approximately 100 lineal feet per square.

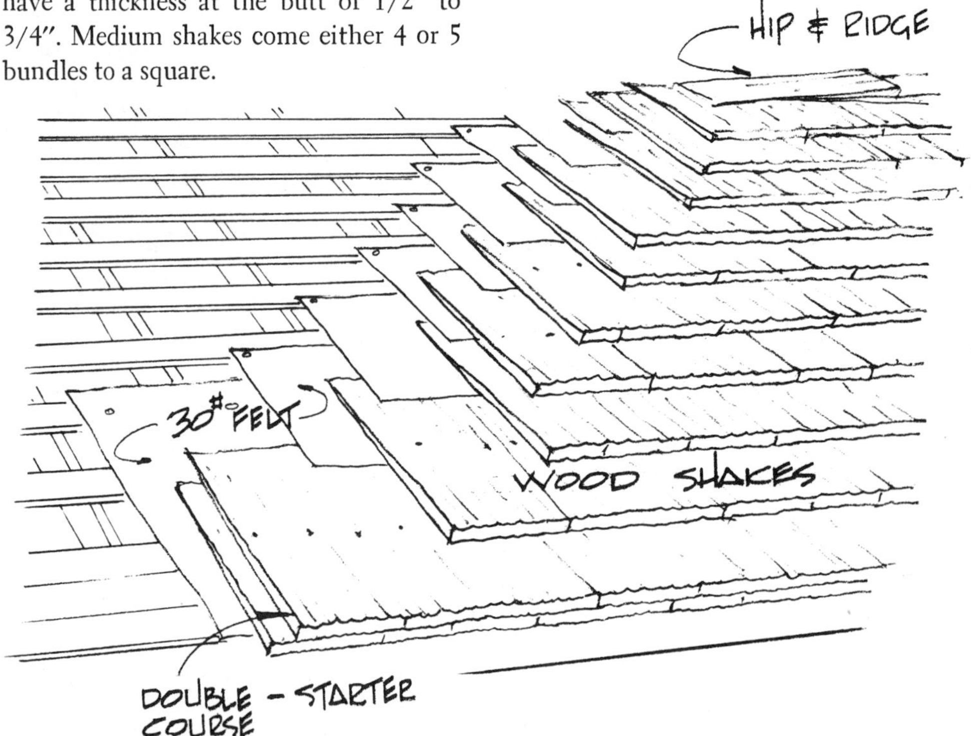

56
Estimating Roofing Materials

The estimator may use two simple methods to find the roof area in estimating roofing materials. The first method can best be used when estimating a simple gable or hip roof. The estimator can scale from an elevation view or a section view the length of the common rafter including the overhang. He can then double the rafter length for both sides of the roof. For those shingle materials that have a double starter course at the eave, he must add 2 more lineal feet. This can then be multiplied by the length of the building, including the overhang on each end of the building, to acquire the square footage of the roof. The total roof area is then divided by 100 for the number of squares of roofing material.

METHOD 1

Example Problem:

Estimate the number of squares of 235-pound asphalt shingle roofing material in the following problem.

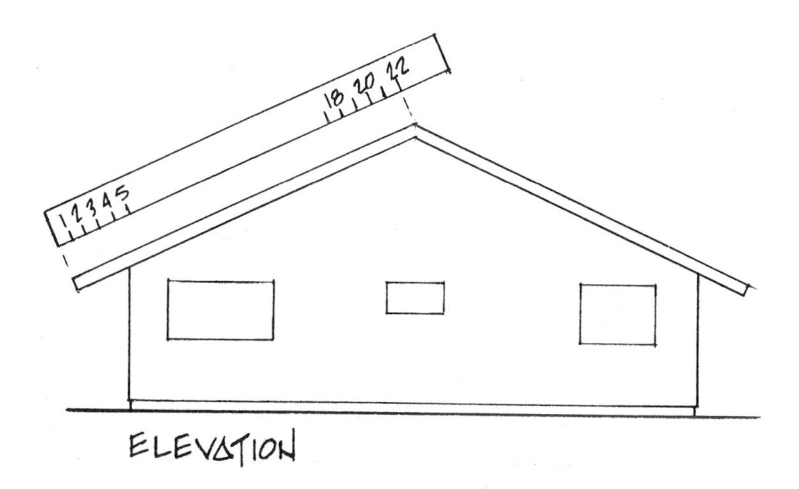

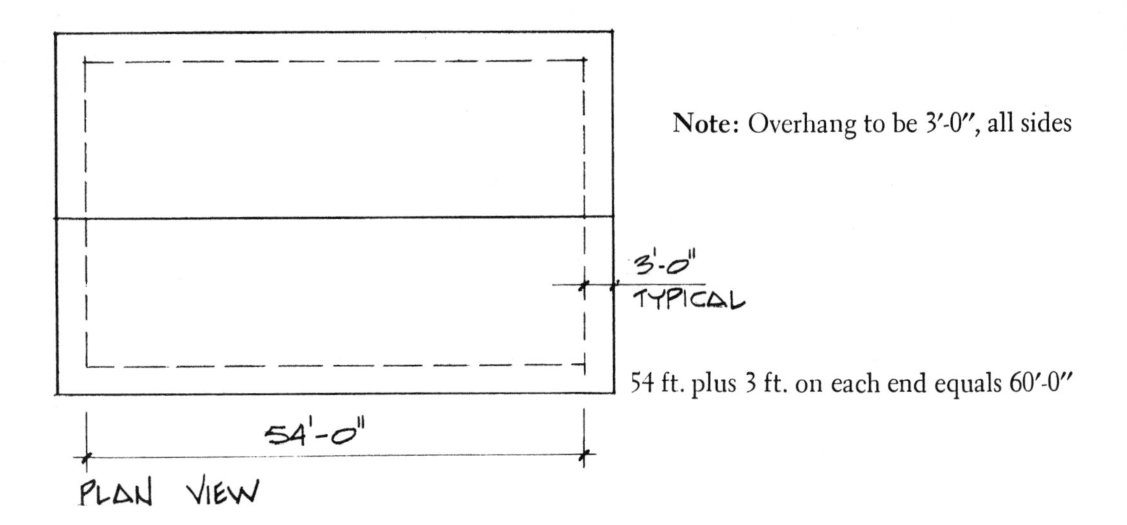

PROCEDURE FOR ESTIMATING ROOFING MATERIAL

1. Scale from an elevation or section view the common rafter length including the overhang: 22′-0″.
2. Double the rafter length for both sides of the roof: $22 \times 2 = 44$ lin. ft.
3. Add 2 more lineal feet for double starter course: $44 + 2 = 46$ lin. ft.
4. Multiply the length above by the length of the building including the gable overhangs: $46 \times 60 = 2{,}760$ sq. ft.
5. Divide the square footage of the roof above (2,760) by 100 to get the number of squares of roofing material:

$2760 \div 100 = 27.60$ or 27-2/3 squares of asphalt shingles

If the roof is a hip and a double starter course of shingles is used, add 2 feet to the scaled rafter width and 2 feet to the length of the roof. This allows 1 foot of extra material around the eave of the hip roof for the double starter course.

Example Problem:

Estimate the number of squares of wood shingles in the following problem.

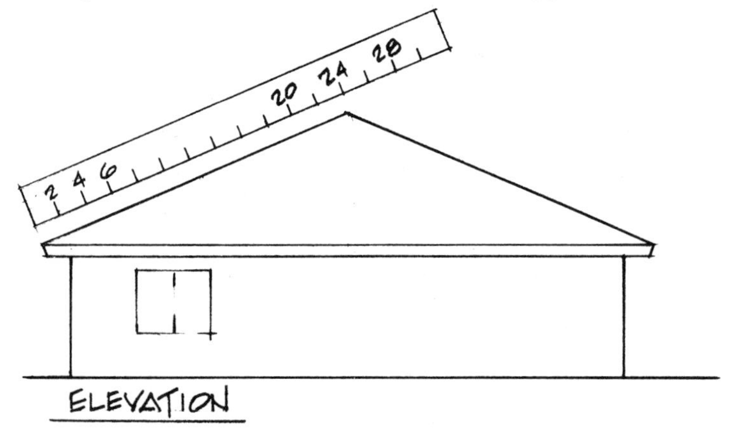

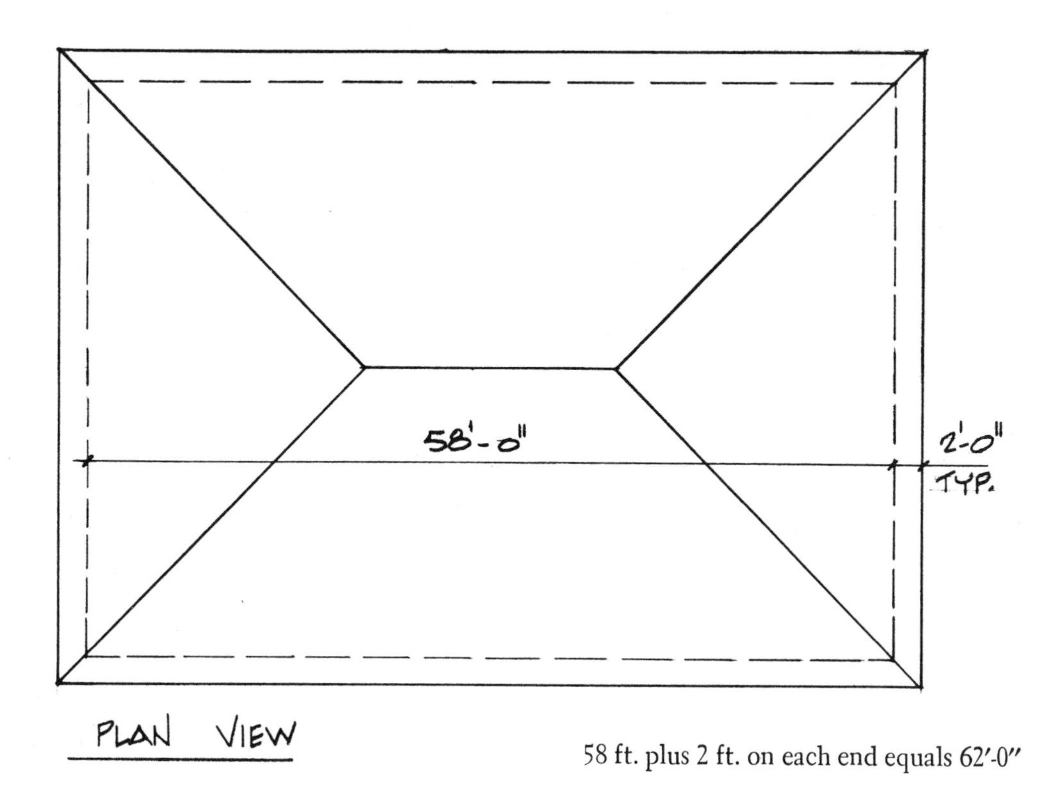

58 ft. plus 2 ft. on each end equals 62'-0"

PROCEDURE FOR ESTIMATING ROOFING MATERIAL

1. Scale from the elevation or section view the common rafter length including the overhang: 24'-0".
2. Double the rafter length for both sides of the roof: $24 \times 2 = 48$ lin. ft.
3. Add 2 more feet for double starter course: $48 + 2 = 50$ lin. ft.
4. Determine the length of the building including the overhang on each end of the building: 58 feet plus 2 feet on each end equals 62'-0".
5. Add 2 more feet to the length for the double starter course: $62 + 2 = 64'$-0" long.
6. Multiply the rafter length (50) by the length of the building (64): $50 \times 64 = 3{,}200$ square feet.
7. Divide the square footage of the roof above (3,200) by 100 to get the number of squares of roofing material:

   $3{,}200 \div 100 = 32$ squares of wood shingles

The second method for estimating roofing material can be used for any shape hip or gable roof, no matter how complicated the design. The square foot area may be found by using Table 56-1.

TABLE 56-1 ESTIMATING MATERIAL BY FINDING SQUARE FOOT ROOF AREA

Slope	*Constant*	*Slope*	*Constant*
4 / 12	1.10	9 / 12	1.31
5 / 12	1.12	10 / 12	1.36
6 / 12	1.17	11 / 12	1.42
7 / 12	1.21	12 / 12	1.50
8 / 12	1.25		

The constant from the table 56-1 that corresponds to the roof slope is multiplied by the horizontal area the roof covers to find the square foot area of the roof. This constant already includes a percentage for shingle waste. The square foot area of the roof can then be divided by 100 for the number of squares of roofing material needed.

METHOD 2

Example Problem:

Estimate the number of squares of wood shingle roofing material in the following problem.

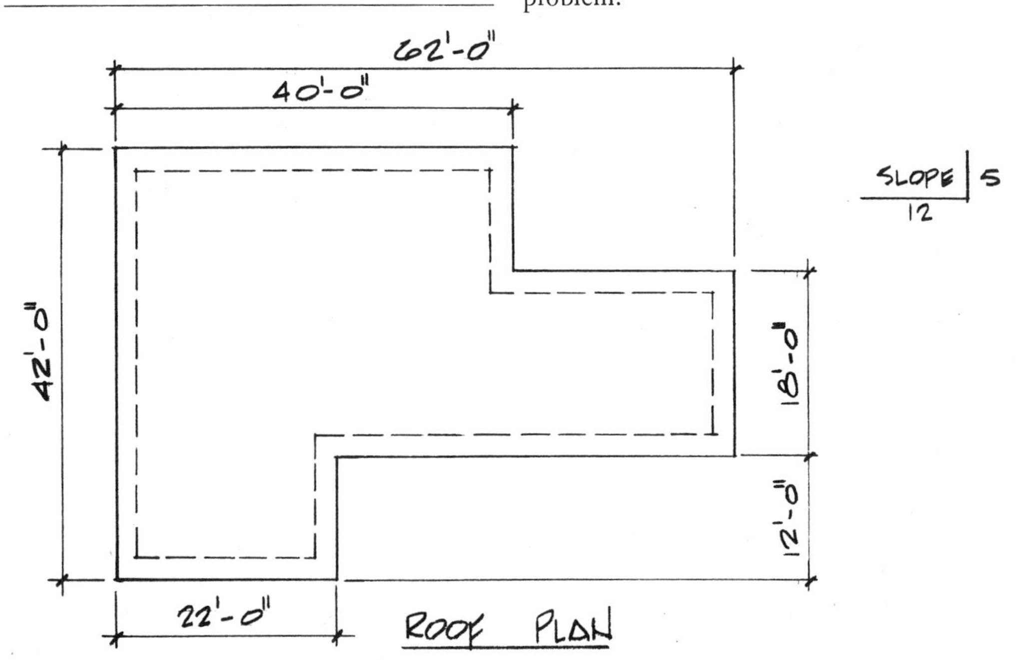

PROCEDURE FOR ESTIMATING ROOFING MATERIAL

1. Determine which constant from Table 56-1 corresponds with the roof slope.

   5 / 12 slope = 1.12 constant

2. Calculate the horizontal square foot area that the roof covers.
3. Multiply the horizontal square foot area (1,860) by the constant that cor-

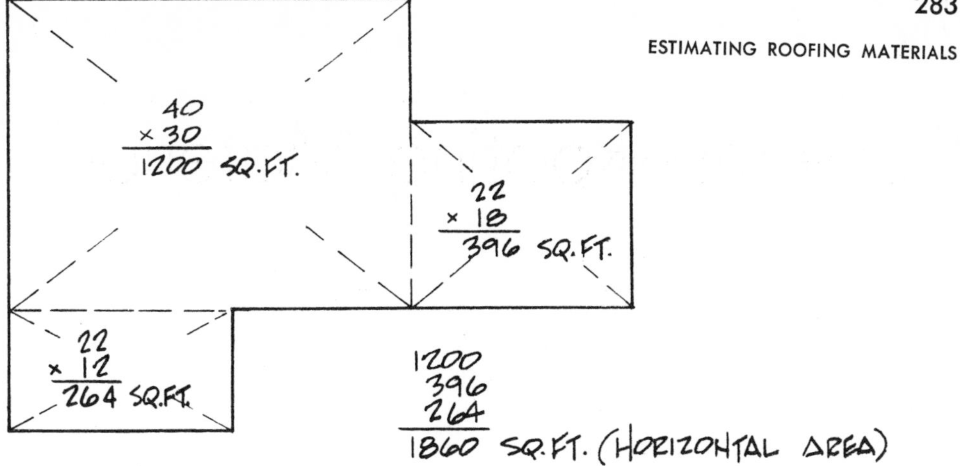

responds to a 5 / 12 roof slope (1.12): 1,860 × 1.12 = 2,083.20 or 2,084 sq. ft. of roof area.

4. Divide the total square footage of roof area (2,084) by 100 to find the number of squares of roofing material:

   2,084 ÷ 100 = 20.84 or 21 squares of wood shingle material

If the roof has a double starter course of shingles, the estimator must calculate the additional amount of material needed. The following information will help the estimator determine the extra material.

COMPOSITION SHINGLES

The starter course for composition shingles is usually a roll of mineral paper 9″ wide and 36 ft. long. To determine the number of rolls needed, divide the lineal footage of the roof eave by 36.

WOOD SHINGLES AND WOOD SHAKES

One square of wood shingles or shakes in field application covers approximately 100 lineal feet. To determine the number of squares needed, divide the lineal feet of roof eave by 100.

57

Problems—Roofing Material

PROBLEM 1

Information needed:

Roof slope 6 / 12
235-lb. composition shingles
15-lb. felt underlayment (320 sq. ft. per roll)

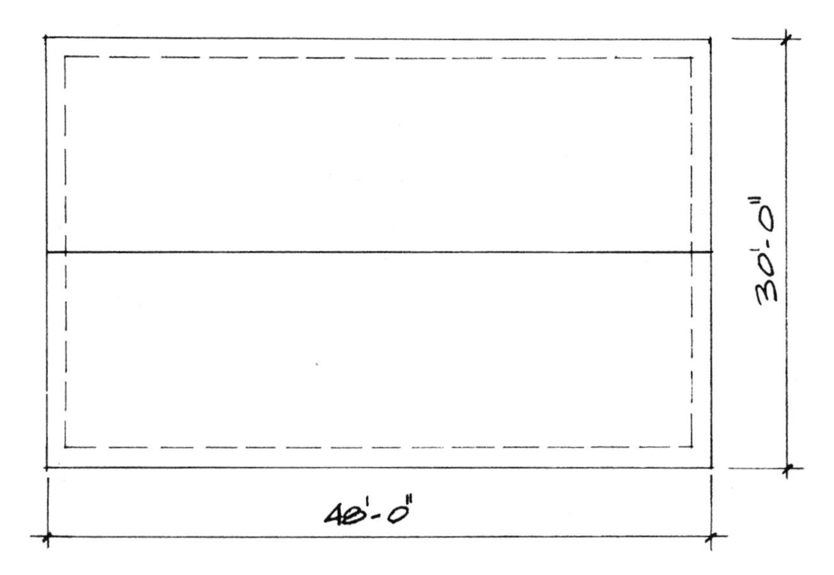

ESTIMATE THE FOLLOWING:

1. Squares of composition shingles __________
2. Bundles or squares of hip and ridge shingles __________
3. Number of rolls of 15-lb felt (coverage 300 sq. ft. per roll) __________

PROBLEM 2

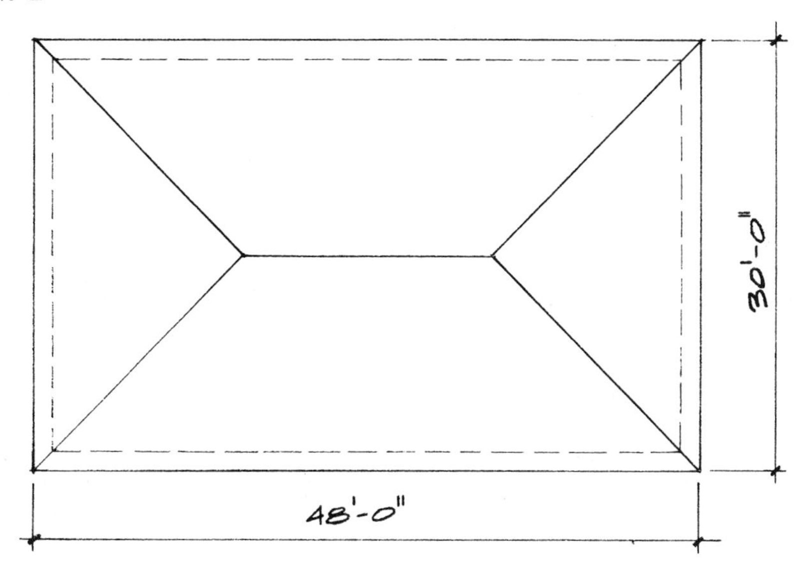

Information needed:

Roof slope 4 / 12
Wood shingles
No felt needed

ESTIMATE THE FOLLOWING:

1. Squares of wood shingles __________
2. Bundles of hip and ridge shingles __________

PROBLEM 3

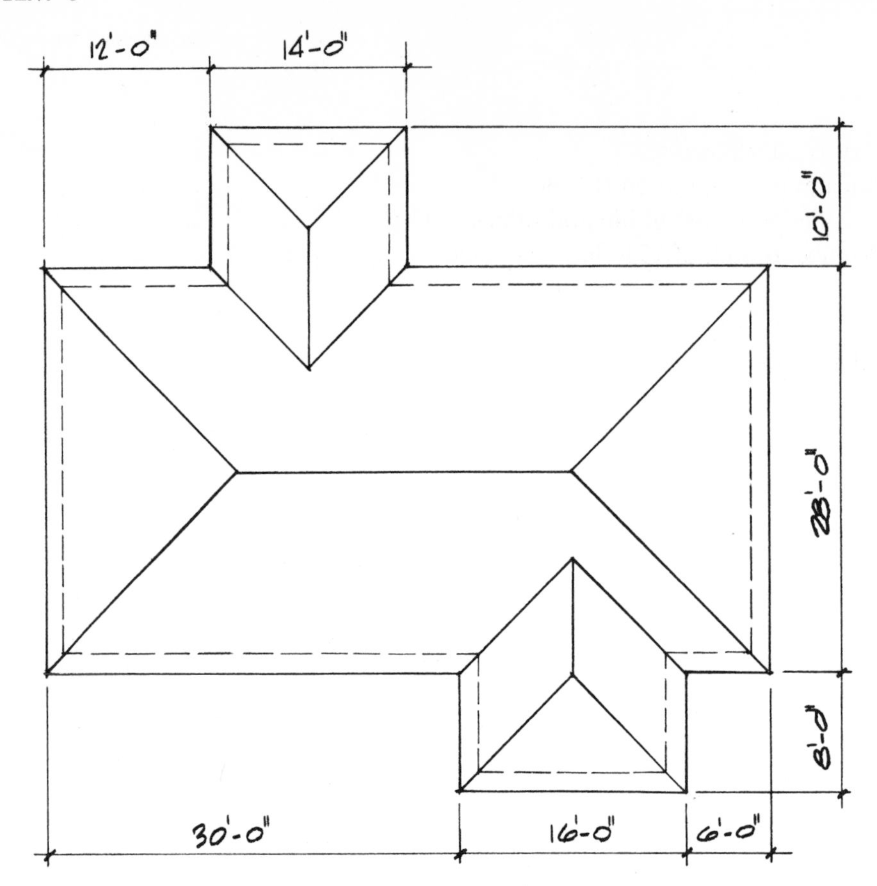

Information needed:

Roof slope 4-1/2 / 12
Heavy shake shingles
30 lb. felt

ESTIMATE THE FOLLOWING:

1. Squares of shake shingles ____________
2. Bundles of hip and ridge shingles ____________
3. Squares of 30-lb felt paper ____________

PROBLEM 4

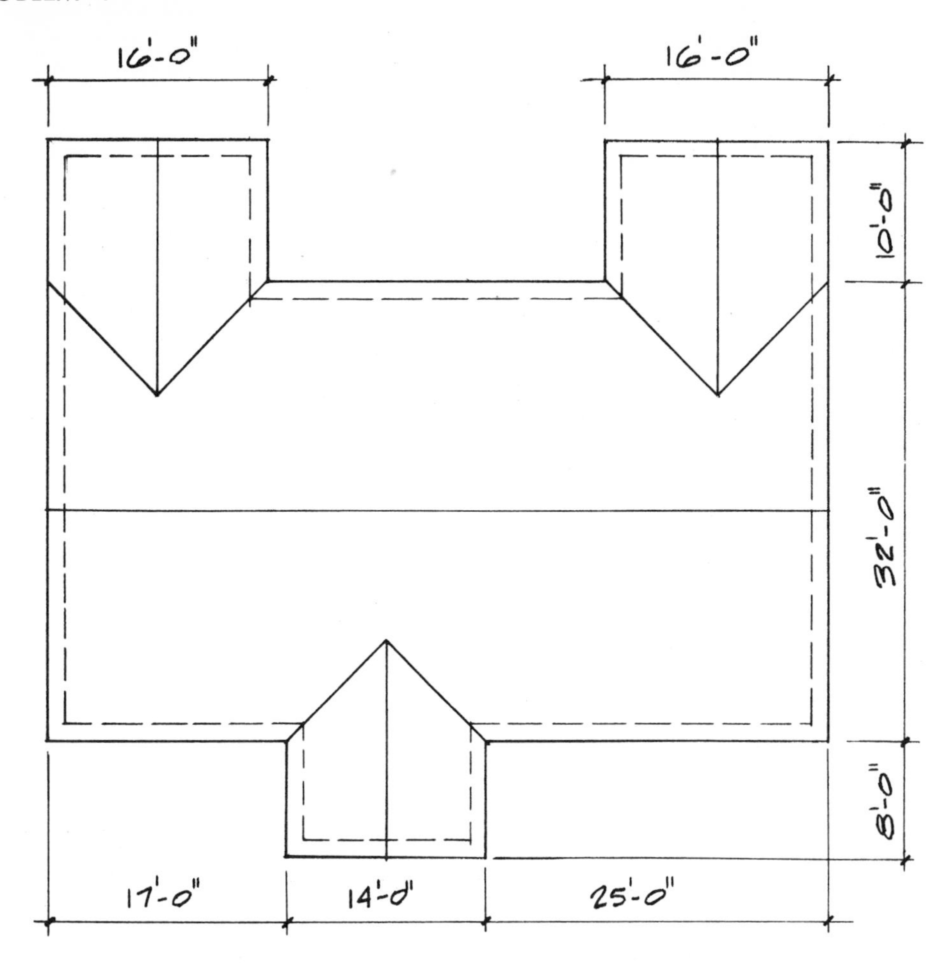

Information needed:

Roof slope 5 / 12
255-lb composition shingles
15-lb. felt underlayment

ESTIMATE THE FOLLOWING:

1. Squares of composition shingles ____________
2. Bundles or squares of hip and ridge shingles ____________
3. Number of rolls of 15-lb. felt (coverage 300 sq. ft. per roll) ____________

PROBLEM 5

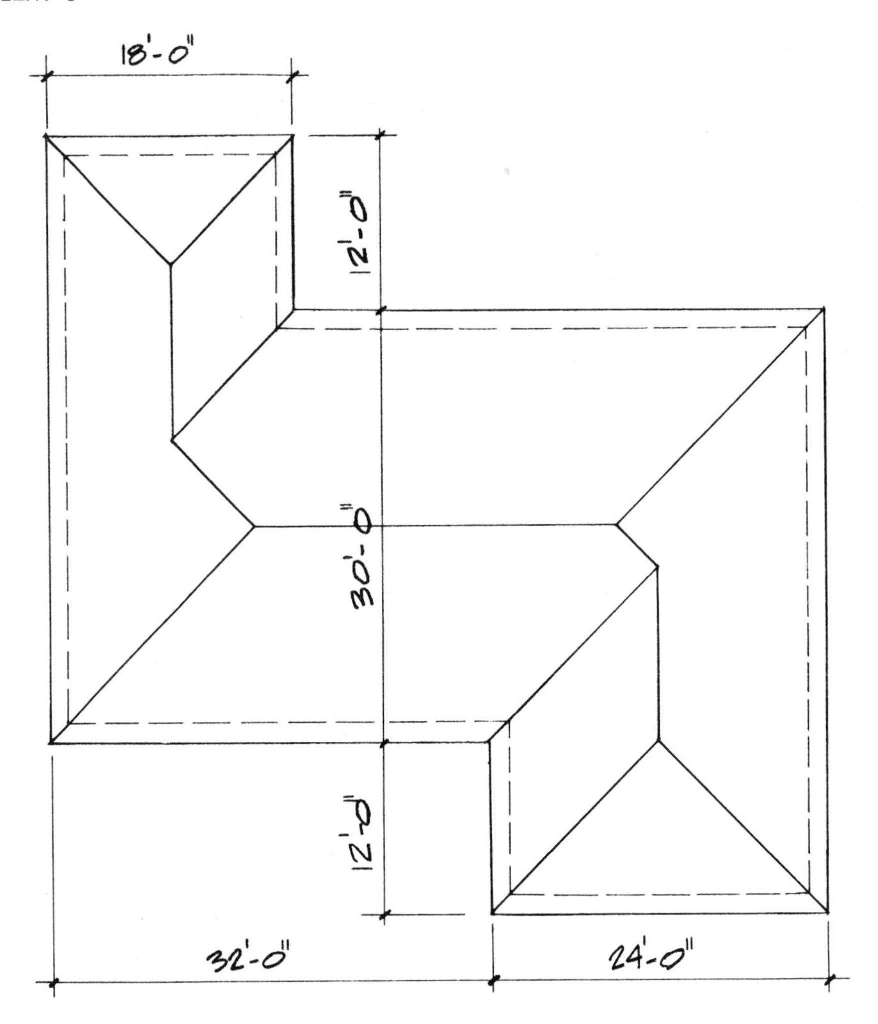

Information needed:

Roof slope 5-1/2 / 12
235-lb. composition shingles
15-lb. felt underlayment

ESTIMATE THE FOLLOWING:

1. Squares of composition shingles ____________
2. Bundles of hip and ridge shingles ____________
3. Number of rolls of 15-lb. felt (coverage 300 sq. ft. per roll) ____________

VIII

EXTERIOR FINISH MATERIAL

58

Estimating Exterior Doors

Exterior doors are usually solid-core flush doors or panel doors 1-3/4″ thick and a minimum of 6′-8″ high. For the sake of economy, a variety of hollow-core flush doors may be used. The usual width for a single exterior entry door is 3′-0″ and for a rear door 2′-8″. These sizes facilitate the moving in and out of furniture and appliances. Following is a variety of common exterior door styles.

SOLID- OR HOLLOW-CORE EXTERIOR FLUSH DOORS

Flush doors are available with a smooth surface or rough-sawn texture. They may also be V-grooved or channel routed for a plank effect.

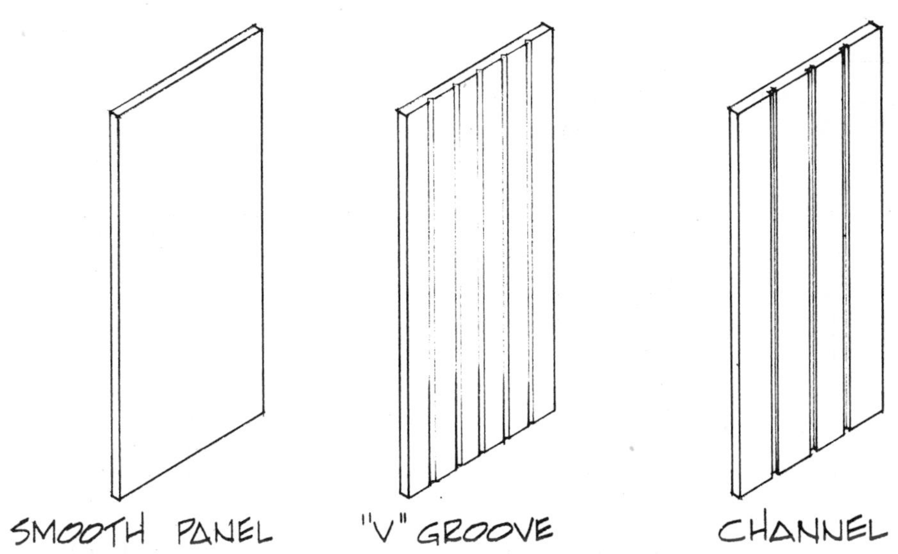

FLUSH DOORS WITH PLANT-ON MOLDING

Various patterns of molding may be applied with glue and nails to smooth- or rough-textured flush doors.

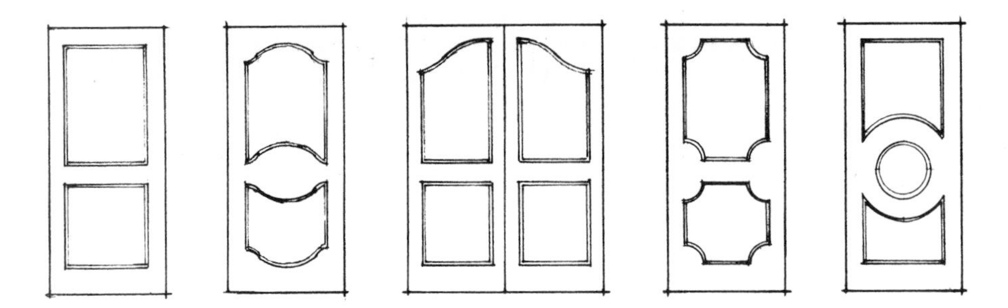

FLUSH DOORS WITH GLAZED LIGHTS

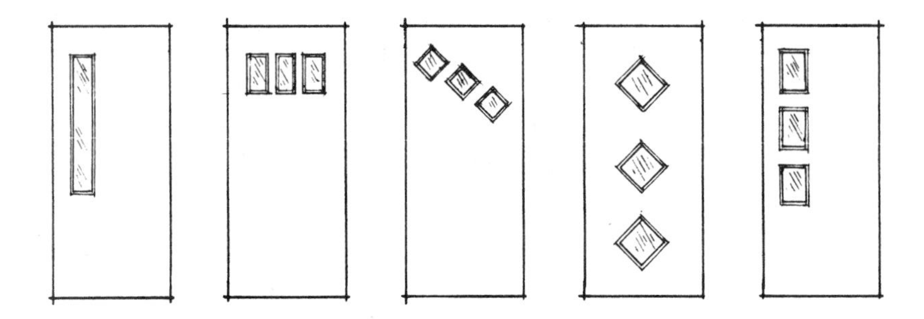

Flush doors with operable windows or louvers furnish light and ventilation. They are usually used as side or rear exit doors.

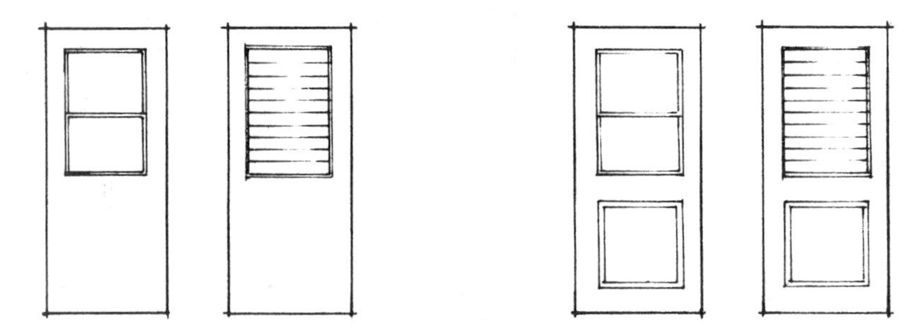

The doors above are also available with a solid panel in the bottom half of the door.

STANDARD SOLID-PANEL DOORS

STANDARD SOLID-PANEL DOORS WITH WINDOW LIGHTS

Exterior doors are usually estimated and listed with the interior doors and trim material. The type, size, and manufacturer are specified on the building plans or in the specifications.

59

Estimating Exterior Door Frames

Exterior door frames are usually installed at the same time as the window frames. They may be ordered assembled or disassembled (K.d. or knocked down). They may be a simple jamb or a very complex window-door unit as illustrated below.

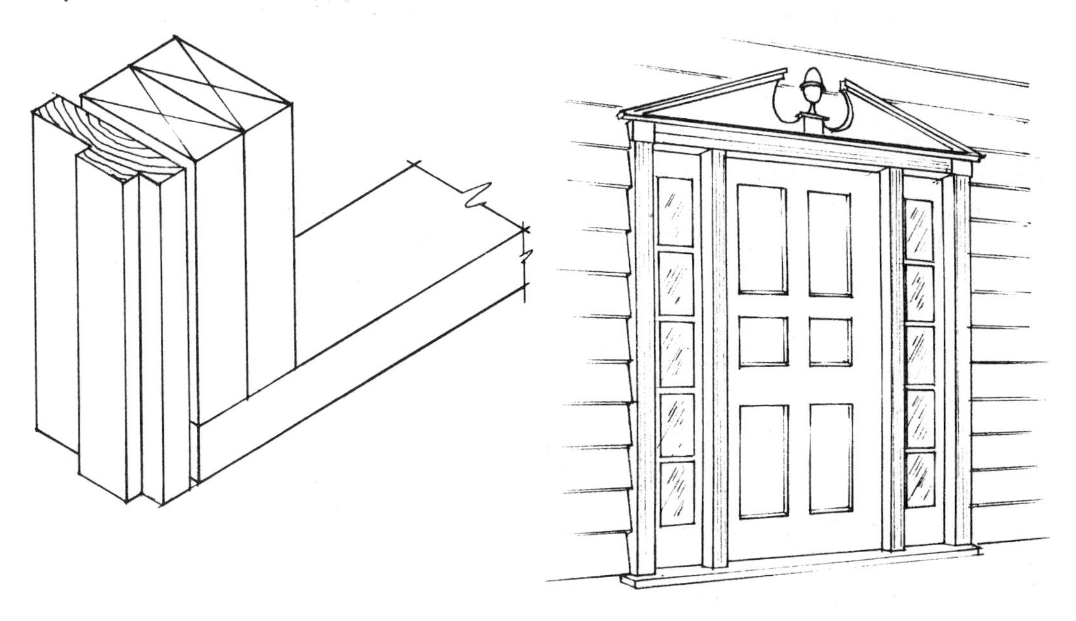

For concrete slab construction, exterior door frames do not require a sill but usually have a wood or aluminum threshold.

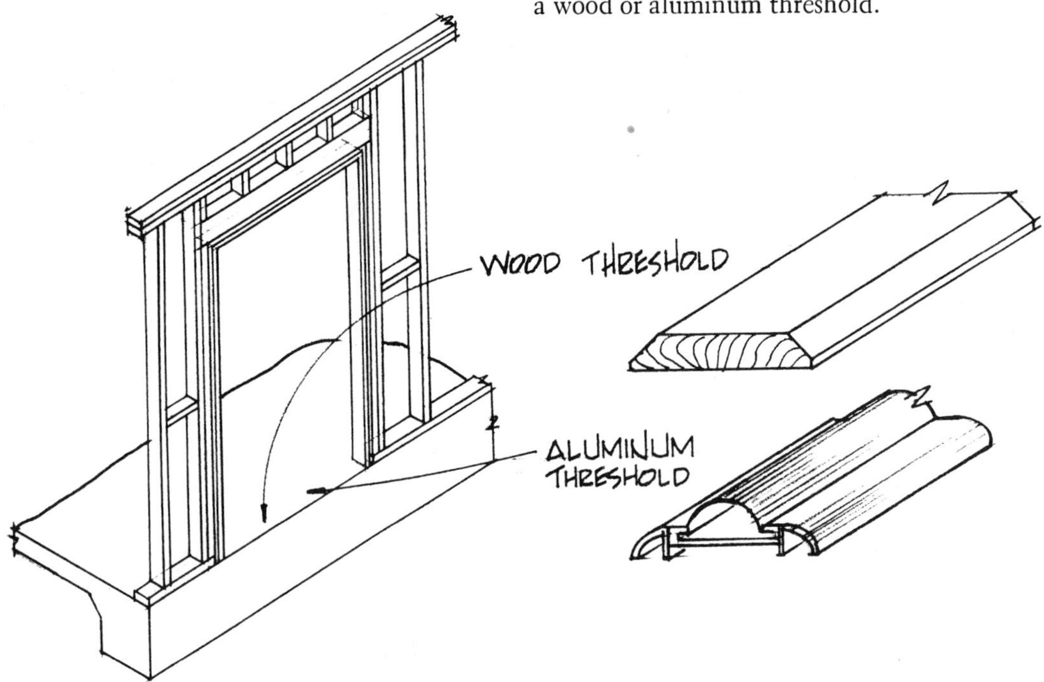

On wood floor construction, a wood still is usually needed. The floor joist is usually notched to receive the sill, and the top of the sill is set so that it will be flush with the entry flooring material. A wood or aluminum threshold is also needed.

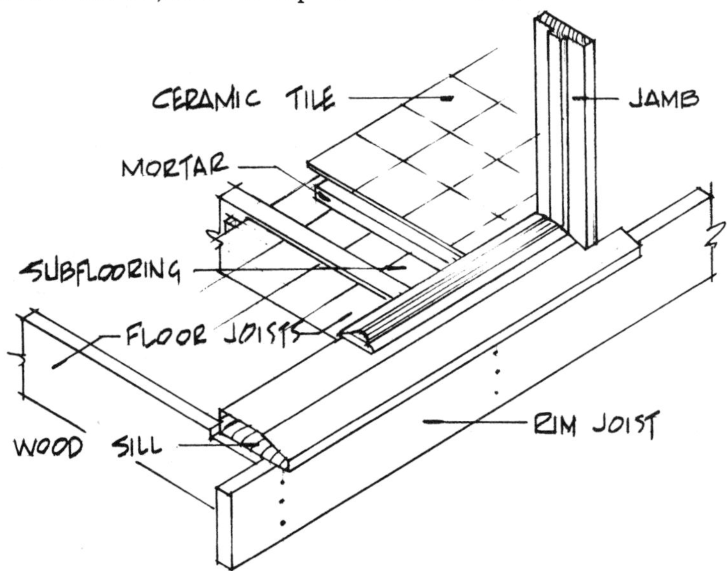

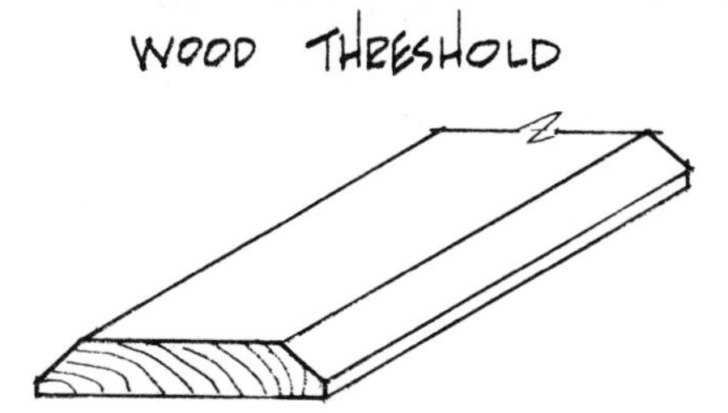

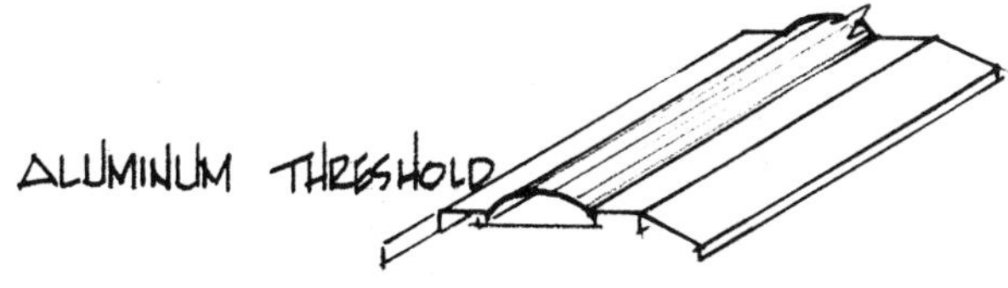

Exterior door frames may also be ordered with vapor barrier flashing paper and stucco or siding mold.

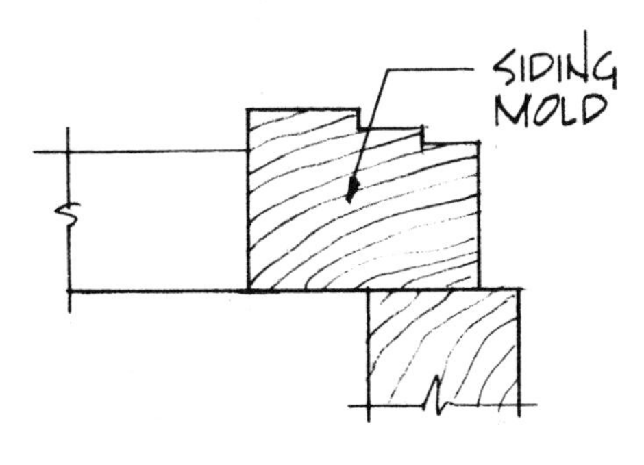

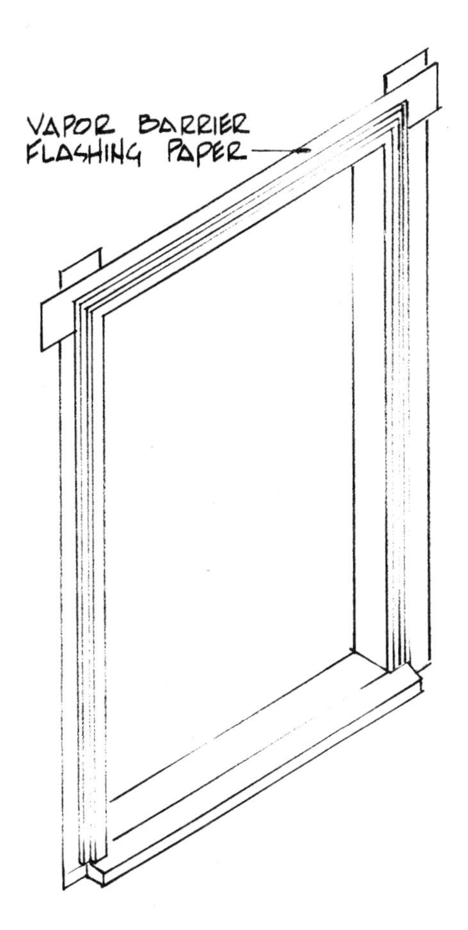

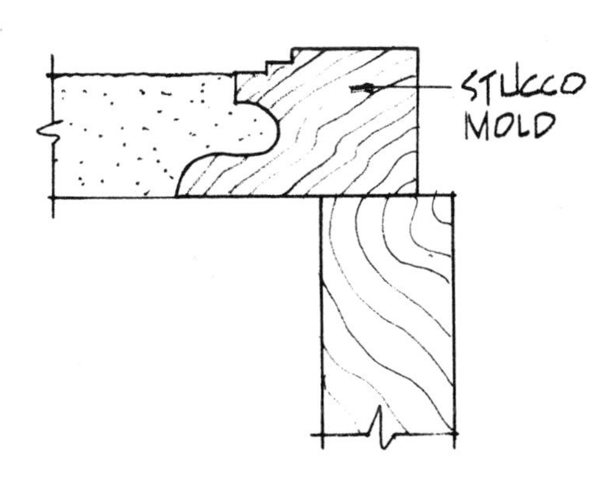

The estimator should keep the following in mind when ordering door frames:

1. Will the door (or doors) swing-in or swing-out?
2. The door width, height, and thickness should be included in the order, the width and height to determine the size of the frame, and the thickness to determine the position of the door stop on the jamb.
3. Is the door frame to come with or without a wood sill?
4. Is the door frame to be preassembled or knocked down and assembled on the job?
5. Sometimes the door jamb width has to be specified. This is determined by the thickness of the interior wall material and the exterior siding used.
6. What type of wood is used for the frame? (This is usually fir or pine.)
7. The style, model, or manufacturer should be included if designated by the architect.

The following description of an exterior door frame includes much of the information asked for above.

1. K.d. door frame, T. M. Cobb Co. 3′-0″ ×6′-8″—1-3/4″ swing-in oak sill, 4″ jamb with siding mold, 1/2″ drywall pine frame.

Exterior door frames are usually included, and delivered with the framing lumber or the window units.

60

Estimating Exterior Window Frames

Window frames are made of one or of a combination of wood, aluminum, steel, or plastic.

All windows may be classified under one of three basic designs.

1. SLIDING WINDOWS

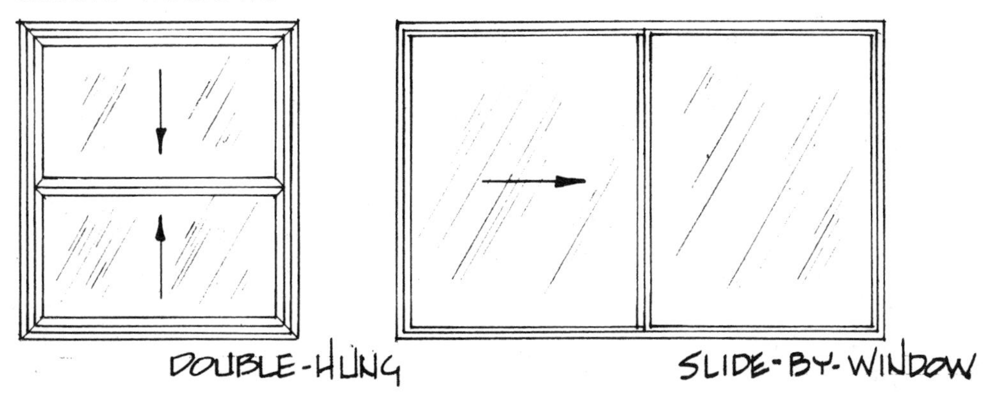

2. SWINGING WINDOWS

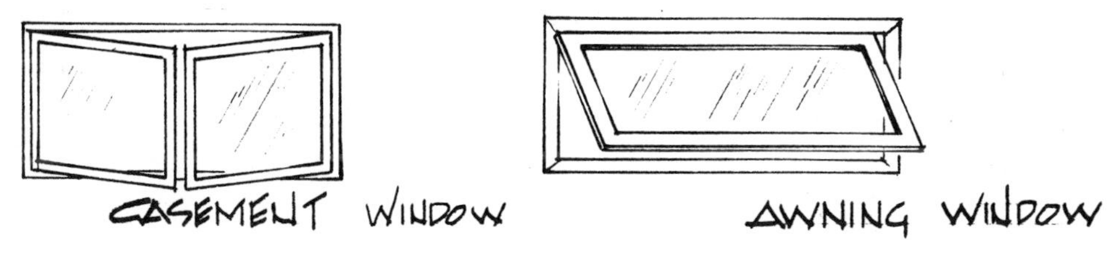

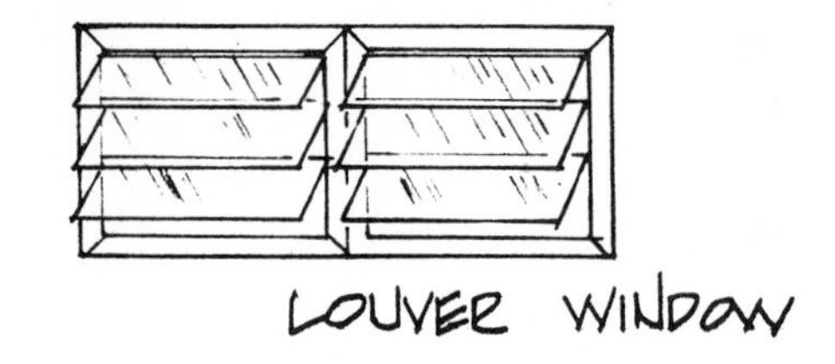

3. FIXED WINDOWS

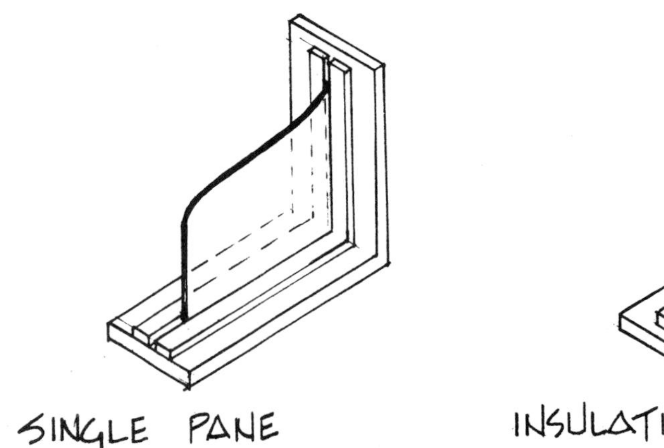

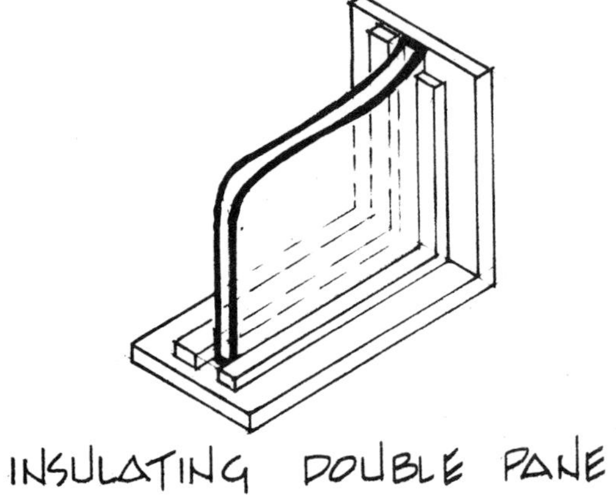

Note: Any window within 18 inches of the floor will require tempered glass.

Window sizes and types are usually found on the floor plan or on a window schedule as illustrated below.

Floor Plan

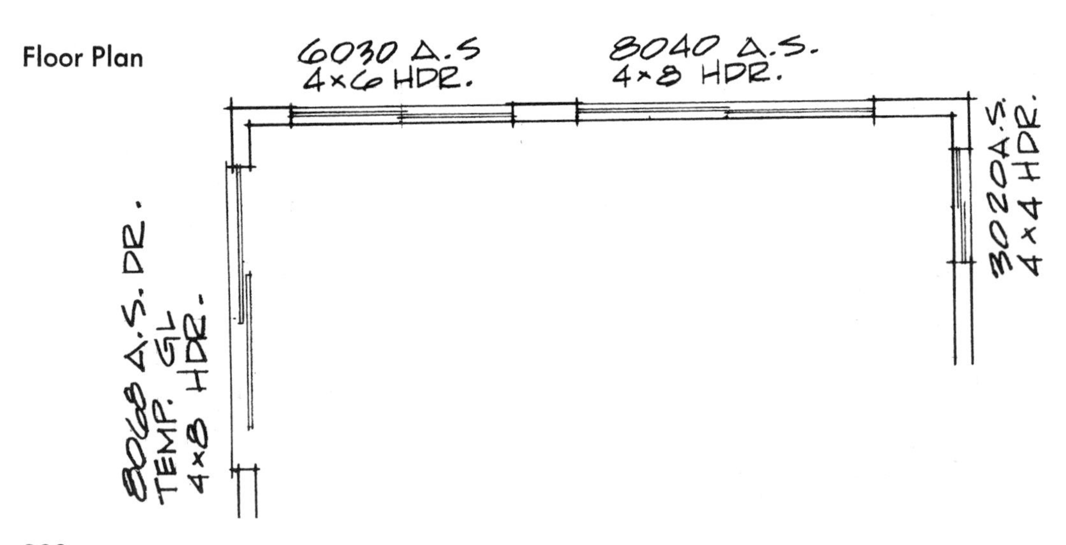

In this case a take-off of windows will have to be made from the floor plan.

Window Schedule

Windows may be designated on the floor plan by a letter or number symbol which refers to the window schedule as illustrated below.

Sliding glass doors will also be designated by a symbol and are found on the window schedule. (Refer to symbol "4" in Table 60-1.)

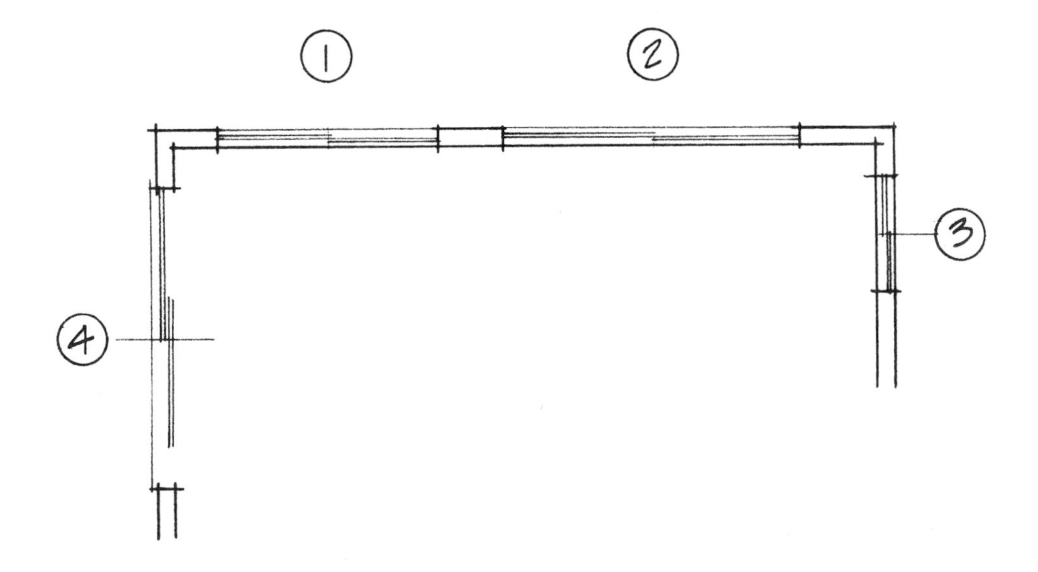

TABLE 60-1 WINDOW SCHEDULE

Symbol	*Window Size*	*No. Required*	*Header Size*	*Window Type*	*Remarks*
1	6030	3	4 × 6	Al.Sl.	
2	8040	1	4 × 8	Al.Sl.	
3	3020	1	4 × 4	Al.Sl.	Obscure glass
4	8068	1	4 × 8	Al.Sl.Dr.	Tempered glass

A specifications schedule such as the one above makes it easy for the estimator to list the materials needed.

61

Flashing Paper

Flashing paper is usually required around all exterior door and window frames. It usually comes in small rolls 6″ wide by 150′ long, or large rolls 6″ wide by 250′ long. The roll or rolls needed may be estimated by taking the appropriate lineal feet around each exterior door and window opening on the plans.

EXAMPLE: If you have an 8040 window, you need approximately 28 lineal feet of flashing paper.

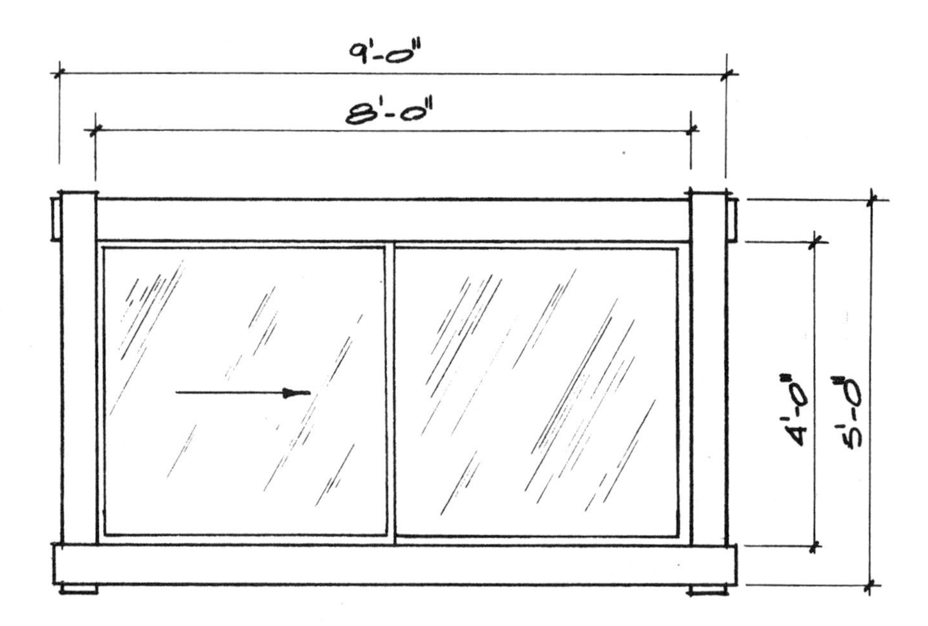

EXAMPLE: If you have a 3068 door, you need approximately 19 lineal feet.

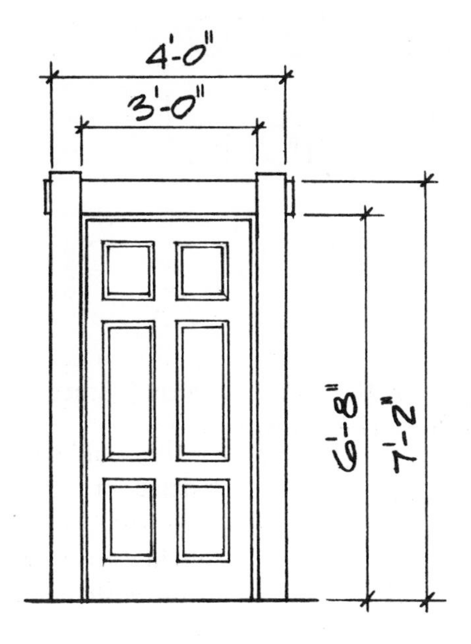

62

Estimating Exterior Cornice Material

The cornice or eave is created by the rafters extending beyond the wall line. The design of the building will determine whether the cornice will be open, closed, or boxed.

OPEN CORNICE

When the cornice is open, the estimator will be concerned with estimating the frieze board or molding and the type of blocking that will be used to close off the area between the rafters. The type and size of blocking and the frieze board or molding will depend on the eave detailing and the type of exterior wall covering as illustrated below.

Detail 1 Stucco

If the building has a hip roof the lineal footage of 2 × 4 material estimated will equal the perimeter of the building. If the building has a gable roof, the lineal footage estimate should equal the lower eave length, excluding the gable rakes.

Detail 2 Stucco

Use the same estimating procedure as in detail 1 with the rafter blocking ordered the same size as the rafter material.

Details 3, 4, 5

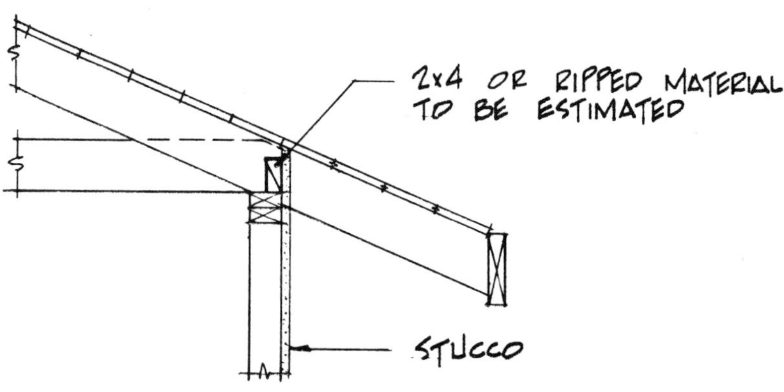

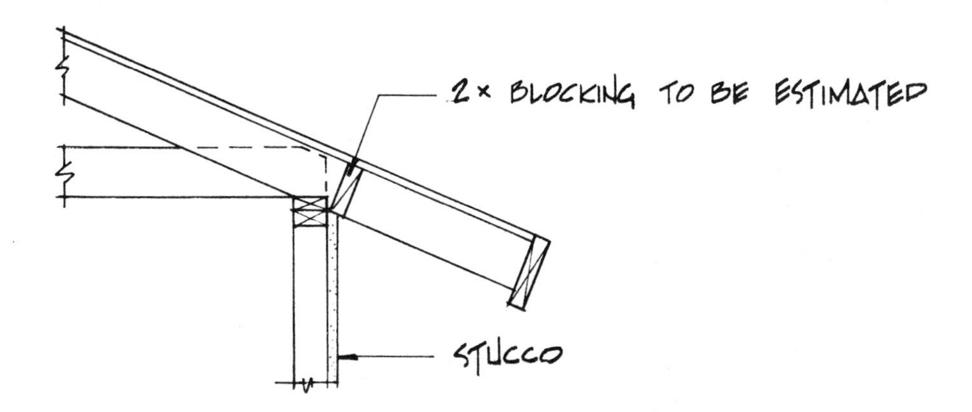

The following three open-eave details use the same procedure for estimating as details 1 and 2 above. The frieze board will also have to be added. The lineal feet of frieze board material will be the same as the lineal footage of blocking material plus 10% for cutting waste.

Detail 3 Horizontal Siding

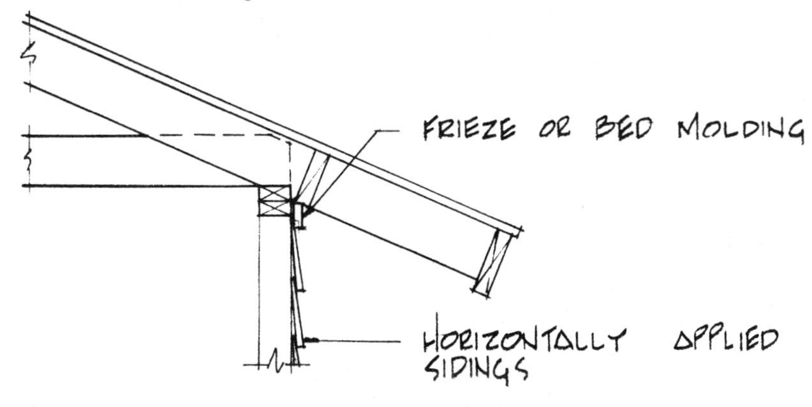

Detail 4 Plywood and Batt

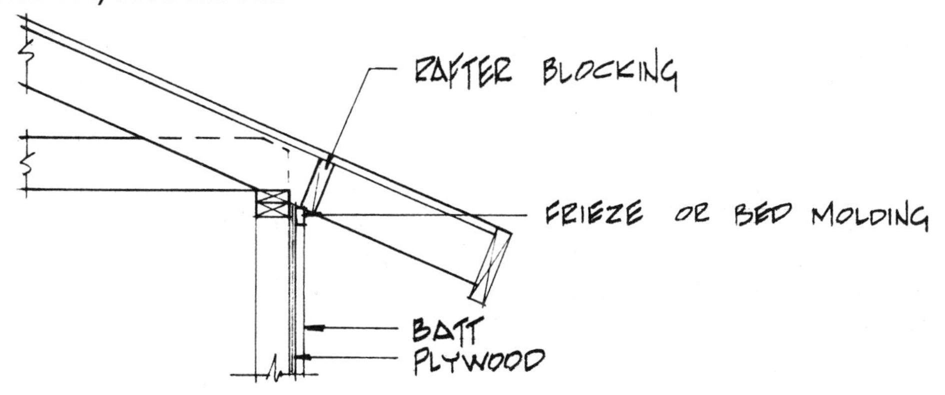

Detail 5 Shingles

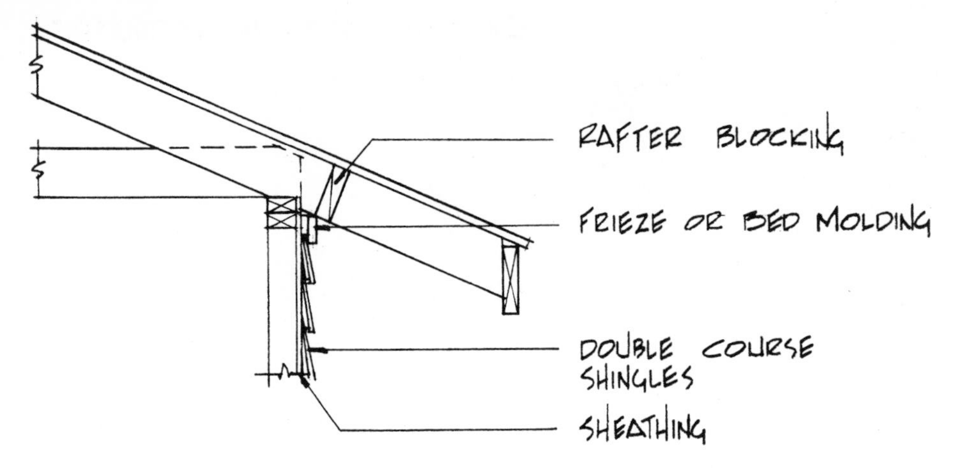

CLOSED CORNICE

In a closed cornice detail, the rafters are cut flush with the wall line, and there is no eave overhang. The estimator in this case is interested in estimating the frieze board and crown molding as illustrated below.

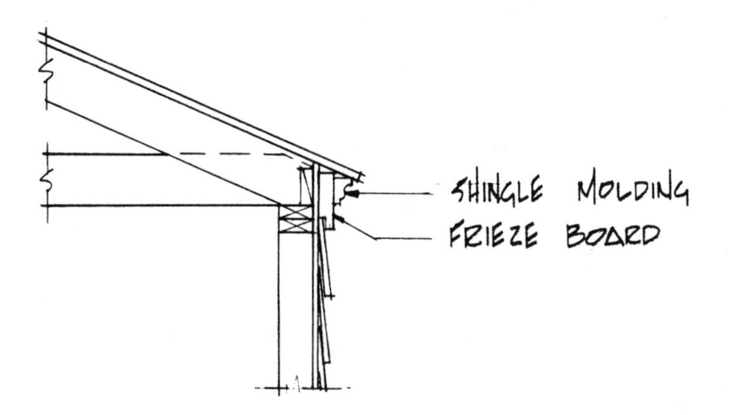

If the building has a hip roof, the lineal footage of frieze board and molding will equal the perimeter of the building. Then 10% should be added for waste. If the building is a gable, the lineal footage should equal the lower eave length excluding the gable end rakes. Again 10% should be added for cutting waste.

Boxing in the cornice is a practice in the northern areas of the country, although it is used occasionally on better quality homes throughout the United States. The boxed soffit was designed to give a structure a more finished look and for lower maintenance in areas where weathering of wood is rather severe and painting is necessary every few years.

The material to be estimated for a boxed cornice includes:

1. The ledger or ribbon, usually 1 × 4 or 2 × 4 material
2. The lookouts, which are usually 2 × 4 material
3. The soffit or plancher material, which may be 1/4- or 3/8-inch plywood or hardboard, or 3/4-inch shiplap or tongue-and-groove boards
4. A frieze board and/or a bed molding.

Following are a few typical boxed soffit details.

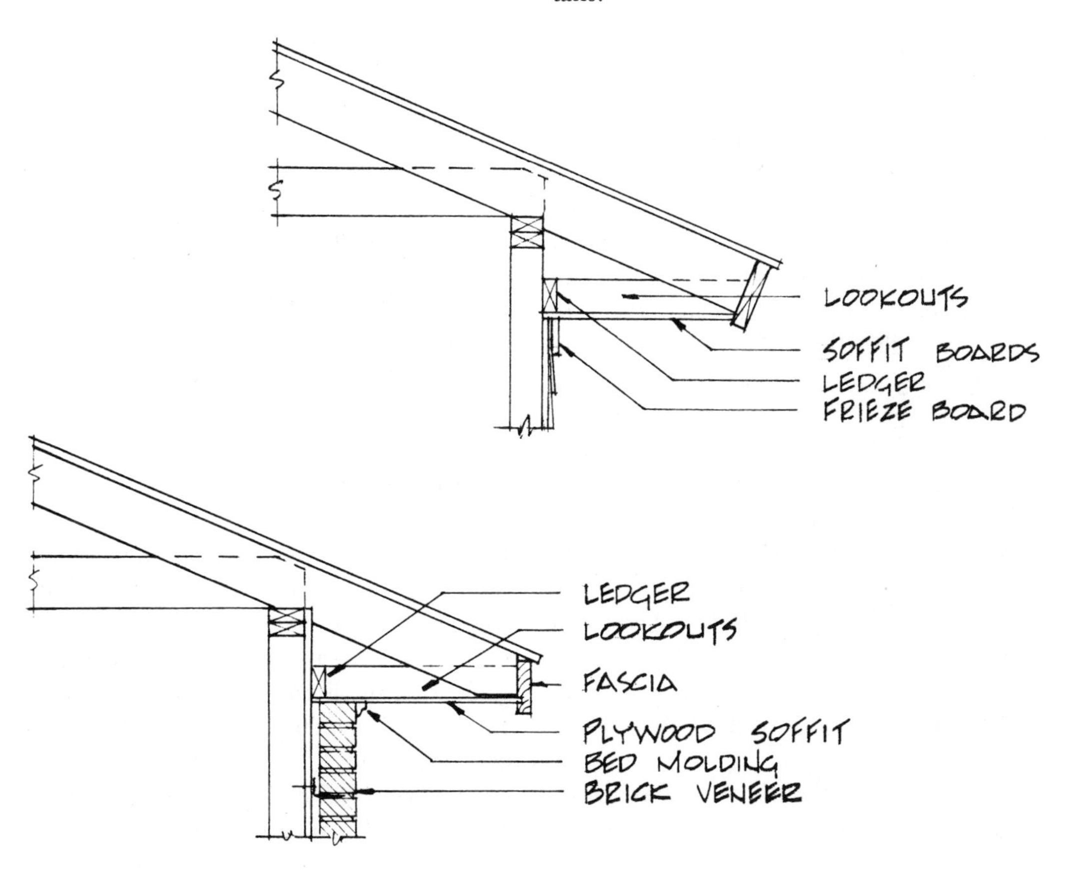

PROCEDURE FOR ESTIMATING MATERIALS FOR A BOX CORNICE

Ledger Material

If a building has a hip roof, the lineal feet of ledger material needed will equal the perimeter of the building. If the building has a gable roof, the lineal footage of ledger estimated should equal the lower eave length, excluding the gable rakes.

Lookout Material

The amount of material needed for lookouts will depend on lookout spacing and width of the overhang. A typical eave overhang width will be 2′-0″ with the lookouts usually spaced 16 or 24 inches on center. To find the number of lookouts needed, the estimator multiplies the lineal feet of box cornice by the spacing constant. If the lookouts are spaced 16″ O.C., the spacing constant is .75. If spaced 24″ O.C., the constant is .50. For waste, 10% should be added. The resulting figure is the number of lookouts needed. The number of lookouts can then be multiplied by the length of each piece for the total lineal footage of material to order.

Soffit or Plancher Material

As previously mentioned, the plancher or soffit material may be plywood or board. If plywood is used, the sheets will be ripped down the center so one sheet will provide two pieces, 2 feet wide by 8 feet long, or enough plywood to cover 16 lineal feet of soffit. The number of sheets of material can then be estimated by dividing 16 into the lineal feet of soffit needed. (The lineal feet of soffit needed should be measured from the edge of the roof line.)

If plywood is used to enclose the gable rake, the lineal footage of the soffit rake should be added to the lineal footage of eave soffit needed. Then divide by 16 for the number of sheets.

If boards are used, the number of boards it will take to cover the 2-foot overhang must be calculated. For example, if 1 × 6 boards are used, the dressed width dimension will be about 5-1/4 inches. The 5-1/4 or 5.25 is then divided into the 2-foot soffit width (24 inches):

$$24 \div 5.25 = 4.57 \text{ or } 5 \text{ boards}$$

The number of boards (5) is then multiplied times the lineal feet of soffit needed. For waste 10% should be added.

Frieze Board or Bed Molding

If the building has a hip roof, the lineal footage of frieze board and/or bed molding will equal the perimeter of the building plus ten percent waste. If the building has a gable roof, the lineal footage of frieze board and/or bed molding will equal the lower eave length plus ten percent waste.

Following are examples and procedures for estimating various soffit materials.

Example Problem:

List the cornice materials needed in the following problem.

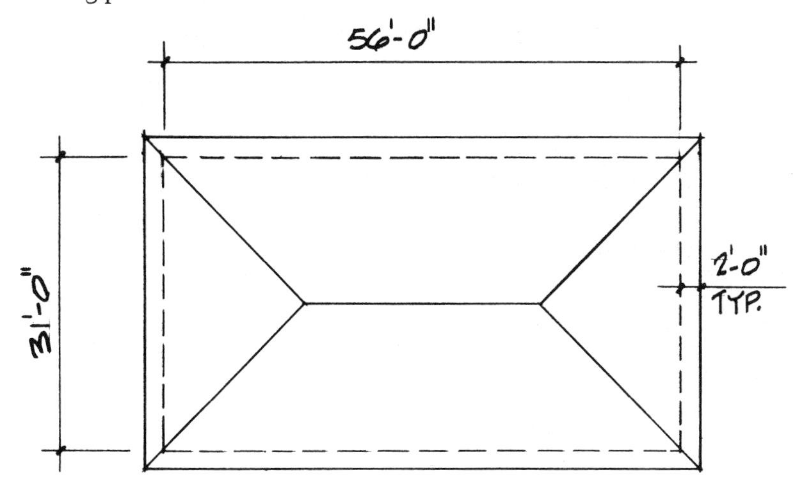

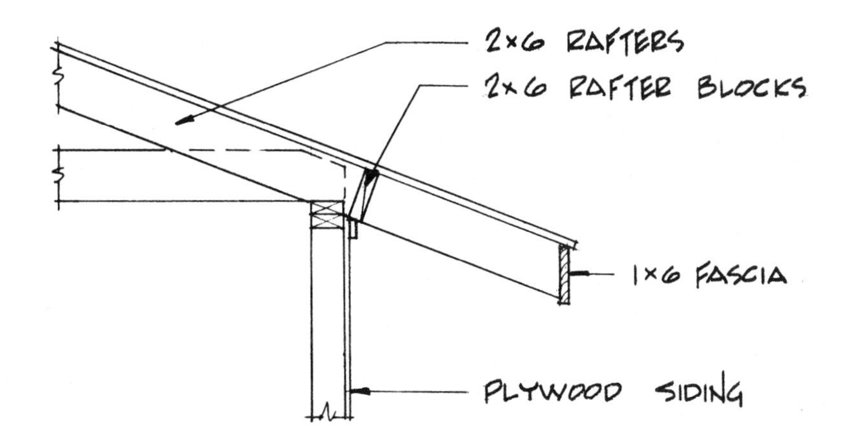

PROCEDURE:

1. Calculate the perimeter of the building: $56 + 56 + 31 + 31 = 174$ lin. ft.
2. The lineal feet of rafter blocks and frieze board will equal the perimeter of the building or 174 lin. ft.
3. Add 10% for waste, $174 \times .10 = 17.4$ or 18 lin. ft. $18 + 174 = 192$ lin. ft. of 2×6 rafter blocks and 192 lin. ft. of 1×2 frieze board.

OPEN CORNICE—GABLE ROOF

Example Problem:

List the cornice materials needed in the following problem.

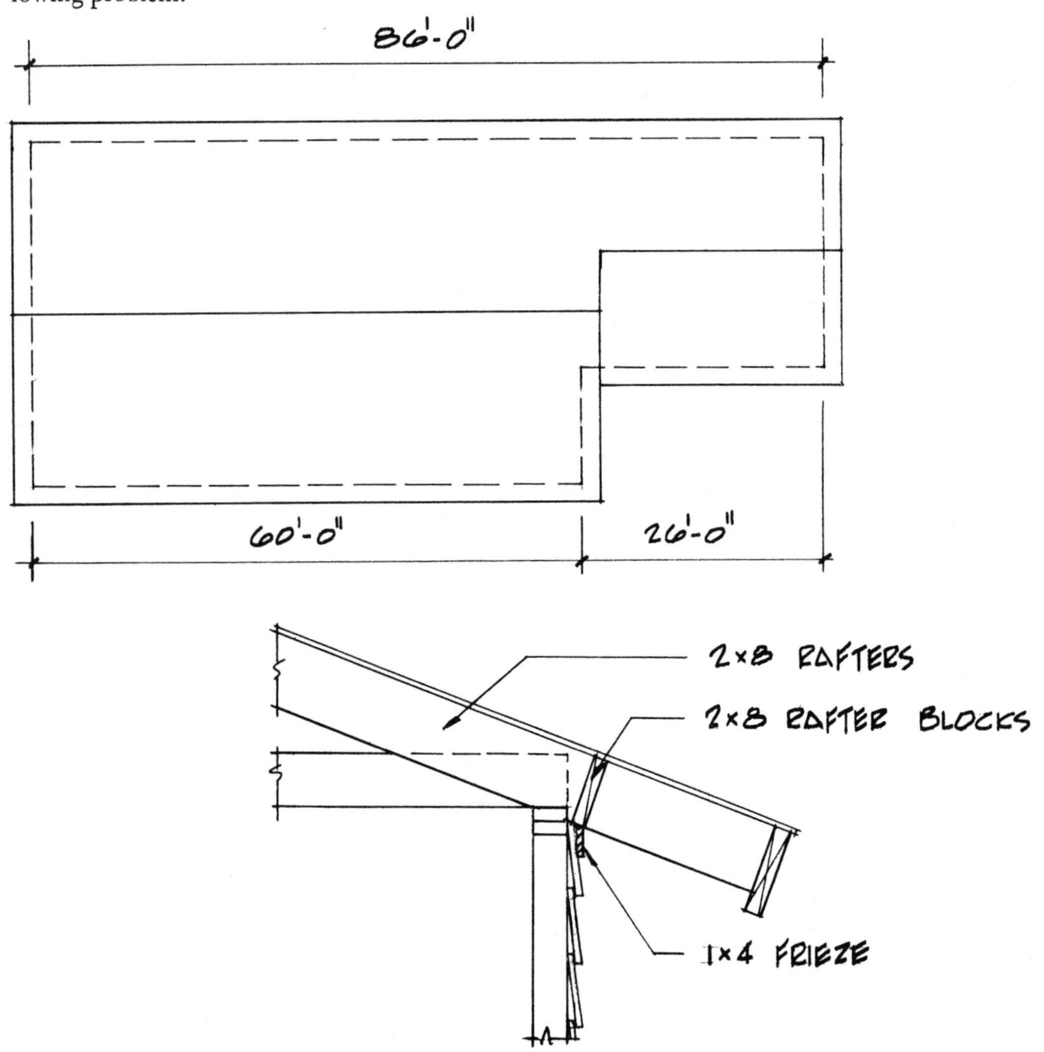

PROCEDURE:

1. Calculate the lower eave length of the building: 86 + 60 + 26 = 172 lin. ft.
2. The lineal feet of rafter blocking and frieze board will equal the lower eave length or 172 lin. ft.
3. Then add 10% for waste to your calculations: 172 × .10 = 17.2 or 18 lin. ft. 18 + 172 = 190 lin. ft. of 2 × 6 rafter blocks and 190 lin. ft. of 1 × 4 frieze board.

BOX CORNICE—HIP ROOF

Example Problem:

List the cornice material needed in the following problem.

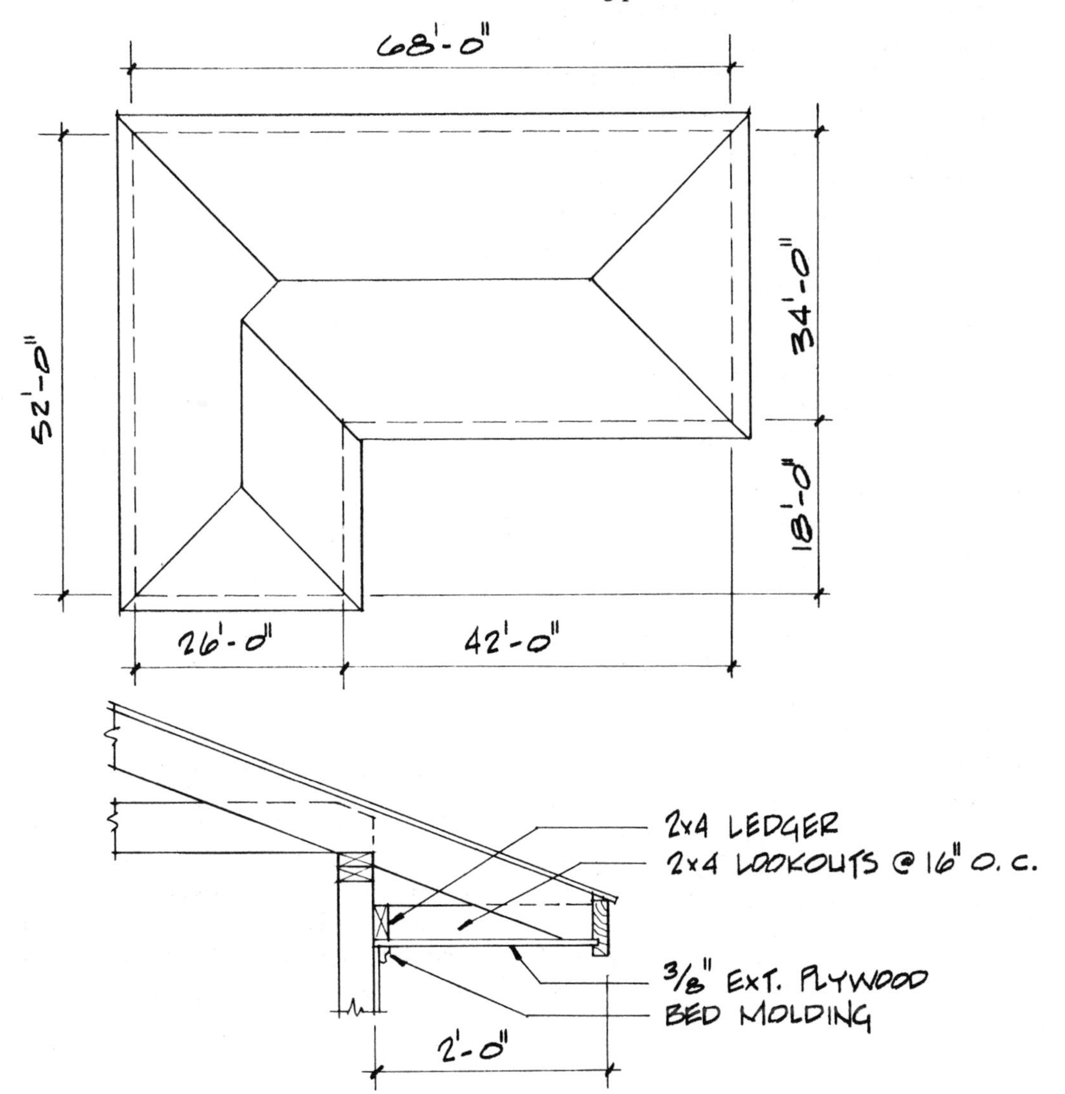

PROCEDURE:

1. Calculate the perimeter of the building: 68 + 52 + 26 + 18 + 42 + 34 = 240 lin. ft.
2. The 240 lineal feet of perimeter plus 10 percent for waste will equal the amount of ledger and bed molding material needed: 240 × .10 = 24 lin. ft. 24 + 240 = 264 lin. ft. of 2 × 4 ledger material and 264 lin. of bed molding.
3. To calculate the number of lookouts, multiply the perimeter of the building (240 lin. ft.) by the spacing constant for 16″ centers (.75) and add 10 percent for waste. The result will be the number of lookouts:

   240 × .75 = 180 lookouts
   180 lookouts × .10 waste = 18
   18 + 180 = 198 lookouts

4. The number of lookouts (198) is multiplied by the length of each lookout or 2′-0″: 2 × 198 = 396 lin. ft. of 2 × 4 lookout material.
5. To estimate the number of sheets of plywood for the soffits, first calculate the perimeter of the roof line: 72 + 72 + 56 + 56 = 256 lin. ft.
6. The roof perimeter (256 lin. ft.) is then divided by the lineal feet of soffit that one 4 × 8 sheet of plywood will cover. (One 4 × 8 sheet cut in half lengthwise will cover 16 lin. ft. of soffit.)

   256 ÷ 16 = 16 pieces of 3/8″ × 4′ × 8′ plywood

BOX CORNICE—GABLE ROOF

Example Problem:

Estimate the cornice material in the following problem.

PROCEDURE:

1. To estimate the ledger material, calculate the lower eave length. Lower eave lengths are illustrated below.

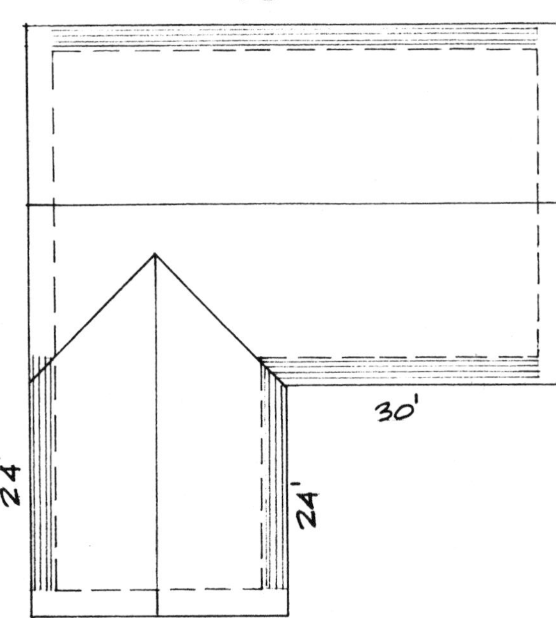

52 + 24 + 24 + 30 = 130 lin. ft.
Add 10% for waste to your total.

2. The 2 × 4 lookout material can also be estimated by multiplying the lower eave length (130) by the spacing constant for 24-inch center-to-center spacings (.5). Add 10% for waste. The result will be the number of lookouts needed:

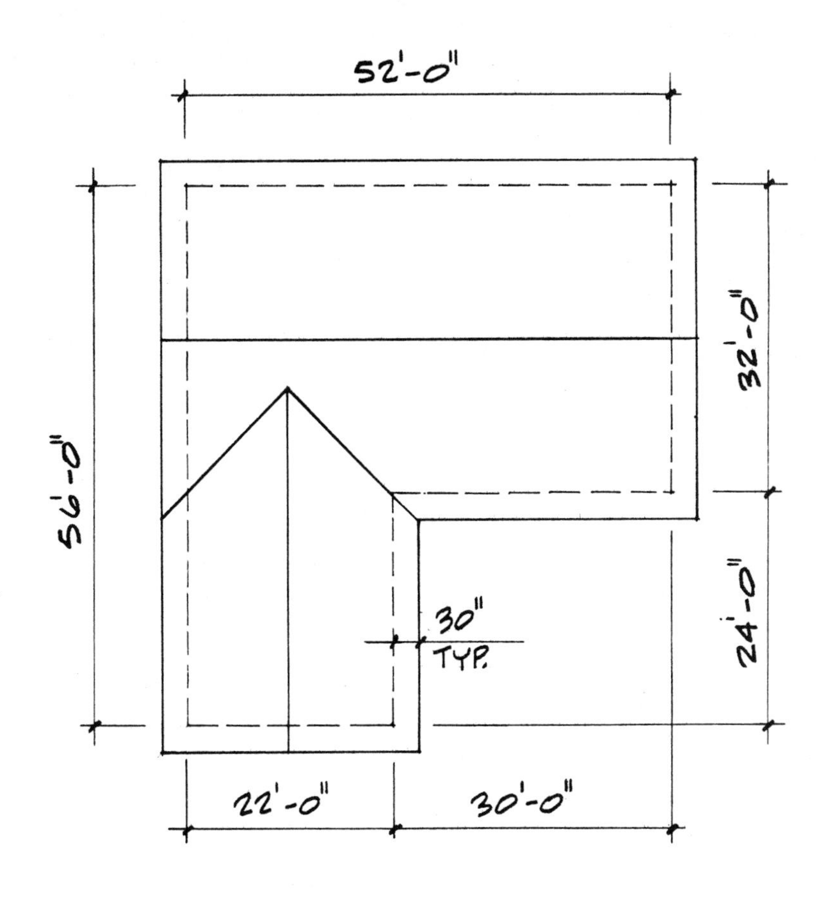
52'-0"
56'-0"
32'-0"
24'-0"
30"
TYP.
22'-0"
30'-0"

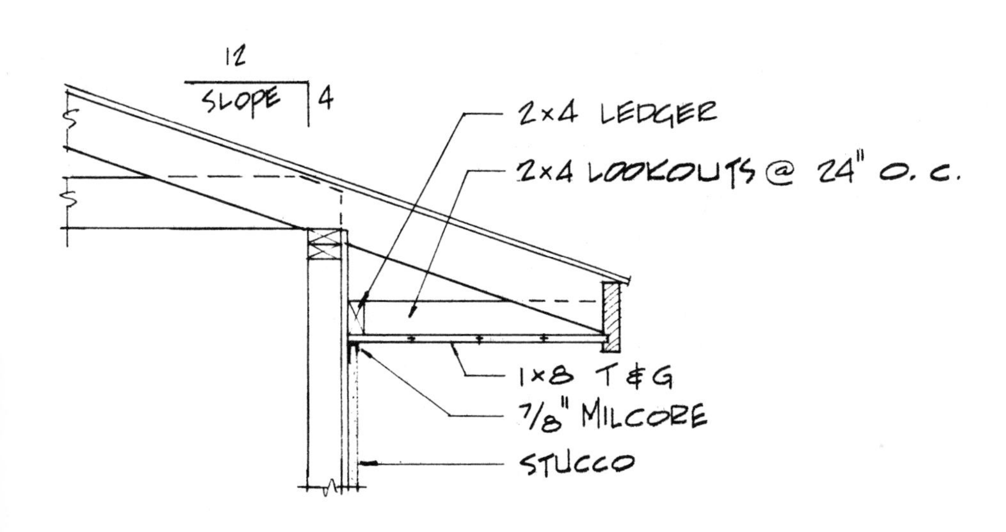
12
SLOPE
4
2×4 LEDGER
2×4 LOOKOUTS @ 24" O. C.
1×8 T & G
7/8" MILCORE
STUCCO

$130 \times .5 = 65$ lookouts
$65 \times .10$ waste $= 6.5$ or 7
$65 + 7 = 72$ pieces

The number of lookouts (72) is then multiplied times the length of each lookout (30″), or if changed to a decimal (2.5).

$72 \times 2.5 = 180$ lin. ft. of 2×4 lookout material

Note: The framing method used and the material needed for supporting the bargeboard and soffit material on a gable roof is discussed in Unit 48 (Fascia and Bargeboard).

3. To estimate the lineal feet of tongue-and-groove boards for the soffit, the estimator must first calculate the perimeter of the roof line. This includes the fascia and barge lengths.

80 lin. ft.—long rafters
32 lin. ft.—short rafters
+ 135 lin. ft.—lower eave line
247 lin. ft.

Next, the number of 1×8 boards to cover a 30″ overhang must be calculated.

$30'' \div 7.25$ (actual width of a 1×8) $= 4.13$ or 5 boards

The 247 lineal feet of eave is then multiplied times the number of boards (5):

$247 \times 5 = 1{,}235$ lin. ft.

Add 10% for waste.

$1{,}235 \times .10 = 124$ lin. ft. of waste
$1{,}235 + 124 = 1{,}359$ lin. ft. of 1×8 tongue-and-groove board.

If the eave is to have a box cornice, some extra barge material is needed for cornice returns as shown in the following diagram.

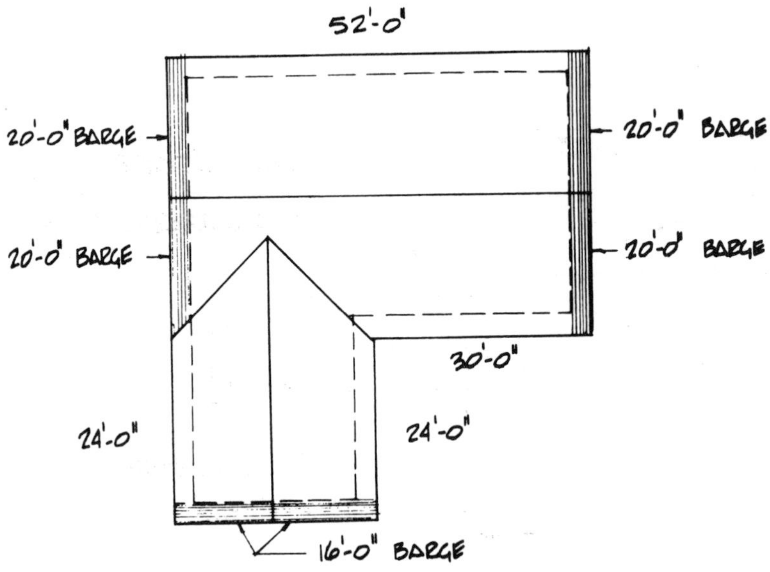

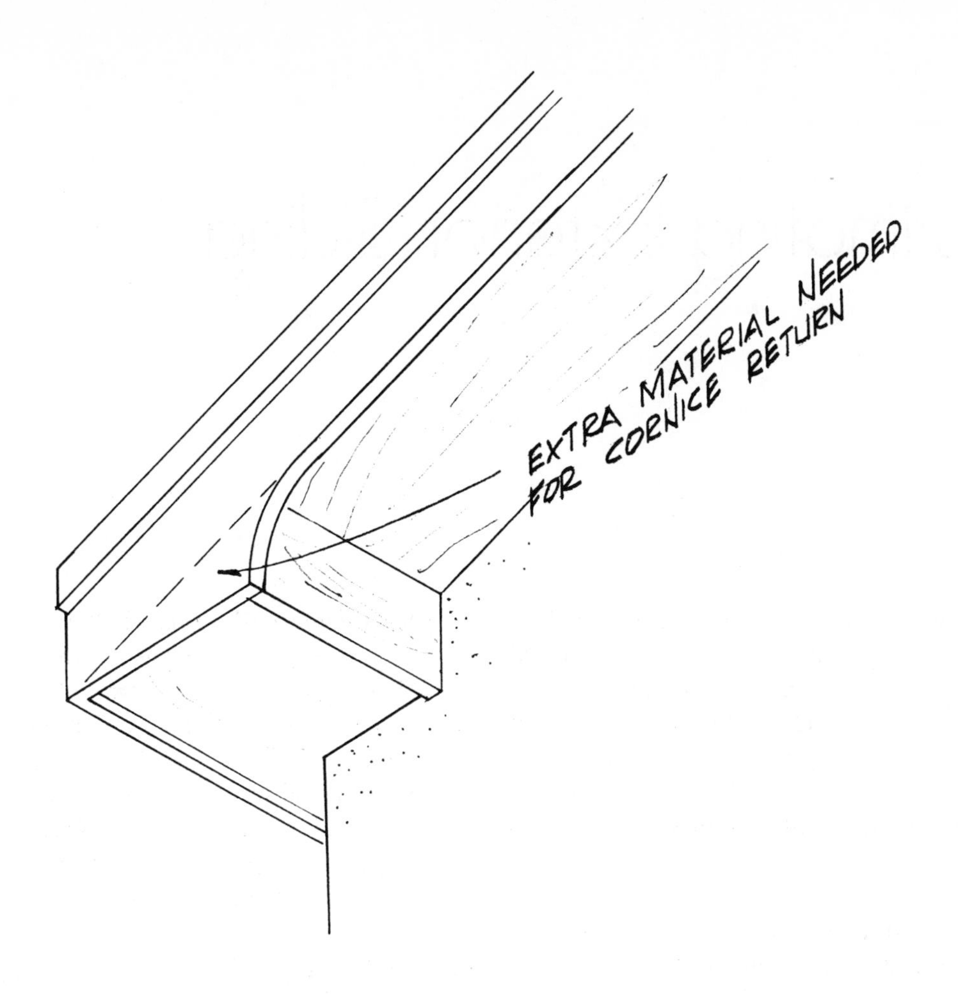
EXTRA MATERIAL NEEDED
FOR CORNICE RETURN

63

Estimating Exterior Siding

There are numerous types of siding and exterior wall-covering materials on the market today. They are made from various materials such as wood, hardboard, composition board, metal, plastic, etc., and they come in a variety of styles, shapes, and textures. These exterior wall coverings come in two forms.

1. As sheet material such as plywood.
2. As strip siding in the form of boards that may be applied vertically, horizontally, or diagonally.

Stucco and shingle siding is also used, but it does not fall into the categories above. These will be discussed later in the unit.

The estimator usually estimates sheet or strip siding using one of the following two methods.

METHOD 1: The estimator lays out by scaling or drawing on the building plans the number of boards to cover the area where siding is used. The boards are counted, and the total lineal footage is calculated. A few extra boards or a certain percentage of the total needed is added to this figure to allow for checked or split material and for end cutting waste. Exterior plywood wall covering can also be estimated by this method. The estimator calculates the number of sheets by checking the floor plan dimensions and elevation views for the areas covered.

METHOD 2: Many estimators prefer the second method for estimating exterior wall materials because any type sheets, boards, shingles, or stucco may be calculated. The square foot area of wall to be covered is calculated, and a percentage is added for lapping of boards, width loss for milling, and cutting waste.

STRIP SIDING

The following illustrations show various applications of strip siding with which the estimator should be familiar.

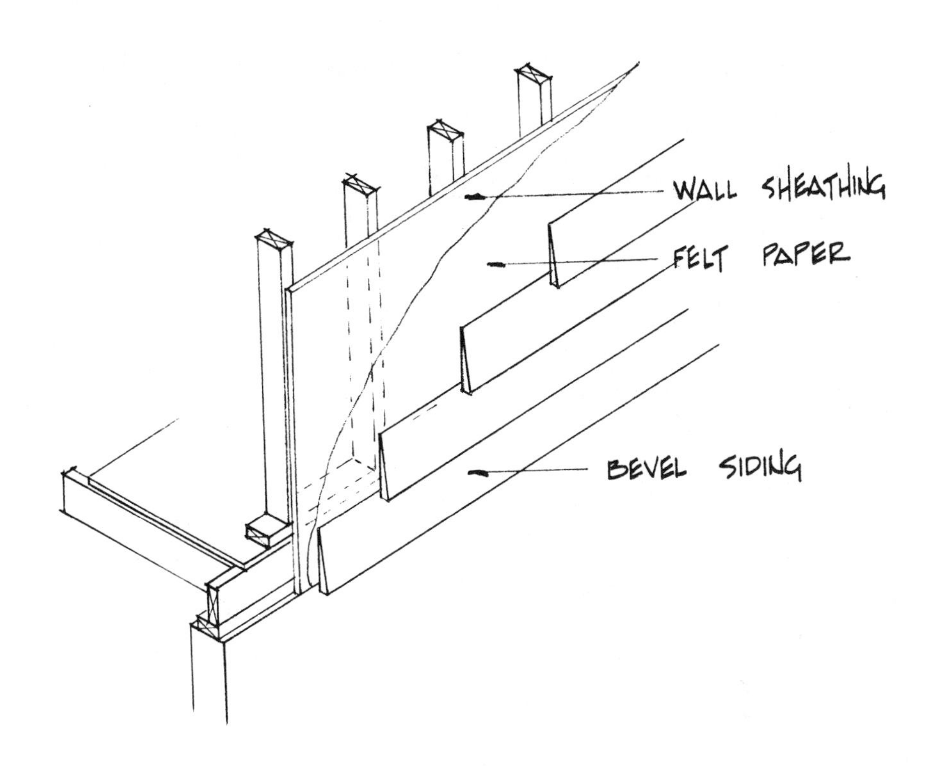

SIDING STYLES

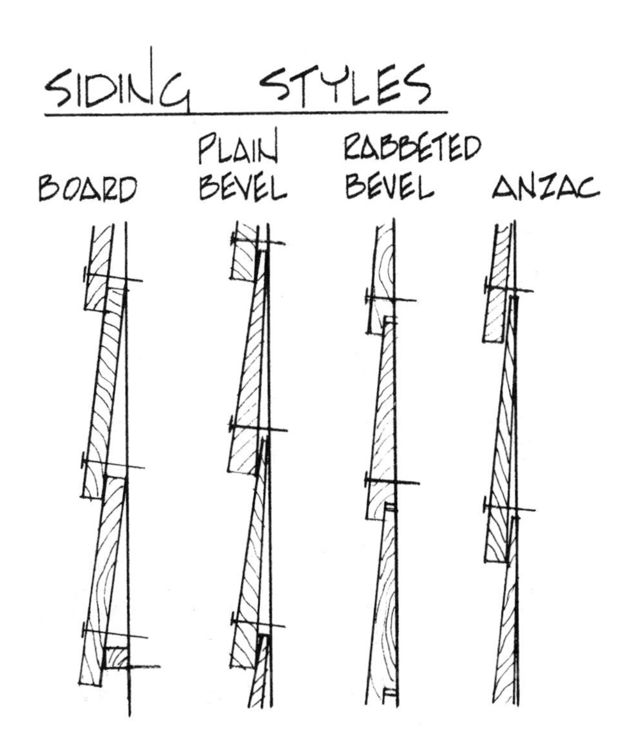

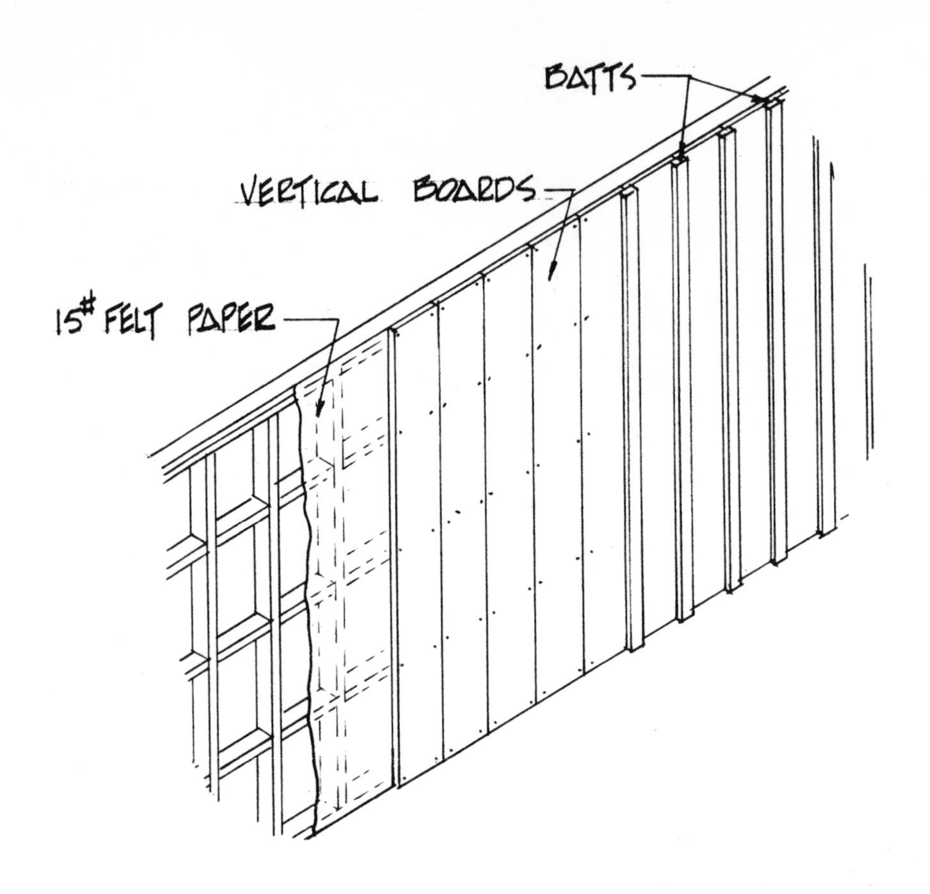
BATTS
VERTICAL BOARDS
15# FELT PAPER

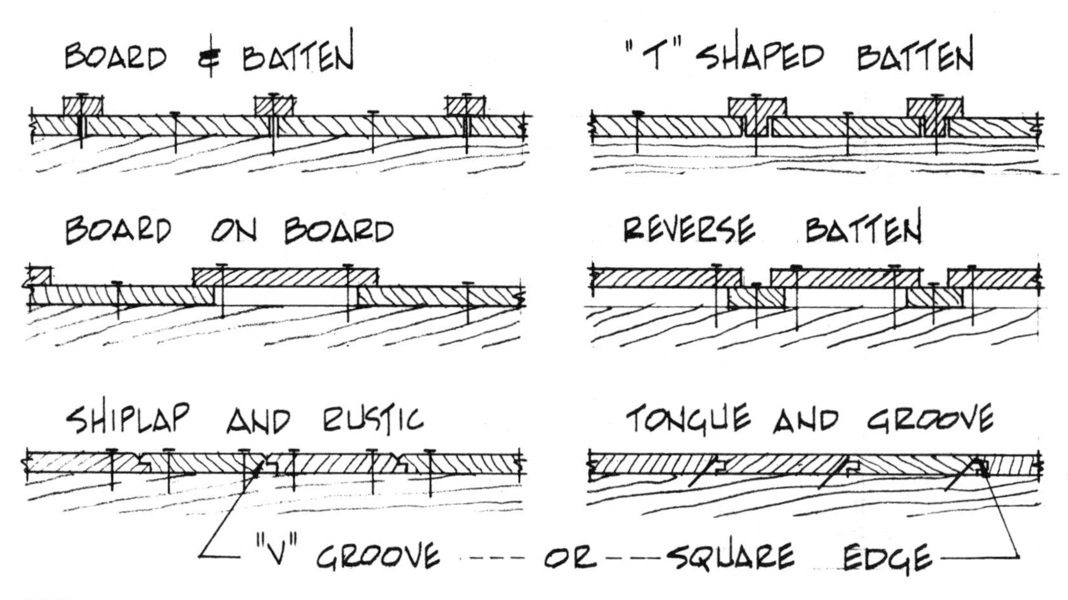
BOARD & BATTEN
"T" SHAPED BATTEN
BOARD ON BOARD
REVERSE BATTEN
SHIPLAP AND RUSTIC
TONGUE AND GROOVE
"V" GROOVE --- OR --- SQUARE EDGE

Strip siding may be ordered in random or specified lengths. Specified lengths, especially over 8 foot, will cost more (sometimes quite a bit more) per board foot. Random lengths are the least expensive per board foot and usually come in a variety of lengths from 6 to 18 feet.

When the siding runs vertically, specified lengths are usually ordered. They look better running full wall height; moveover, it's hard to weather seal board joints that run horizontally.

When siding is applied horizontally, random lengths are usually ordered because pieces will have to be joined on long walls. These joints are not visually unpleasing, and if properly fitted will be weather tight.

PROCEDURE FOR ESTIMATING STRIP SIDING

1. Calculate the square foot area the siding is to cover by multiplying the wall height by the wall width or, in some cases, by the perimeter of the building; then substract the square foot area of the windows and doors. For these openings the estimator rounds off to the nearest foot. For example, a 3068 exterior door would be considered 3′ × 7′ or 21 square feet.
2. Multiply the square foot area to be covered by factor A from the following tables, according to the particular type and size of siding to be used. The resulting amount is the needed siding in square feet. A certain amount of cutting waste must be added to this figure. Factor B adds a standard additional 20% cutting waste to factor A. This is used by many estimators and contractors.

   The actual amount of cutting waste depends on the carpentry crew applying the siding. If they are experienced and use the various lengths of material carefully, the standard waste factor may be cut to 5 or 10%. But, if 20% cutting waste seems to be appropriate for your needs, then multiply the square footage to be covered by the factor in column B. If 20% is too much waste, multiply the square footage by the factor in column A, then add the appropriate percentage of waste.
3. Siding may be ordered from the lumberyard by square foot or lineal foot measure. When the siding is applied vertically and specified lengths of material are to be ordered, the estimator must convert from square foot measure to lineal foot measure. To convert square feet to lineal feet, multiply the square footage needed by the conversion factor in column C for the type and size of siding used. The lineal feet of material may then be divided by the specified length of board needed. The result will be the number of boards necessary for coverage.

Following are various types of siding and the waste factors which take into account the lapping of boards, width loss for milling, checking, splitting, and end cutting waste.

Note: All the following sidings require a layer of 15-lb. felt paper between the studs or wall sheathing and the siding.

TABLE 63-1 SQUARE EDGE S4S STRIP SIDING

Size	(A) Width Loss, & Milling Factor	(B) A Factor Plus 20% Waste	(C) Conversion: Sq. Ft. to Lin. Ft.
1 × 4	1.14	1.37	3
1 × 6	1.10	1.32	2
1 × 8	1.09	1.31	1.33
1 × 10	1.08	1.30	1.20
1 × 12	1.07	1.29	1.0

TABLE 63-2 PLAIN BEVEL STRIP SIDING

Size	(A) Lapping, Width Loss, & Milling Factor	(B) A Factor Plus 20% Waste	(C) Conversion: Sq. Ft. to Lin. Ft.
1/2 × 4	1.60	1.92	3
1/2 to 1 × 6	1.34	1.61	2
1/2 to 1 × 8	1.28	1.54	1.33
1/2 to 1 × 10	1.20	1.44	1.20
1/2 to 1 × 12	1.17	1.40	1.0

TABLE 63-3 RABBETED BEVEL STRIP SIDING

Size	(A) Lapping, Width Loss, & Milling Factor	(B) A Factor Plus 20% Waste	(C) Conversion Sq. Ft. to Lin. Ft.
5/8 × 4	1.28	1.54	3
5/8 or 3/4 × 6	1.17	1.41	2
5/8 or 3/4 × 8	1.13	1.36	1.33
3/4 × 10	1.10	1.32	1.20
3/4 × 12	1.08	1.30	1.0

TABLE 63-4 Shiplays Strip Siding (Straight or V-Groove)

Size	(A) *Lapping, Width Loss, & Milling Factor*	(B) *A Factor Plus 20% Waste*	(C) *Conversion: Sq. Ft. to Lin. Ft.*
3/4 or 1 × 4	[N.A.]	[N.A.]	[N.A]
3/4 or 1 × 6	1.19	1.43	2
3/4 or 1 × 8	1.16	1.39	1.33
3/4 or 1 × 10	1.13	1.36	1.20
3/4 or 1 × 12	1.10	1.32	1.0

TABLE 63-5 T & G Strip Siding

Size	(A) *Lapping, Width Loss, & Milling Factor*	(B) *A Factor Plus 20% Waste*	(C) *Conversion: Sq. Ft. to Lin. Ft.*
1 × 4	1.26	1.51	3
1 × 6	1.16	1.39	2
1 × 8	1.14	1.37	1.33
1 × 10	1.11	1.33	1.20
1 × 12	1.09	1.31	1.0

TABLE 63-6 Channel Rustic Strip Siding

Size	(A) *Lapping, Width Loss, & Milling Factor*	(B) *A Factor Plus 20% Waste*	(C) *Conversion: Sq. Ft. to Lin. Ft.*
1 × 6	1.20	1.44	2
1 × 8	1.15	1.38	1.33
1 × 10	1.12	1.34	1.20
1 × 12	1.09	1.31	1.0

TABLE 63-7 Anzac Strip Siding S4S

Size	(A) *Lapping, Width Loss, & Milling Factor*	(B) *A Factor Plus 20% Waste*	(C) *Conversion: Sq. Ft. to Lin. Ft.*
1 × 8	1.28	1.54	1.33
1 × 10	1.22	1.46	1.20
1 × 12	1.18	1.42	1.0

Example Problem:

Estimate the square foot amount of 1 × 6 plain bevel siding needed to cover 1,292 square feet of wall area. Twenty percent waste is needed.

PROCEDURE:
Multiply the 1,292 square feet of wall by the factor from Table 63-2 column B for plain bevel siding.

```
   1,292 sq. ft.
×   1.61 factor for 1 × 6
         plain bevel
         siding
  ------
    1292
   7752
  1292
  ------
```

Coverage needed for 2,080.12 sq. ft.

Convert the square foot amount in the problem above to lineal feet.

PROCEDURE:
Multiply the 2,080 square feet from the problem above by the lineal foot conversion factor from column C Table 63-2 for 1 × 6 siding (2).

```
   2,080
×      2
  ------
   4,160 lin. ft. of material
```

PLYWOOD SIDING

Typical plywood application

Plywood siding may be estimated by either of the two methods previously mentioned:

1. By scaling the plan and counting the sheets.
2. By calculating the number of sheets from the square foot area to be covered.

Plywood siding usually comes in 4′ × 8′, 4′ × 9′, or 4′ × 10′ sheets, and in thicknesses from 3/8 to 3/4 inch.

Before the sheets are applied, a layer of 15-lb. felt should cover the studs or wall sheathing.

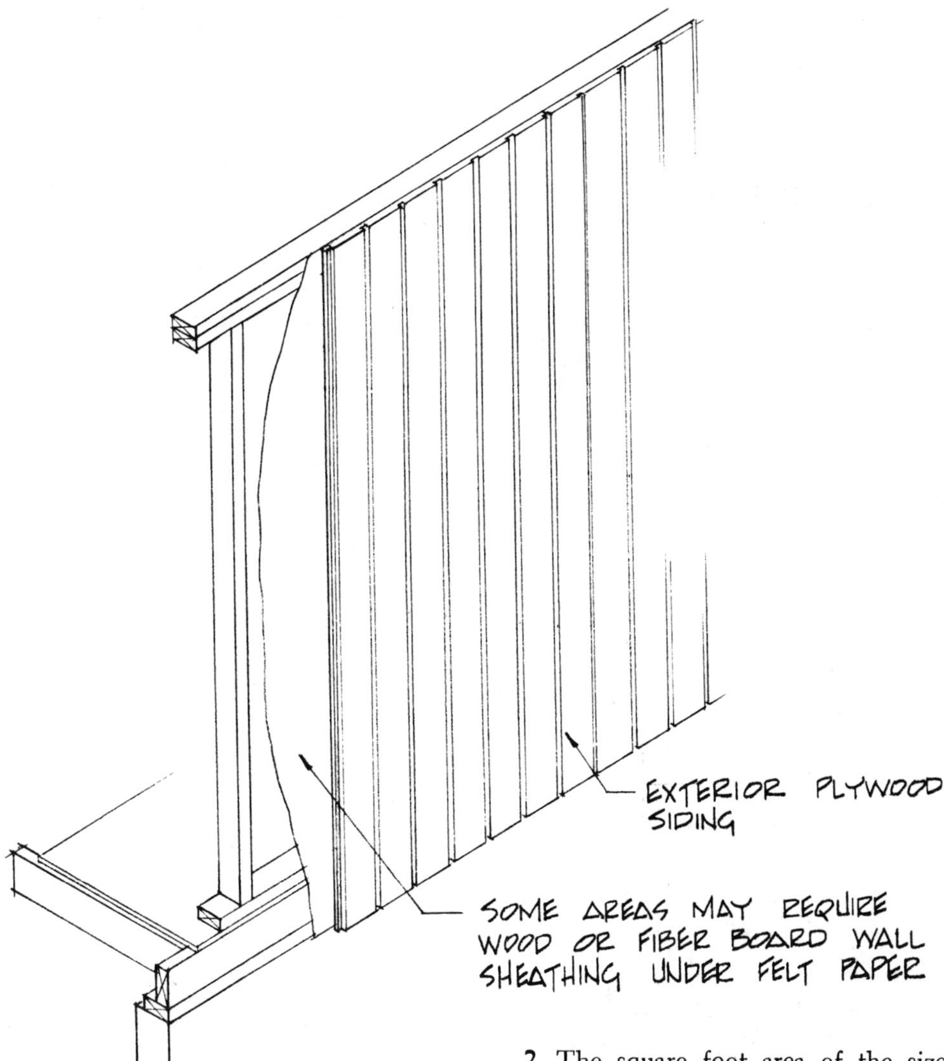

PROCEDURE FOR ESTIMATING PLYWOOD SIDING

1. When estimating plywood siding, calculate the square foot area of the walls covered, and include the window and door areas.
2. The square foot area of the size of sheet used, 4 × 8 = 32 sq. ft., 4 × 9 = 36 sq. ft., 4 × 10 = 40 sq. ft., is divided into the total square foot area to be covered. The result will be the number of sheets needed.

Note: The estimator may find it a wise practice to add one extra sheet for any unforeseen problem on the job, such as a defaced sheet, cutting mistake, theft, etc.

PLYWOOD AND BATT APPLICATION

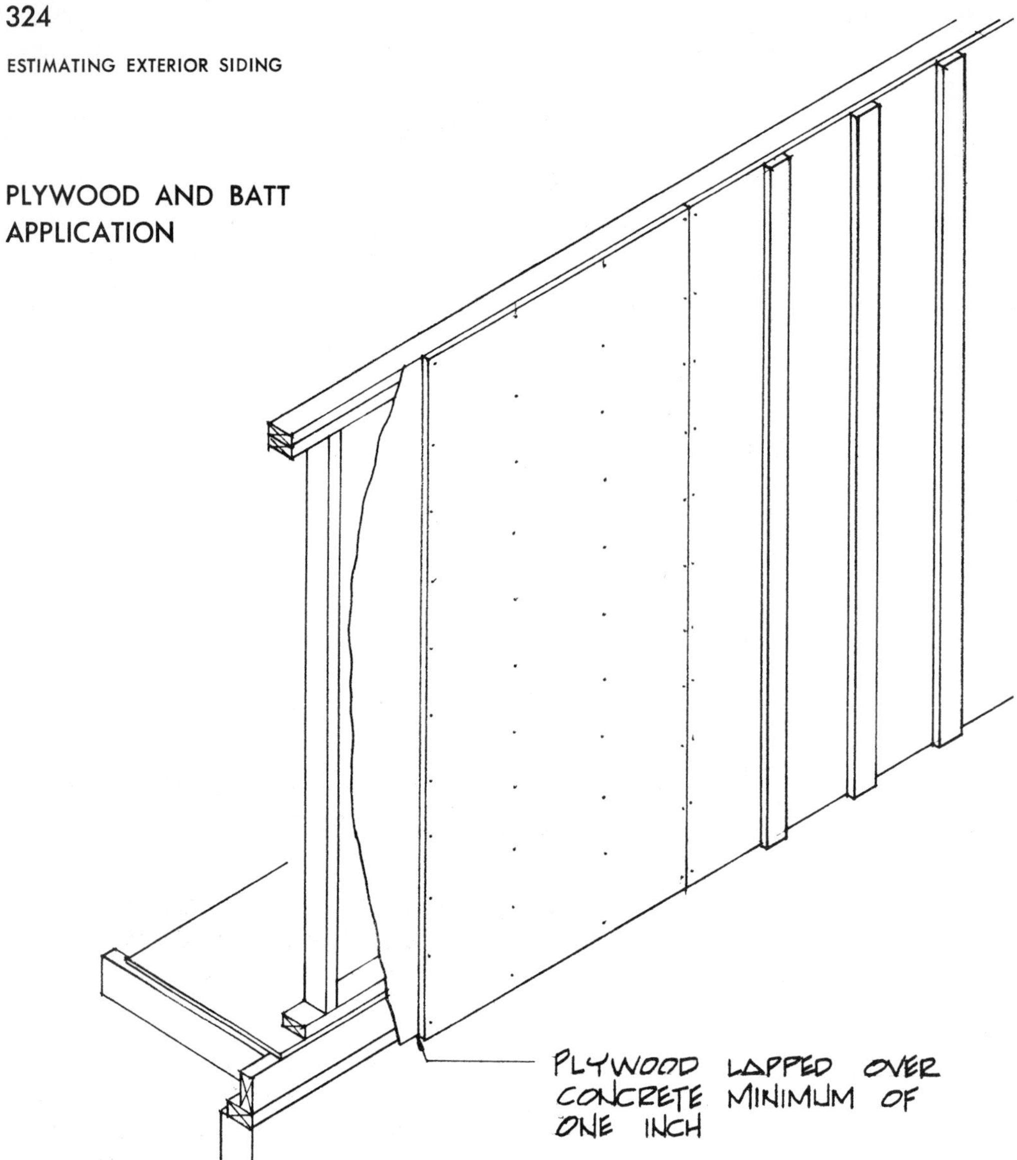

The easiest way for the estimator to calculate the number of batts needed is to estimate the number of sheets of plywood and figure how many batts per sheet are needed. If the batts are spaced 1′-0″ on center, 4 batts per sheet are needed. If the batts are spaced 16″ on center, 3 batts per sheet are needed. Batt material is usually ordered in full height pieces but may be spliced. Waste is seldom added because there will be some extra material from the window and door openings.

ESTIMATING SHAKE AND WOOD SHINGLES FOR WALLS

Typical shake and wood shingle applications

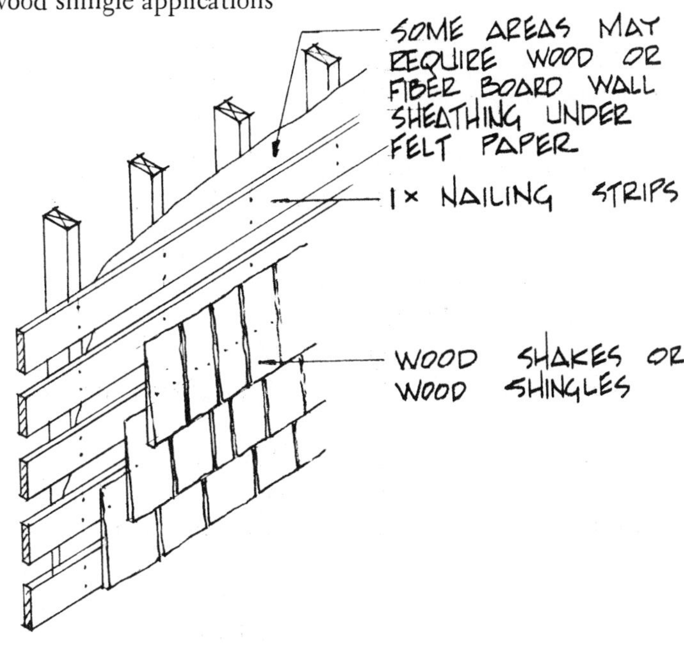

WOOD SHAKE SHINGLES

Wood shake shingles 24″ long with a 10″ exposure to weather are usually used for wall application. Wall shakes may be estimated in the same manner as roof shakes. The square foot area to be covered is calculated (wall area minus the window and door areas), then divided by 100 to get the number of shingle squares. One extra bundle should be added to insure coverage. If the exposure is less than 10″, more shake material will be needed. The amount should be adjusted according to the percentage factor in the following table.

TABLE 63-8 Coverage for 24″ Wood Shake Shingles

Weather Exposure	*Exposure Factor*
10″	1.00%
9-1/2″	1.05%
9″	1.10%
8-1/2″	1.15%
8″	1.20%

Example Problem:

How many squares of shake shingles are needed to cover 8,400 sq. ft. of wall area at a 9-1/2″ exposure?

PROCEDURE:

1. Divide 8,400 sq. ft. by 100 to get the number of shingle squares: 8,400 ÷ 100 = 84 squares.
2. Multiply the 84 squares by the percentage factor from Table 63-8 for the 9-1/2″ shingle exposure (1.05):

$84 \times 1.05 = 88.20$ (to the nearest part of a square) or 88-1/4 squares
88-1/4 squares plus one extra bundle = 88-1/2 squares

Note: Before wood shake shingles are applied, a layer of 15-lb. felt paper should be used to cover the studs or wall sheathing.

WOOD SHINGLES

Wood shingles 16″ long are usually used for wall siding. The shingle exposure usually runs from 5″ to a maximum allowable of 7-1/2″. Wall shingles may be estimated in the same manner as roof shingles, considering that one square of shingles at 5″ exposure covers 100 sq. ft. If the weather exposure is greater than 5″, then fewer shingles are needed.

The following table gives the percentage factor for various exposure conditions.

TABLE 63-9 COVERAGE FOR 16″ WOOD SHINGLES

Weather Exposure	*Exposure Factor*
5″	1.00
5-1/2″	.95
6″	.90
6-1/2″	.85
7″	.80
7-1/2″	.75

Example Problem:

How many squares of wood shingles are needed to cover 5,800 sq. ft. of wall area at a 6-1/2″ exposure?

PROCEDURE:

1. Divide 5,800 sq. ft. by 100 to get the number of shingle squares: 5,800 ÷ 100 = 58 squares.
2. Multiply the 58 squares by the percentage factor from Table 63-9 for the 6-1/2″ shingle exposure (.85).

$58 \times .85 = 49.30$ or (to nearest part of a square) 49-1/2 squares
49-1/2 squares plus one extra bundle = 49-3/4 squares

Note: Before wood shingles are applied, a layer of 15-lb. felt paper should be used to cover the studs or wall sheathing.

Wood shingles are sometimes applied with a double course of shingles as illustrated below.

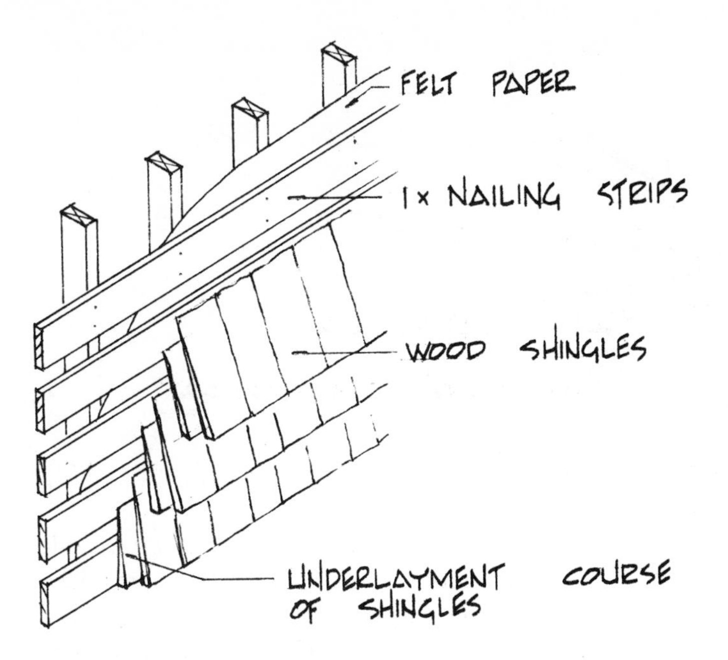

This is known as a *shadow* or *underlayment* course. A lesser grade of shingle, #3 or #4 backing shingle, is used for the first course. If a double layer of shingles is used, the same number of squares of underlayment material should be ordered as was ordered for the top course.

ESTIMATING STUCCO

Typical stucco applications:

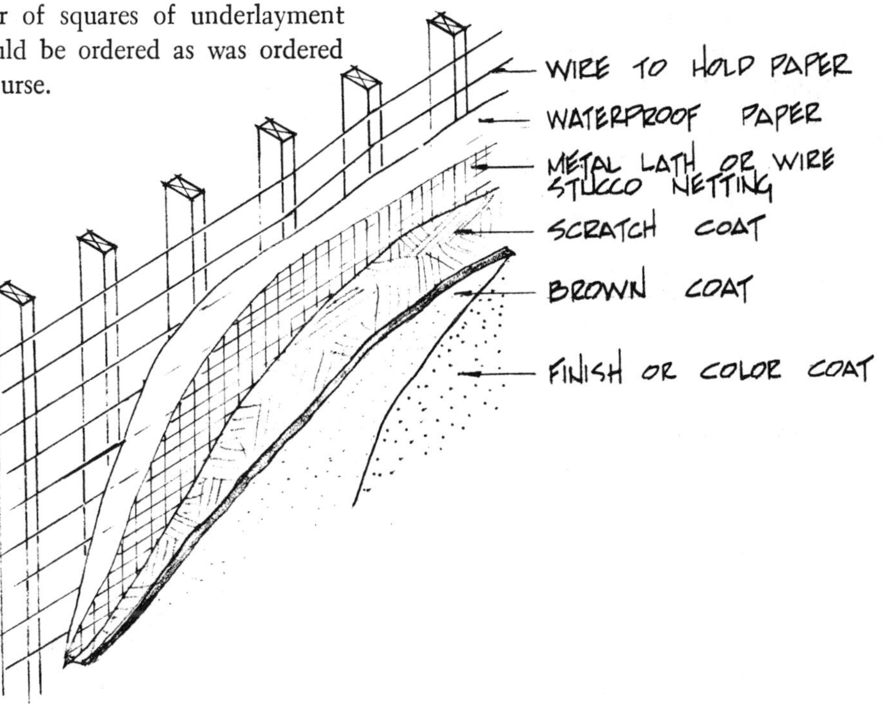

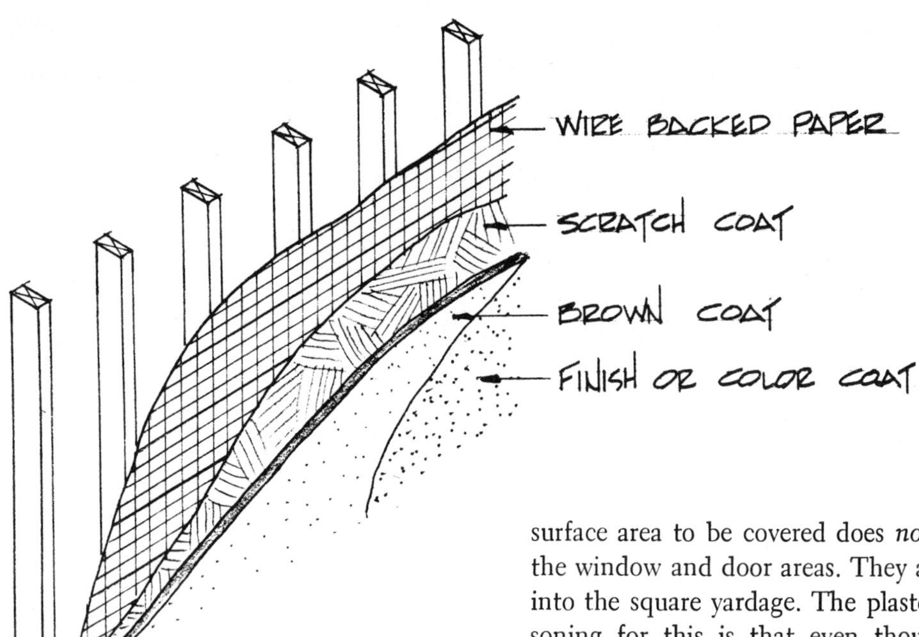

The plasterer estimates stucco by the square yard of coverage. A square yard is 3 ft. by 3 ft., or 9 sq. ft. The cost per square yard includes the materials and labor (including cleanup). The measurement for wall surface area to be covered does *not* exclude the window and door areas. They are added into the square yardage. The plasterer's reasoning for this is that even though there is no material used in the window and door openings, there is an off-setting labor cost for trimming around these openings. On most single-story stucco homes (either open eave or box cornice), the height of the wall will be approximately 9′-0″. This means that there is one square yard for every lineal foot of building perimeter that has stucco.

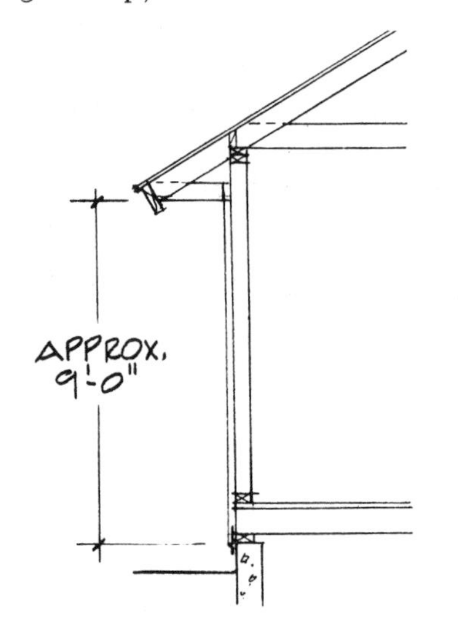

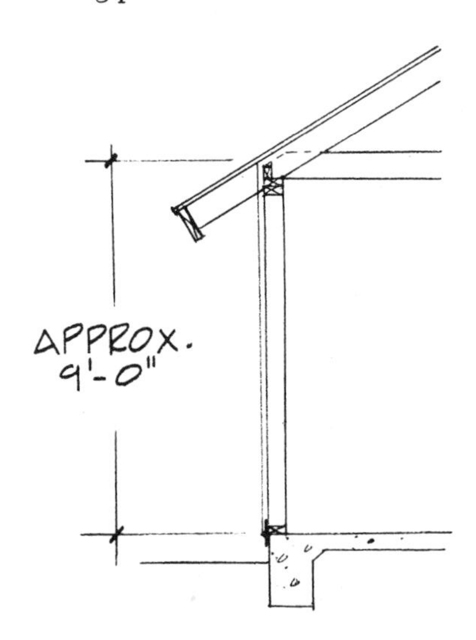

Example Problem:

Estimate the square yards and the cost of stucco for the following building:

Note: Stucco four sides. Cost per sq. yd. $7.50.

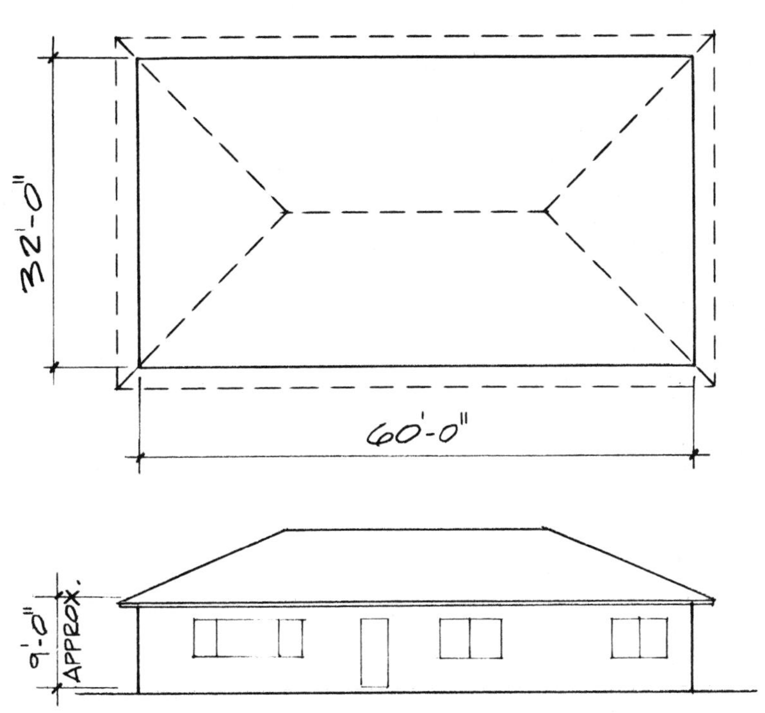

PROCEDURE:

1. Calculate the building perimeter in lineal feet: 184 lin. ft.
2. Calculate the building height to be stuccoed: 9'-0".
   This means for every lineal foot of perimeter, there is one sq. yd. or 184 sq. yds.
3. Multiply the square yards (184) by the cost per square yard ($7.50).

	184 sq. yd.
	× 7.50 per sq. yd.
	000
	920
	1288
Total	$1,380.00

When the building has a gable roof, the area of the gables will have to be calculated and added to the wall area. To calculate the additional square yardage, the estimator must check the floor plan and the elevation views for the following information:

1. The building run to the nearest foot.
2. The total rafter rise to the nearest foot. (This may be scaled from the elevation views or calculated by multiplying the building run by the unit rise per foot of run.)

Example Problem:

Find the square yardage on the gable ends of the following building.

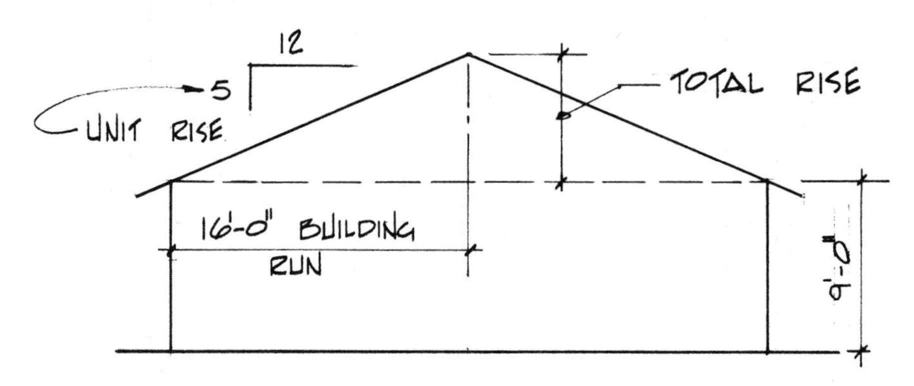

PROCEDURE:

1. Calculate the total rise by multiplying the building run 16′-0″ by the unit rise of 5″: $5 \times 16 = 80''$. Take total rise to the nearest even foot or 7′-0″.

2. To find the square foot area of the gable, multiply the building run 16′-0″ by the total rise of 7′-0″:

$$\begin{array}{r} 16 \\ \times\ 7 \\ \hline 112 \text{ sq. ft.} \end{array}$$

3. The 112 square feet should then be multiplied by the number of gable ends of similar size (in this case 2).

$$\begin{array}{r} 112 \\ \times \quad 2 \\ \hline 224 \text{ sq. ft.} \end{array}$$

4. The square footage is then converted to square yards by dividing the 224 sq. ft. by 9:

$$\begin{array}{r} 24.9 \text{ or } 25 \text{ sq. yds.} \\ 9\,/\,\overline{224.0} \\ \underline{18} \\ 44 \\ \underline{36} \\ 8.0 \end{array}$$

The 25 square yards is then added to the wall square yardage and multiplied by the cost per sq. yd.

Note: For a 2-story building, estimate the square foot area to be covered (including window and door areas), then divide by 9 to get the number of square yards. The cost per square yard for two-story structures is higher because of scaffolding needs and handling of materials.

IX

INTERIOR FINISH MATERIAL

64

Estimating Interior Doors

INTERIOR PASSAGE DOORS

Many types of interior doors are available to the builder today. Three of the most basic types are illustrated below.

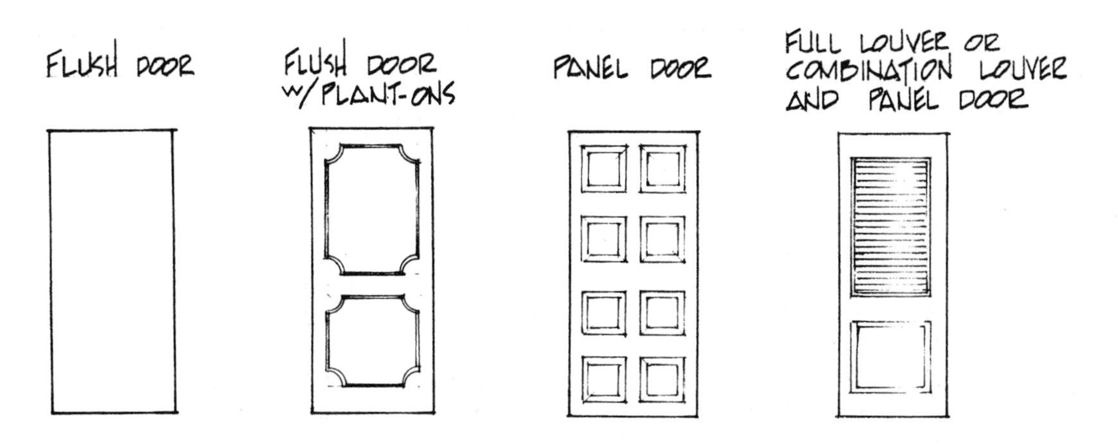

Standard interior doors are usually 1-3/8″ thick and 6′-8″ high. The minimum widths and heights for various interior doors are listed below:

Bedrooms	2′-6″ × 6′-8″
Bathroom	2′-0″ × 6′-8″
Single Door Closets	2′-0″ × 6′-8″
Hall Doors	2′-8″ × 6′-8″

Door sizes and types are usually found on the floor plan or on one of two types of door schedules as illustrated below.

Type 1:

Doors are designated on the floor plan with a letter or number symbol which refers to a door schedule as illustrated below.

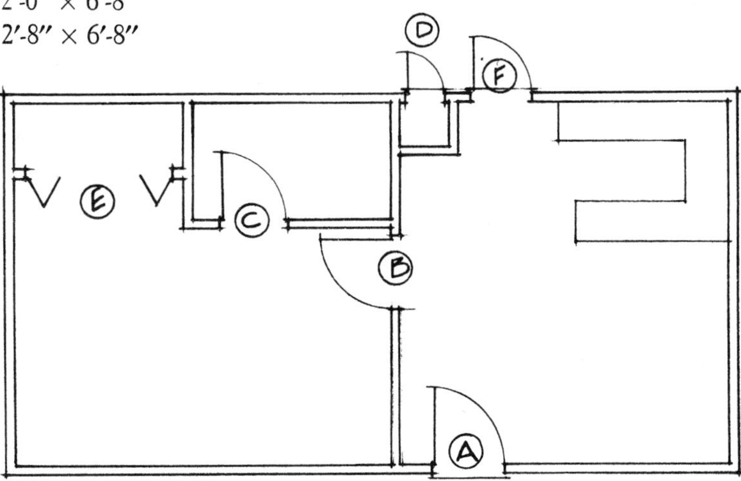

TABLE 64-1 Door Schedule

Symbol	*Size*	*Number Required*	*Material*	*Remarks*
A	3068 × 1-3/4	1	Beech	Paint grade self-closing
B	2868 × 1-3/8	1	Beech	Paint grade
C	2668 × 1-3/8	1	Beech	Paint grade
D	2068 × 1-3/8	1	Hardboard	Vent top and bottom
E	8068 × 1-3/8	1	Hardboard	4-door bi-fold unit
F	2868 × 1-3/4	1	Hardboard	Belair

A door schedule makes it easy for the estimator to list the doors needed.

Type 2:

The second type of door schedule also uses the letter or number symbol on the floor plan to designate each door but refers to a door schedule detail drawing as illustrated below.

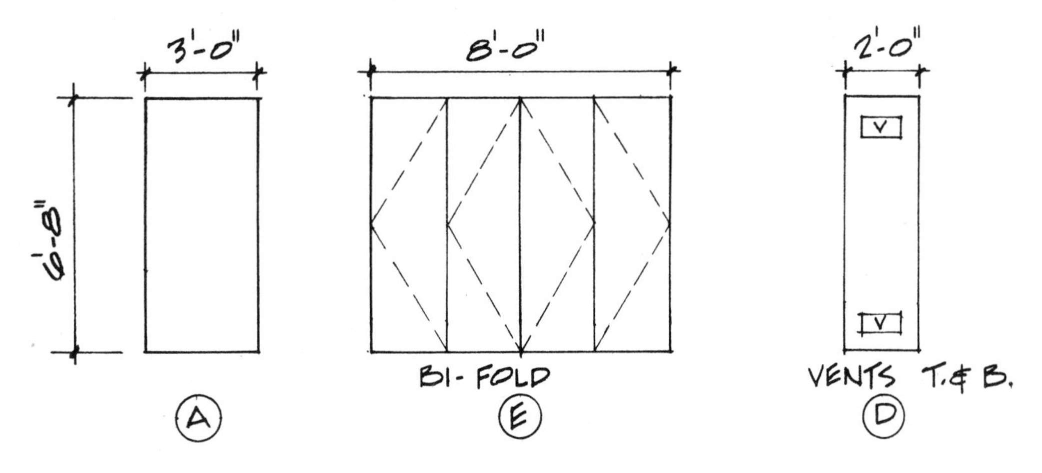

CLOSET DOORS

Closet doors are usually ordered in pairs or as a unit, depending on the type of door. The exposed surface of closet doors may be wood for painting or staining, smooth- or pattern-textured hardboard, vinyl coated, or mirrored. Illustrated below are the common closet and closet door units used in residential construction.

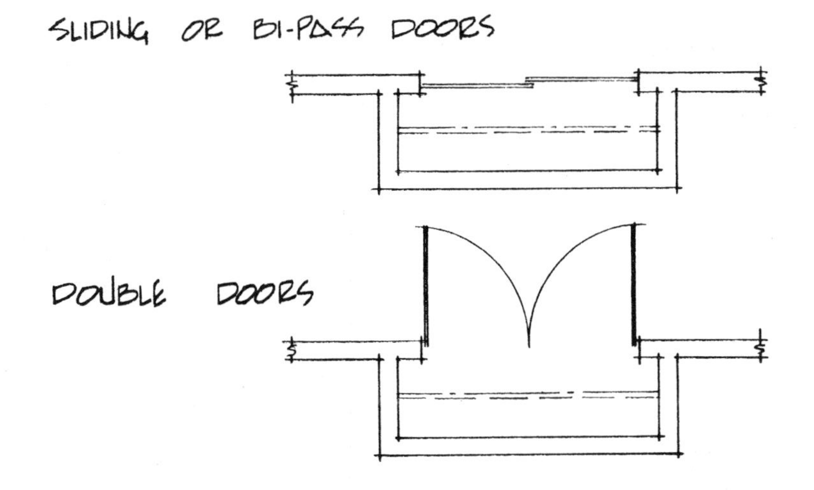

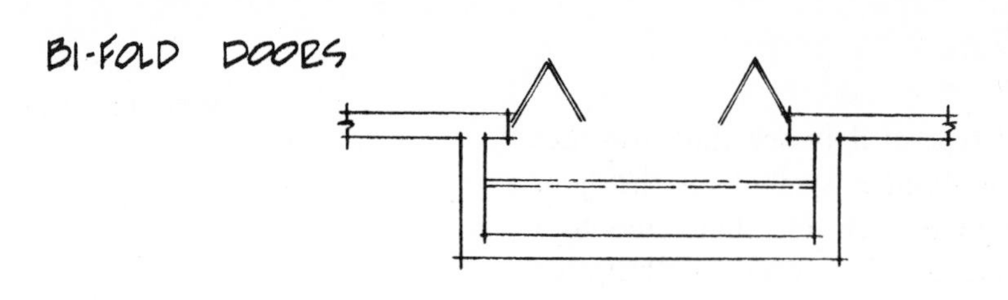

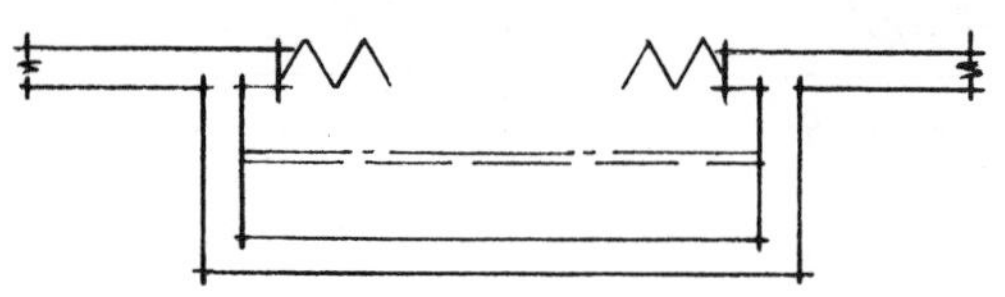

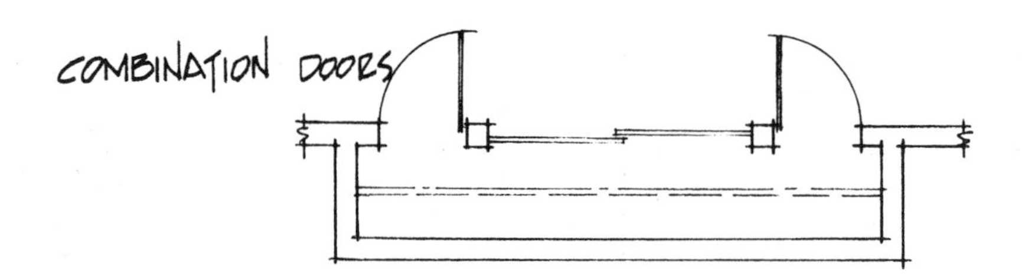

POCKET DOORS

Pocket doors are used when it is inconvenient for a door to swing into either room. The pocket doorjamb unit is purchased partially assembled. The carpenter must finish assembling and set the unit. The roller hardware for the door usually comes with the jamb unit. The door does not come with the jamb unit and should be ordered with the other doors in the building.

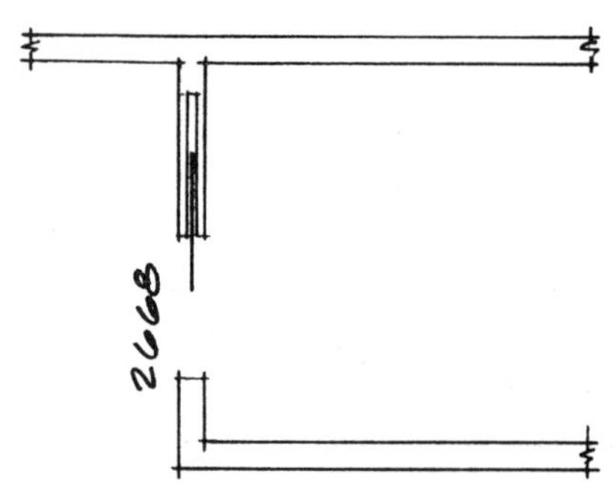

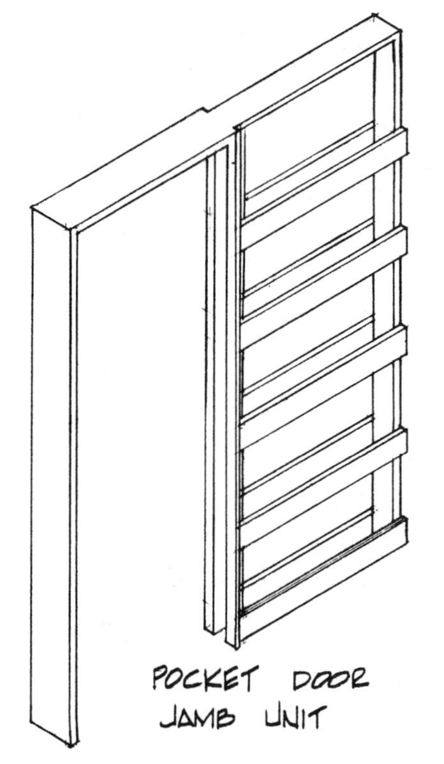

PROCEDURE FOR ESTIMATING INTERIOR DOORS

If a door schedule is available, the door sizes and types may be acquired, but the number of doors of each size, if not on the schedule, must be found by studying the floor plan. The estimator usually lists the passage doors first, including the pocket doors, and then the closet door units. If the exterior doors have not already been ordered, they should also be added to the list at this time. If a door schedule is not available, the estimator must make up his list from the floor plan.

65

Estimating Door Frames (Jambs)

The frame of the interior door trims off the rough door opening. It is made up of three pieces, two side jambs and a head jamb. Various types of jambs available to the builder are illustrated below.

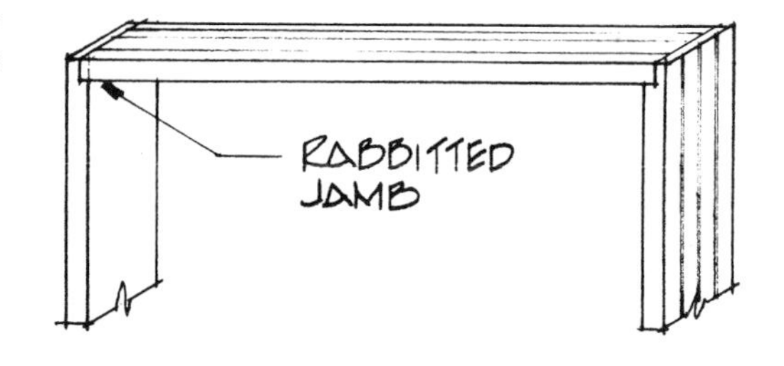

COMMON ONE-PIECE JAMB

SOLID JAMB

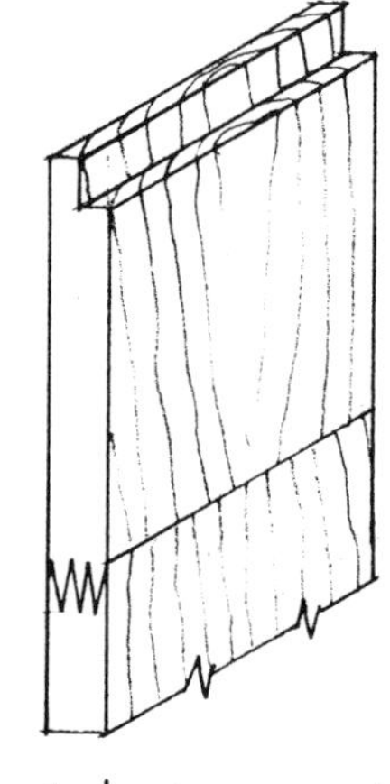

FINGER JOINTED JAMB

TWO-PIECE JAMB

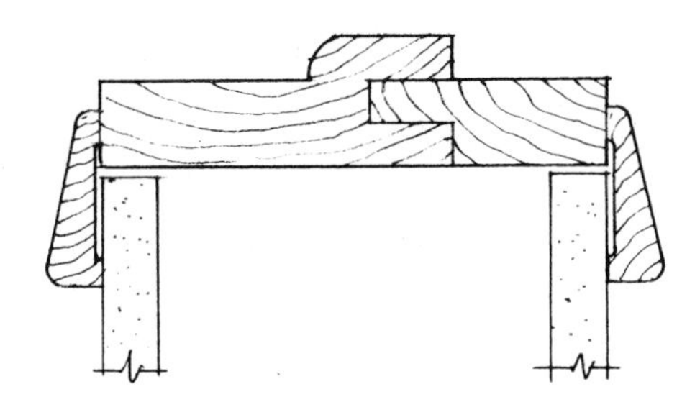

THREE-PIECE JAMB

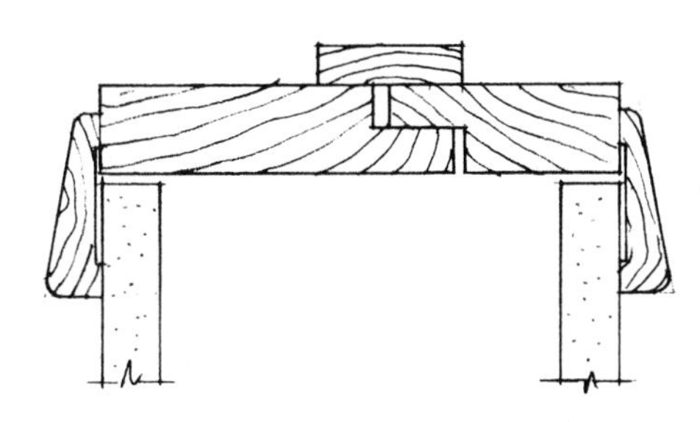

ASSEMBLED JAMB UNIT

This unit has the casings and doorstop stapled or glued to the jamb. It is made up of three pieces, two side jambs and a head jamb, and the hinges and strike latch are prenotched.

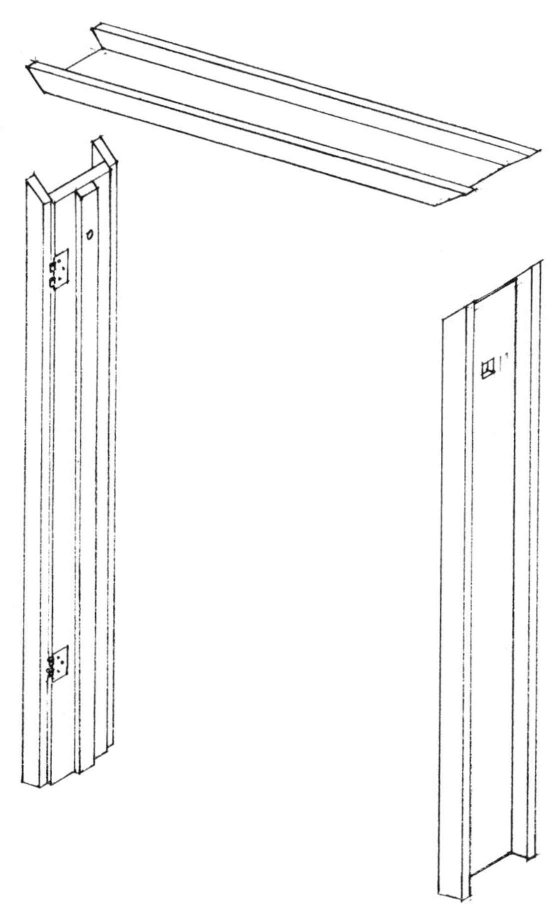

Doors may be provided with these units, or they have to be ordered separately from a door manufacturer. The hinge gain and the hole for the doorknob must align with the assembled jamb unit.

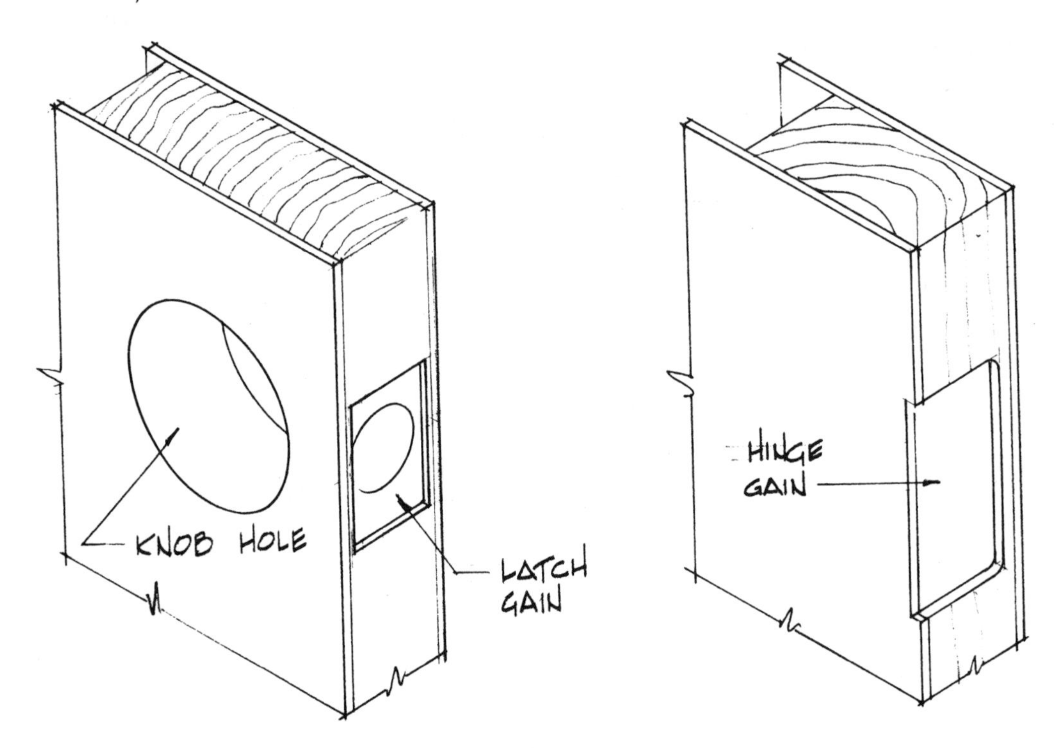

The common one-piece jamb is the most widely used. It is usually ordered in sets, a set consisting of three pieces, two side jambs and a head jamb. The width of the jamb material depends on the type of interior wall covering used. For 1/2″ drywall the jamb width will be 4-1/2 inches and for plaster 5-1/4 inches.

The length of the side jamb pieces for passage or closet doors is usually the same, but the head jamb varies in length, depending on the width of the opening. Doors up to 3′-0″ wide use a standard-stocked 3′-3/4″ long head jamb and may be ordered as a standard jamb set. Cased openings and closet doors wider than 3′-0″ should be listed separately. For these openings the lumberyard will provide a piece of head jamb stock of appropriate length.

PROCEDURE FOR ESTIMATING INTERIOR DOOR FRAMES

The estimator must study the floor plan to make a list of doorjamb sets. The standard jamb sets are usually estimated first, followed by the cased openings and closet jamb sets. Doorjambs should be ordered according to the opening sizes listed opposite:

12 sets doorjamb for 3068 opening Douglas fir 4-1/2 wide solid stock;
1 set doorjamb for 8068 opening Douglas fir 4-1/2 wide solid stock;
3 sets doorjamb for 6068 opening Douglas fir 4-1/2 wide finger joint;
2 sets doorjamb for 4068 opening Douglas fir 4-1/2 wide finger joint.

66

Estimating Door Casings

Door casings are applied to the top and sides of a door to cover the space between the doorjamb and the trimmer stud. Nailing the casing to the doorjamb and through the wallboard into the trimmer stud helps to make the doorjamb rigid.

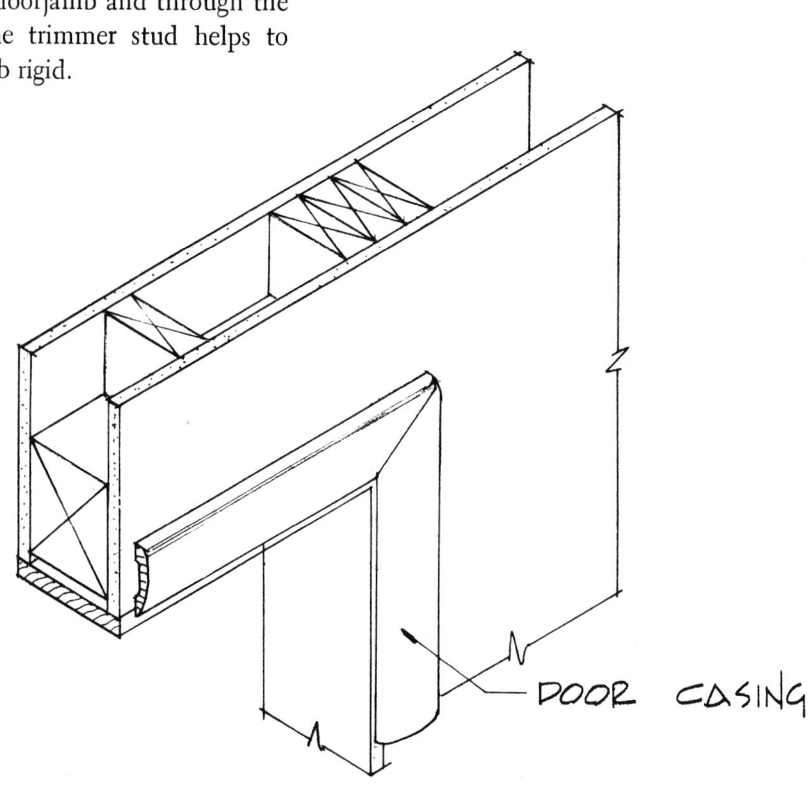

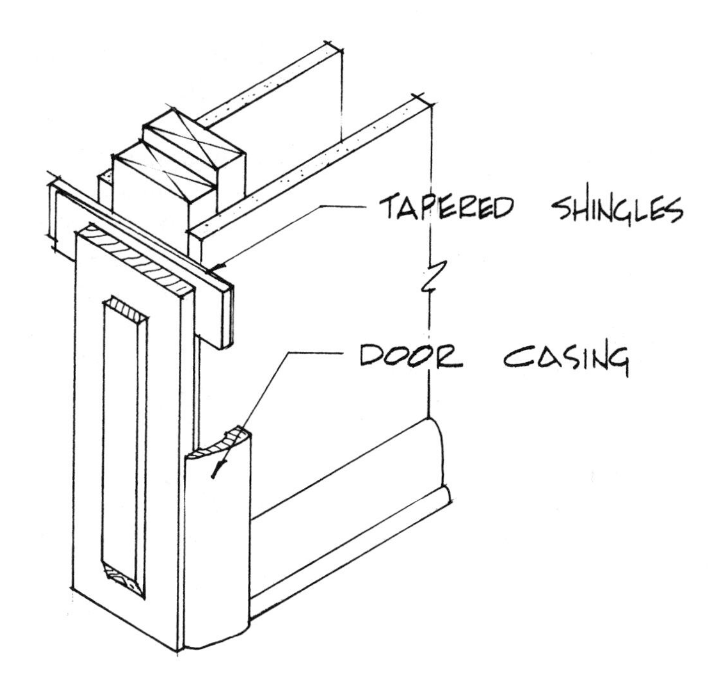

Door casings come in a variety of sizes and shapes as illustrated below.

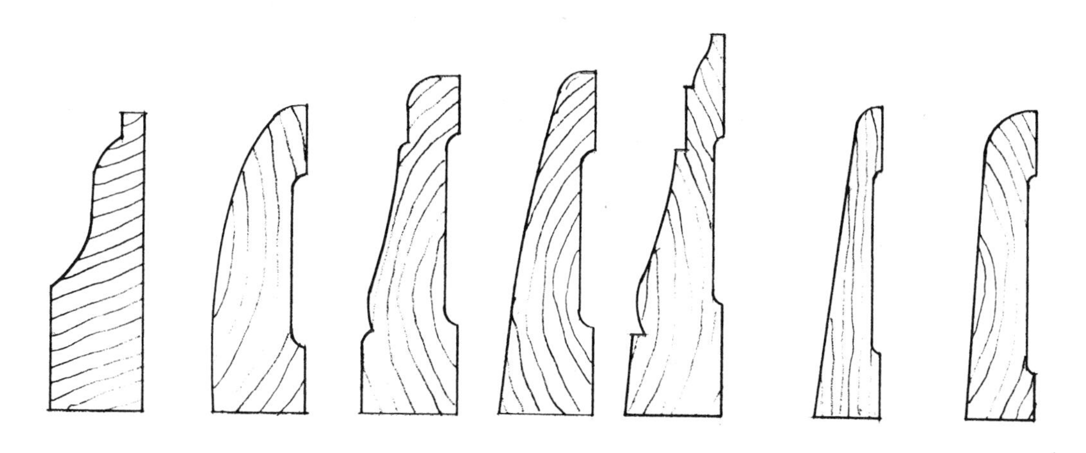

Door casings are usually ordered by the set; one set consisting of two side casings and one head casing. The set cases one side of an interior door opening. A typical set of casings for an interior door up to a width of 3′-0″ consists of two pieces of side casing 7′-0″ and a head casing approximately 3′-4″ long. Openings wider than 3′-0″ should be listed separately; for these openings the lumberyard will provide a piece of head casing stock of appropriate length.

PROCEDURE FOR ESTIMATING DOOR CASING

The estimator must list the number and size of door openings to be cased. He must then calculate the number of casing sets, two sets for each interior door opening and one set for each exterior door opening such as front and rear doors. Casings for standard passage doors and closet doors should be ordered by the size of the opening as listed below:

24 sets #720 streamline pine casing 1/2 × 1-5/8 for up to 3068 opening;
2 sets #720 streamline pine casing 1/2 × 1-5/8 for 8068 opening;
6 sets #720 streamline pine casing 1/2 × 1-5/8 for 6068 opening;
4 sets #720 streamline pine casing 1/2 × 1-5/8 for 4068 opening.

67

Estimating Doorstops

The doorstop is nailed to the surface of the doorjamb and serves as a stop when the door swings closed. Doorstops are ordered in standard sets, a standard set consisting of two 7′-0″ side stops and a 3′-0″ head stop. Doorstops are usually not used on closet openings unless the doors are of the swinging type.

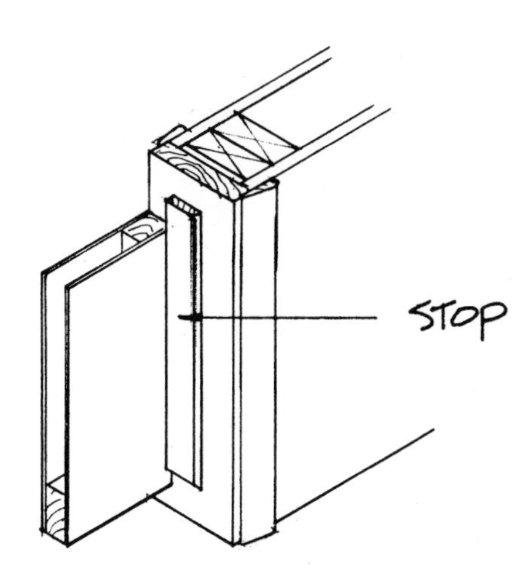

STANDARD DOOR STOP MATERIAL

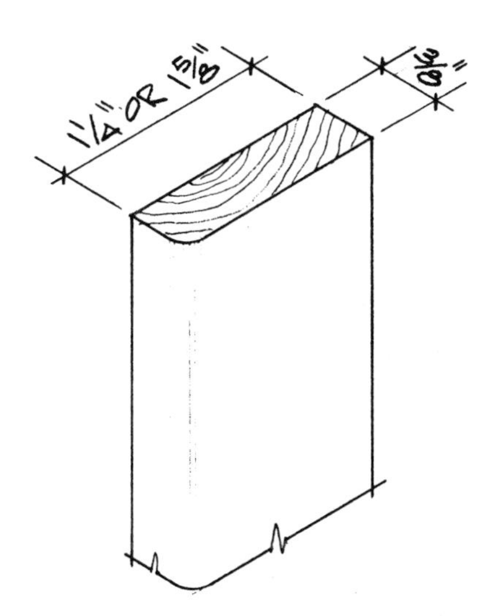

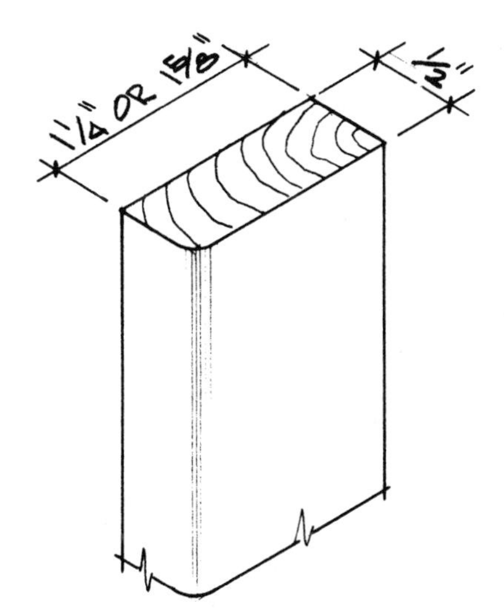

PROCEDURE FOR ESTIMATING DOORSTOPS

The estimator must count the number of passage doors from the floor plan and allow one set of doorstops for each door. For example, they would be listed: 12 sets #755 pine doorstop 3/8″ × 1-1/4″.

68

Estimating Window Trim

Various methods are used to trim a window. The estimator must check the plan details, plan notes, or specifications in order to determine which method to use before making a material take-off:

1. The window may be trimmed completely with plaster or wallboard, thus eliminating any wood trim estimate.

2. The window may have a wood sill (stool) and apron to be estimated, with the remainder of the window trimmed with plaster or wallboard.

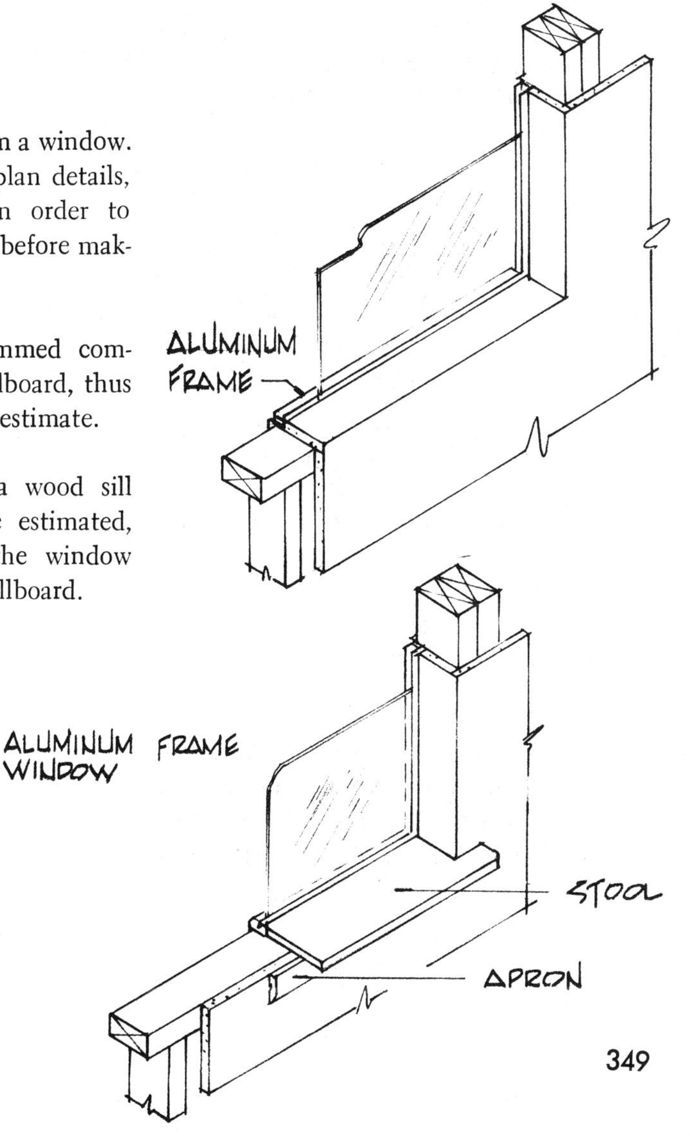

3. The window may have a wood sill (stool), apron, wood jamb, and casing to be estimated.

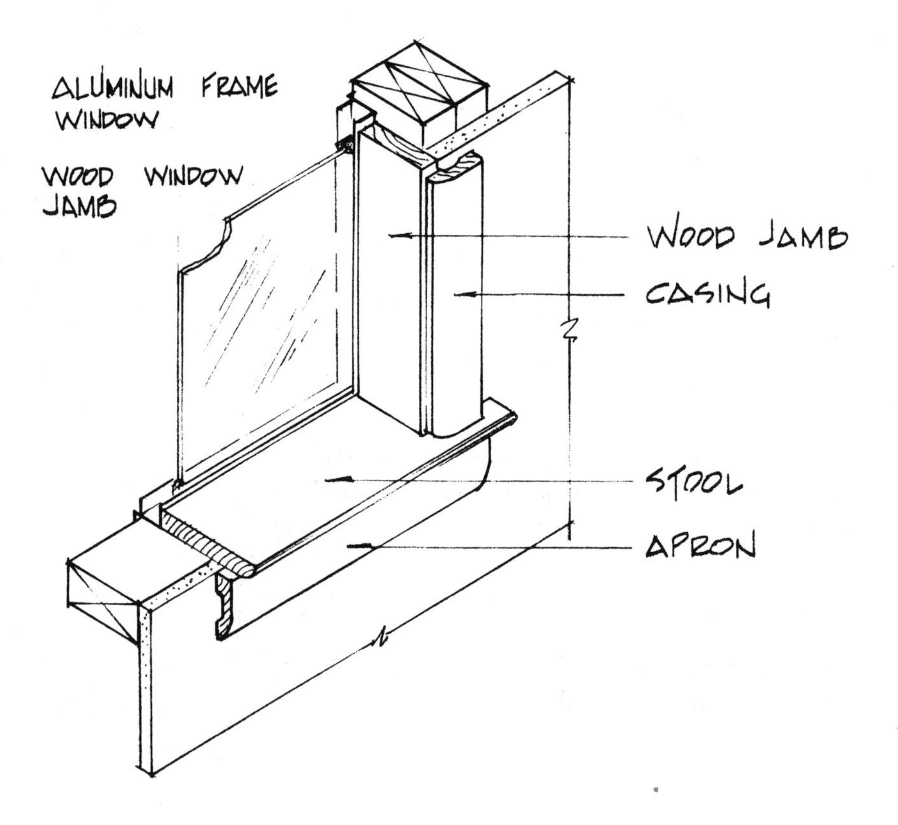

4. The window may be double-hung. Then the stool, apron, and window casing material must be estimated. The other members are already included in the double-hung window unit.

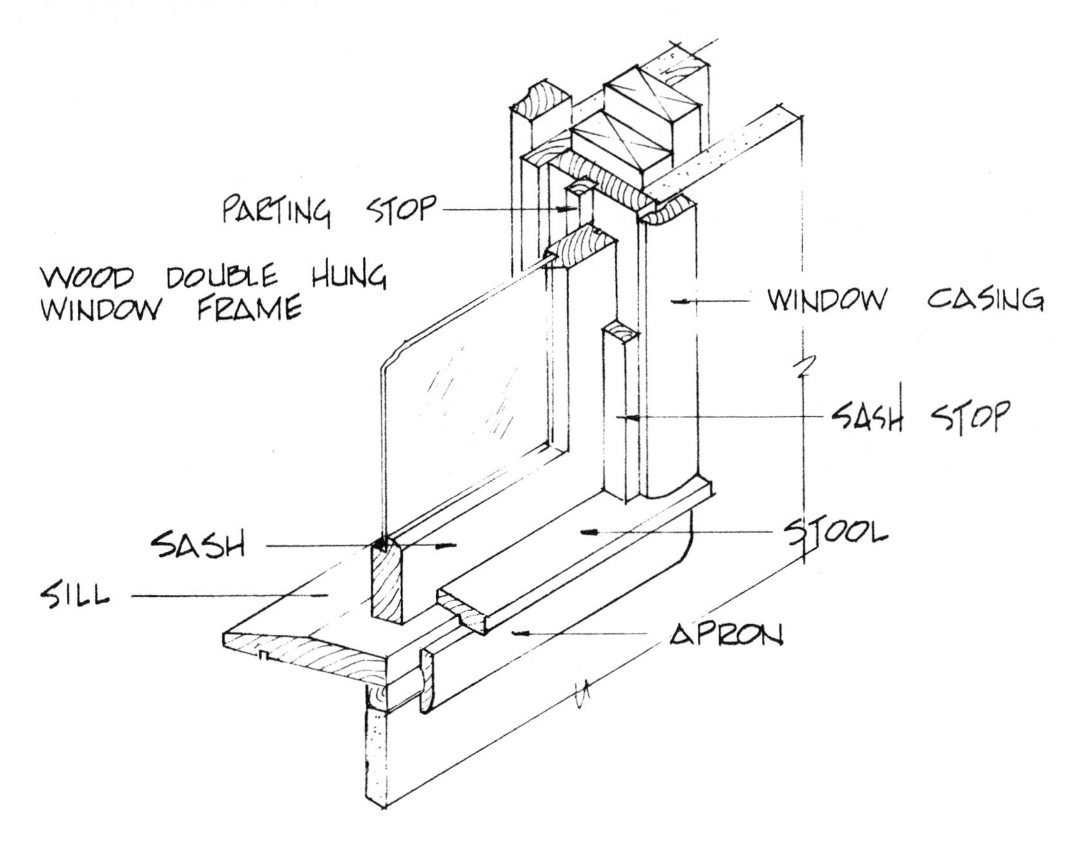

5. The window may be fixed glass with wood stops, window jamb, and casings, all of which must be estimated.

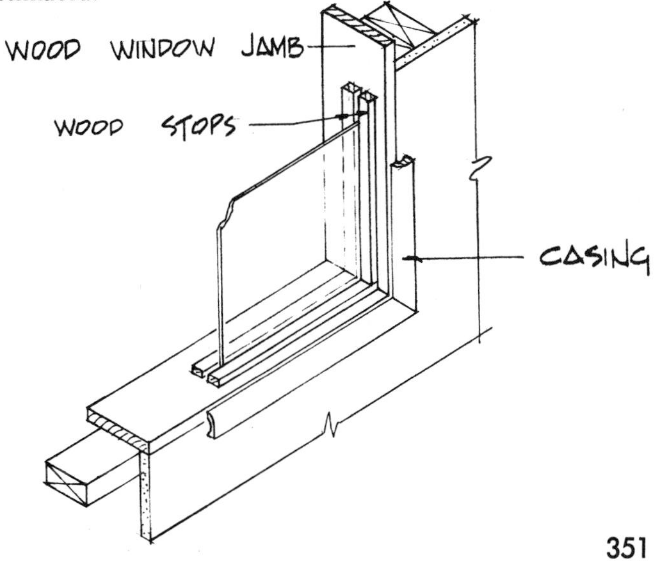

PROCEDURE FOR ESTIMATING WINDOW TRIM

The estimator must determine which method of window trim is to be used and then list the material for each window. The material may then be combined into convenient lengths for ordering.

69

Estimating Closet Trim

The material needed to trim out a closet consists of the following members:

1. Closet ribbon (1″ × 4″ board), upper shelf ribbon (1 × 2 board)
2. Closet shelf (1″ × 12″ board)
3. Wood closet pole (1-3/8″ diameter)
4. Wood or metal pole bracket—needed approximately every 4 to 5 feet
5. Wood or plastic closet pole rosettes.

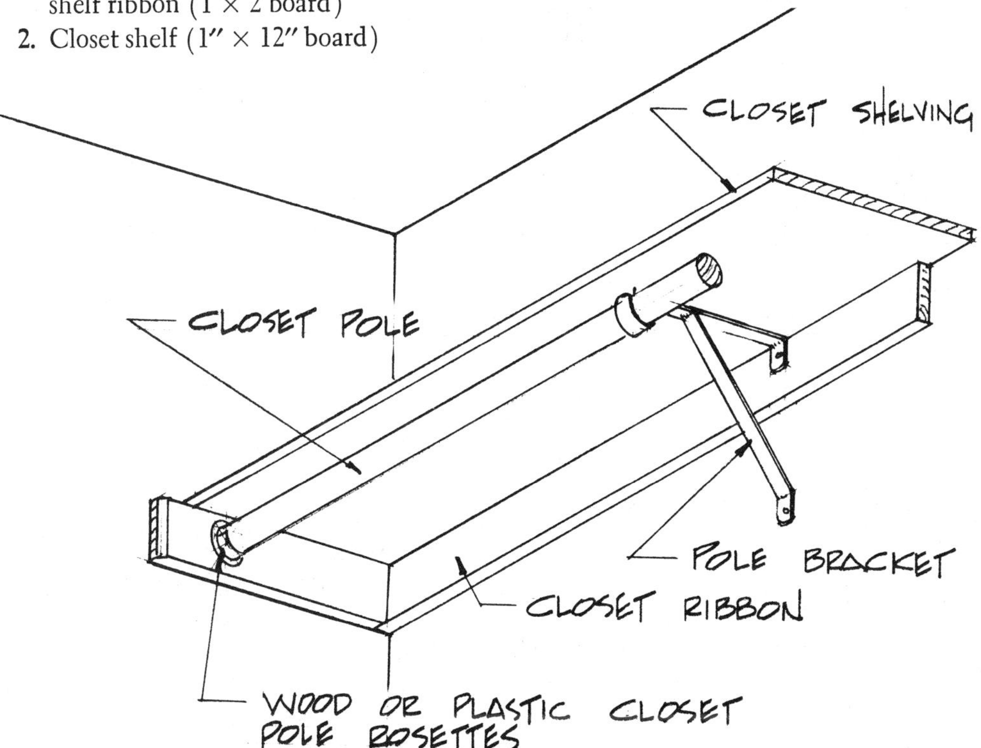

The number of closet shelves, poles, and other materials is usually found on a floor plan, such as the one illustrated below.

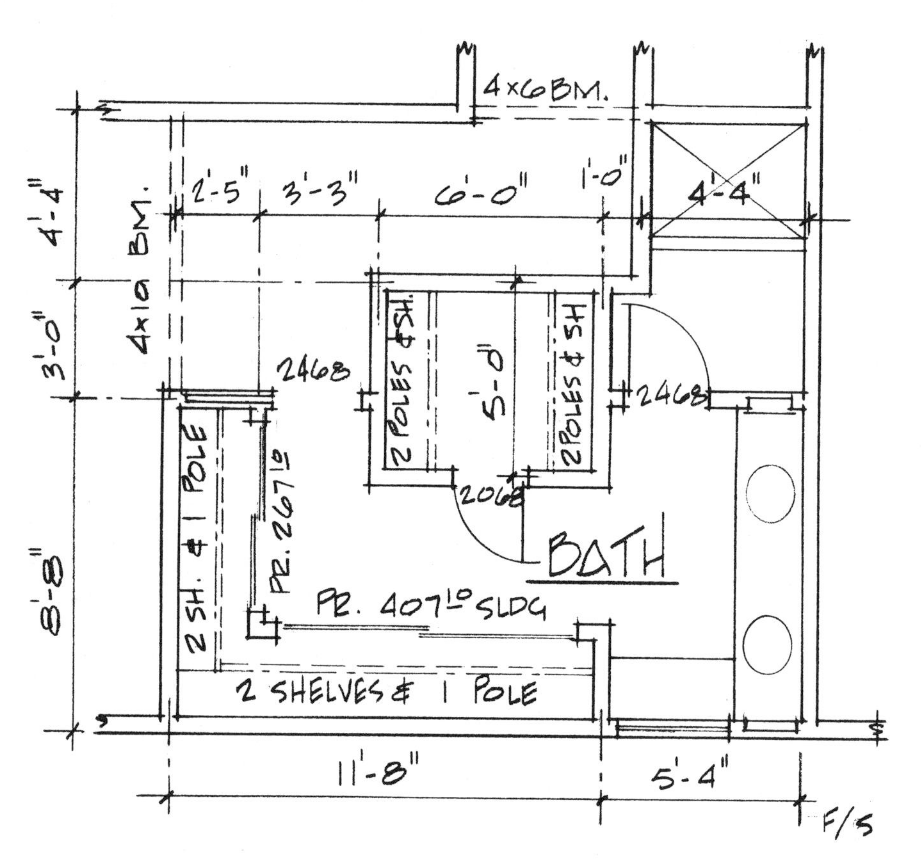

PROCEDURE FOR ESTIMATING CLOSET TRIM

To estimate closet trim, scale from the floor plan the material needed for each closet; then combine the materials into convenient lengths for ordering. The following list of closet material is for the closet illustrated above. Keep in mind that some of the lengths of material have been combined.

Shelves:

- 2 pcs. 1 × 12 × 10′
- 2 pcs. 1 × 12 × 12′
- 2 pcs. 1 × 12 × 9′

Poles:

1 pc. 1-3/8″ round × 12′
1 pc. 1-3/8″ round × 9′
2 pcs. 1-3/8″ round × 10′

Closet Rosettes 6 pair
Metal Closet Pole Bracket 7 each

Closet Ribbon:

Walk-in closet

2 pcs. 1 × 4 × 8′
2 pcs. 1 × 4 × 10′

Bottom Shelf
(L-shaped closet):

1 pc. 1 × 4 × 12
1 pc. 1 × 4 × 14

Upper Shelf
(L-shaped closet):

1 pc. 1 × 2 × 12
1 pc. 1 × 2 × 14

70

Estimating Baseboard and Base Shoe

BASEBOARD

The baseboard runs continuously around the perimeter of each room and covers the space between the plaster or wallboard and the floor. Baseboard is made from softwood (pine), from hardwood (ash, oak, birch), from plastic, and from plastic or wood covered with a wood-grain, vinyl, print.

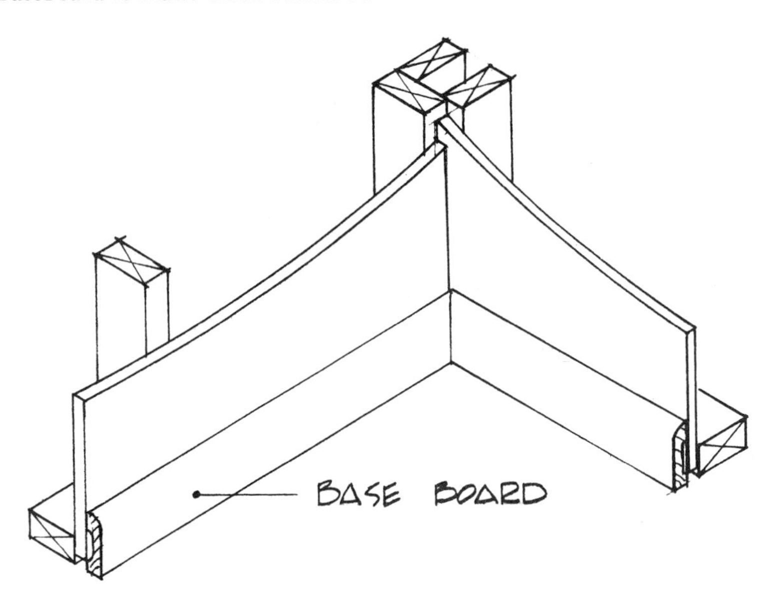

Various sizes and shapes of baseboard are available to the builder as illustrated below.

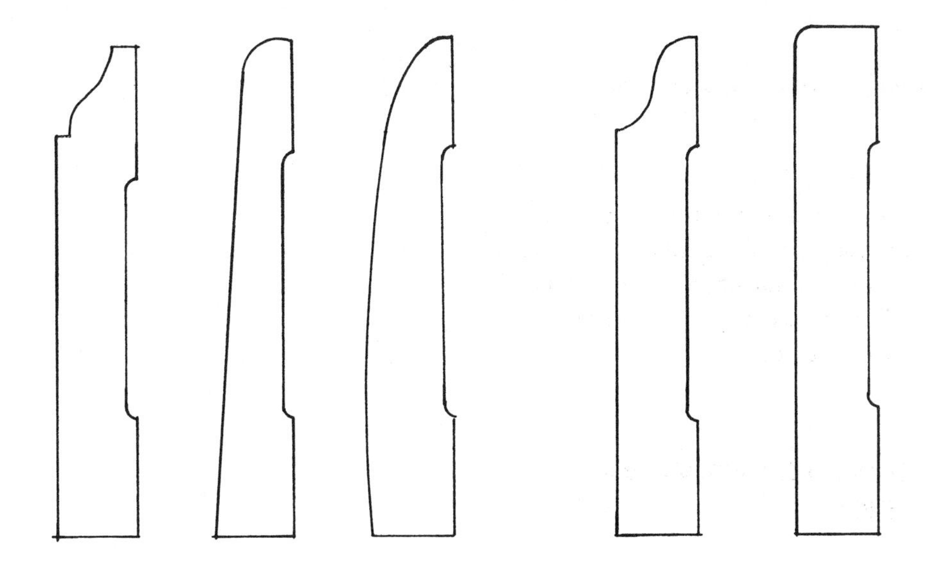

PROCEDURE FOR ESTIMATING BASEBOARD

To estimate the amount of baseboard needed, the estimator scales from the floor plan the approximate lineal footage needed in each room. Then he totals the lineal footage and adds 5% for waste. (Baseboard is not usually used around kitchen cabinets or bathroom pullmans.)

BASE SHOE

Base shoe is used to close the joint between baseboard and the finished floor or around

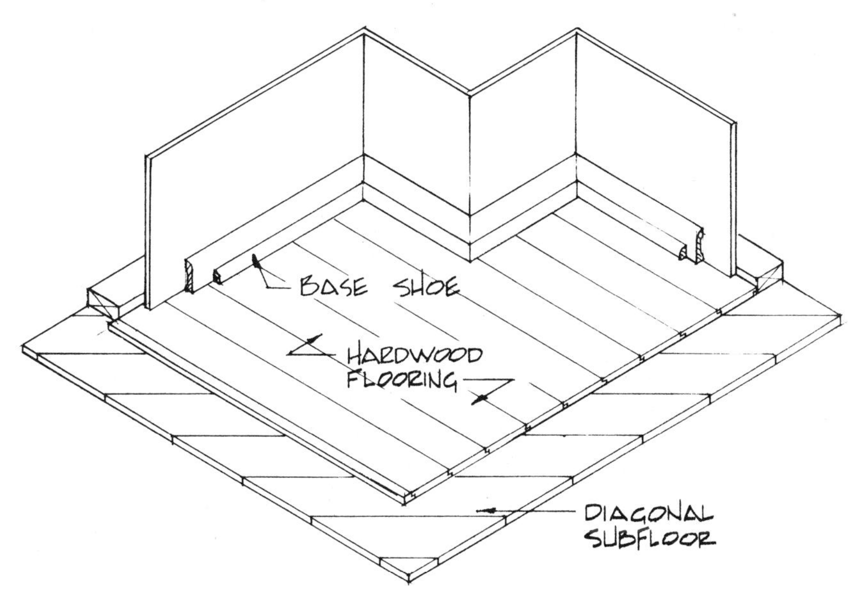

bathroom pullmans and kitchen cabinets. It is usually used when the floors are made of wood or when resilient flooring material is used. It is not necessary to install base shoe when carpet is used.

PROCEDURE FOR ESTIMATING BASE SHOE

If base shoe is required, the lineal footage may be scaled from the floor plan like the baseboard.

71

Estimating Interior Wall Paneling

BOARD PANELING

Interior wall paneling may consist of 1″ × 4″ to 1″ × 12″ wide boards with tongue-and-groove, shiplap, or square edges. Most board paneling is applied vertically, but it may also be applied horizontally or diagonally as illustrated below.

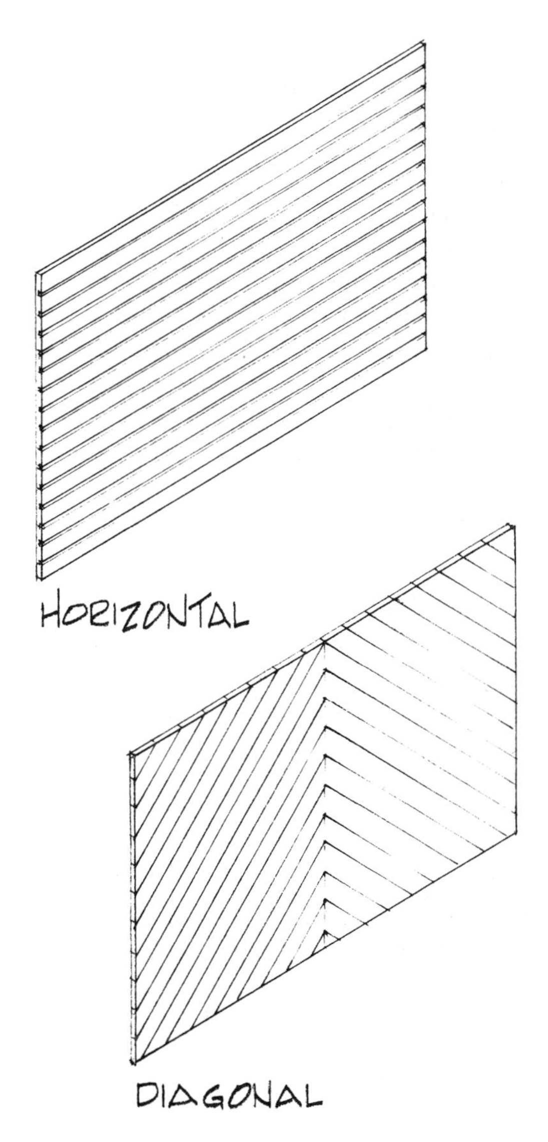

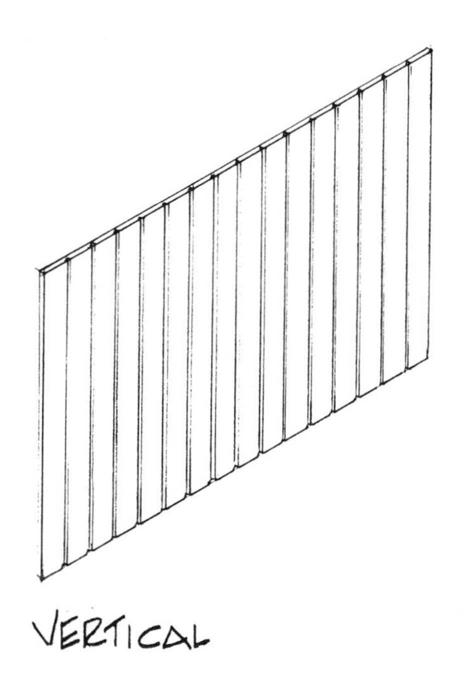

The boards may be of the same width or of random widths, and they may have a smooth or rough-sawn surface.

ESTIMATING BOARD PANELING

The estimator may confront various situations when estimating interior board paneling. If the boards are of the same width, he may divide the total number of lineal inches of wall to be covered by the paneling by the actual coverage width of one board. A few extra boards may be added for waste. The result will be the number of boards needed. The estimator, depending on the situation, may also choose to subtract some material for window and door openings.

Example Problem:

How many pieces of 1″ × 6″ tongue-and-groove rough cedar board applied vertically are needed to cover a wall 8′-0″ high by 49′-0″ long.

PROCEDURE:

1. Change the 49 lineal feet to inches: 49 × 12 = 588 inches.
2. Divide the total inches (588) by the width coverage of a 1 × 6 tongue-and-groove board (5-1/4 or 5.25).

```
          1 12 boards
5.25 / 588.00
       525
        630
        525
        1,050
        1,050
```

3. Add a few extra boards for waste. In this case 4 extra boards: 112 + 4 = 116 pieces of 1″ × 6″ × 8′ T&G resawn cedar.

ESTIMATING RANDOM BOARD PANELING

If random-width boards are used, for example 6″, 8″, and 10″ boards, combine the actual total width in inches of the three boards (5-1/4″ plus 7-1/4″ plus 9-1/4″ or 21-3/4″) and divide that total width into the total number of lineal inches to be covered by the paneling. Add a few extra boards for waste. The result will be the number of boards needed. The estimator, depending on the situation, may choose to subtract some material for window and door openings.

Example Problem:

How many pieces of 1″ × 6″, 1″ × 8″, and 1″ × 10″ knotty pine T&G board applied vertically are needed to panel a wall 8′-0″ high by 76′-0″ long.

PROCEDURE:

1. Change the 76 lineal feet to inches: 76 × 12 = 912″.
2. Next calculate the actual total coverage

of the three boards by combining the actual widths of the nominal sizes:

5-1/4″(6″) + 7-1/4″(8″) + 9-1/4″(10″) = 21-3/4″

3. Then, divide the 912 inches by the actual width (21-3/4 or 21.75) of the three boards: $912 \div 21.75 = 41.9$ or 42 boards.
4. Finally, add a few boards for waste (in this case two extra boards of each size): $42 + 2 = 44$ boards

Answer:
44 pieces 1″ × 6″ × 8′ T&G knotty pine
44 pieces 1″ × 8″ × 8′ T&G knotty pine
44 pieces 1″ × 10″ × 8′ T&G knotty pine

INTERIOR BOARD PANELING

Interior board paneling may also be estimated by the square foot as in exterior siding. The estimator must calculate the square foot area the paneling is to cover, subtract the window and door areas, and then multiply the result by the constant from Table 71-1. To convert the square footage of paneling material to lineal feet, first multiply by the following factors:

Size	*Factor*
1 × 6	2
1 × 8	1.33
1 × 10	1.20
1 × 12	1.0

The constant in Table 71-1 takes into account both the size differential of the boards and waste. The result will be the number of square feet of board paneling needed.

The constants for estimating diagonally applied paneling from the table below were taken for walls 8′-0″ high. These constants will vary somewhat, depending on the height and width of the walls and whether the siding is to be applied in full-height pieces or end spliced.

TABLE 71-1 ESTIMATING WALL PANELING SQUARE FOOTAGE

Panel Size	*Factors for Paneling Applied Vertically or Horizontally*		*Factors for Paneling Applied Diagonally*	
	Square Edge	*T&G or Shiplap*	*Square Edge*	*T&G or Shiplap*
1 × 6	1.15	1.25	1.33	1.35
1 × 8	1.17	1.25	1.33	1.35
1 × 10	1.17	1.25	1.33	1.35
1 × 12	1.17	1.25	1.33	1.35

PLYWOOD OR HARDBOARD PANELING

When 4′ × 8′ sheets of plywood or hardboard paneling are used, the estimator may calculate the square foot area of wall to be covered, including the window and door openings, and divide it by 32, the number of square feet in a 4′ × 8′ panel. Any fractional part of a panel should be rounded off to the next full sheet. If there are any large openings in the walls, the estimator, with careful study, may be able to reduce the number of sheets needed.

Another method of estimating sheet paneling is to calculate the lineal feet of wall the paneling is to cover and divide it by the width of a sheet of paneling 4′-0″. A reduction in the number of sheets of paneling may also be made for any large openings that exist.

X

HARDWARE

72

Hardware—General Information

To estimate the building hardware for a job, the estimator must carefully study the building plans, notes, and specifications. Certain types of hardware will be found on the building plan and notes, and other types just in the specifications. On a complete set of building plans there may be a hardware schedule. A schedule of this sort will usually include only the interior finish hardware such as doors, windows, closets, and cabinets. The architect will usually specify the manufacturer's name, the name of the product, the type or style, an identifying number or symbol, and the finish desired.

An estimator or contractor with a lot of experience adds a lump sum of money, sometimes known as a hardware allowance, to the bid. The owner will then use this allowance to pick out the hardware he wants. If the owner spends more than the allotted hardware allowance, the contractor is entitled to the overage. If the owner adds any extra hardware, other than that normally specified, the contractor is also entitled to a fee for installation, overhead, and profit.

A minimum set of building plans may not call out any of the hardware needed. If this is the case, the estimator will have to know not only what hardware is available but what hardware is needed to meet local code requirements. In other words, the estimator may have to read all kinds of things into the plans.

Many estimators break hardware down into two categories, rough and finish. But in recent years many changes in construction framing methods and building code requirements have brought about an increase in the use of building hardware. For this reason our discussion of hardware is broken down into various categories.

73

Framing Hardware

Framing hardware includes concrete and wood framing, hangers, anchors, and connectors. These hangers, anchors, and connectors will usually be called out on the building plan section views and structural details. The hardware will be designated either as approved hanger, in which case the estimator will have to know what hanger and size to use, or specified by the name, size, and manufacturer's symbol or number, such as Simpson LUP–26 joist hanger. Framing hardware most commonly used for residential construction is illustrated below.

CONCRETE HOLD-DOWN ANCHORS

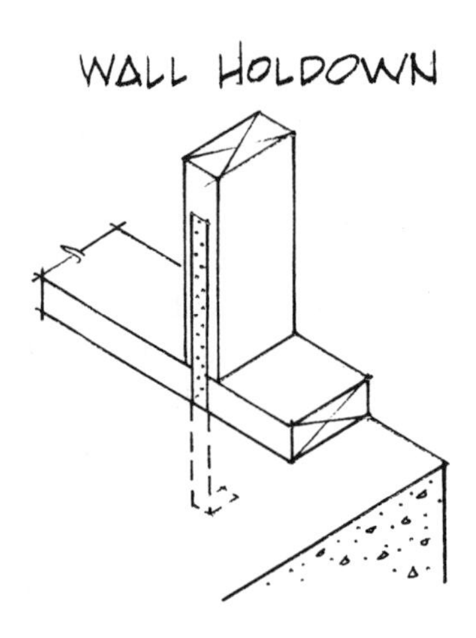

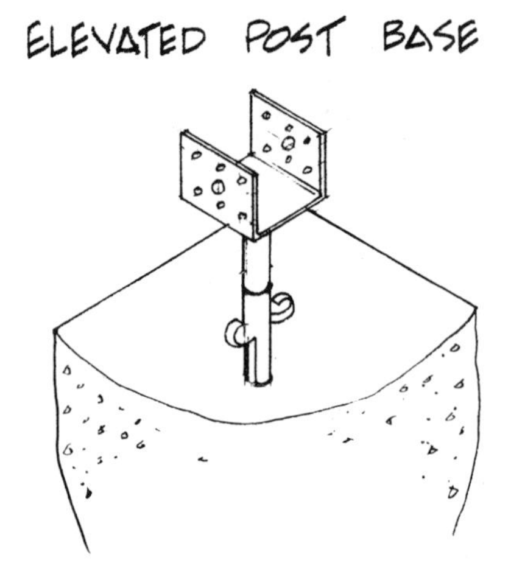

POST BASE

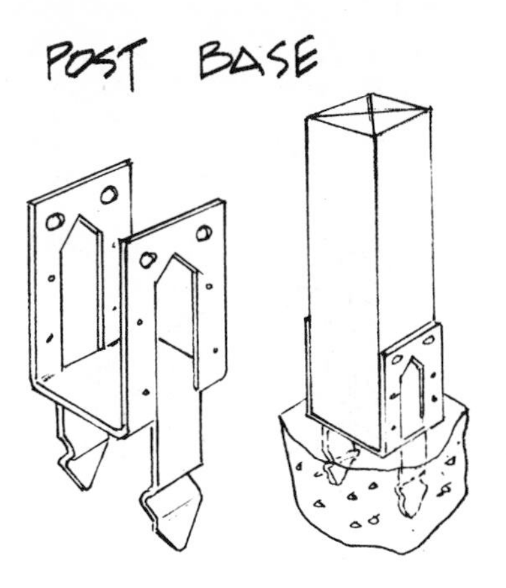

ADJUSTABLE POST BASE

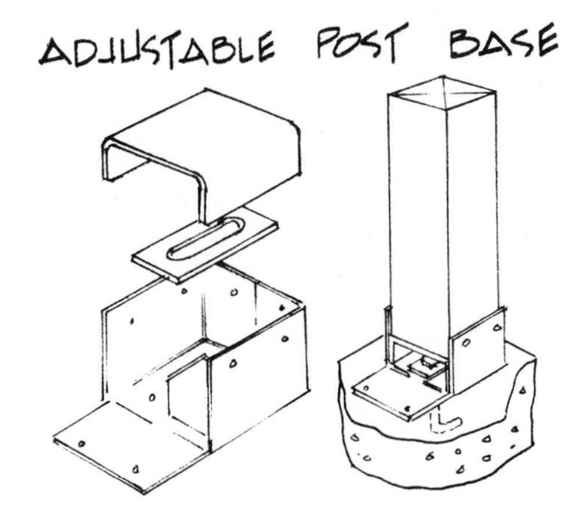

MUDSILL ANCHOR

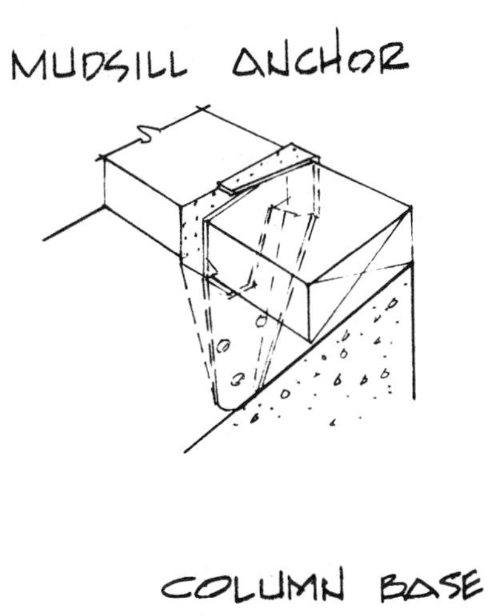

PURLIN ANCHORS

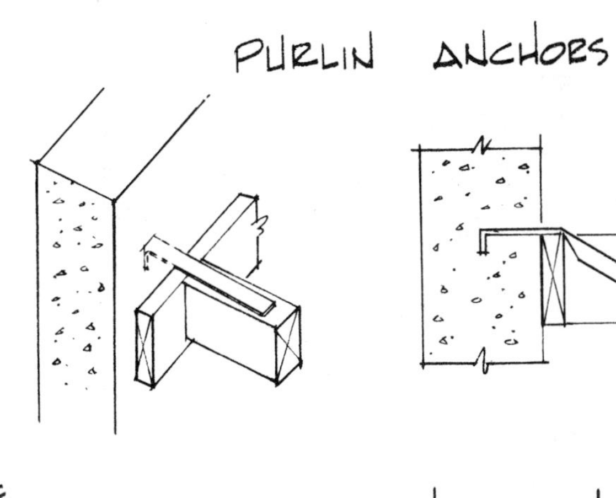

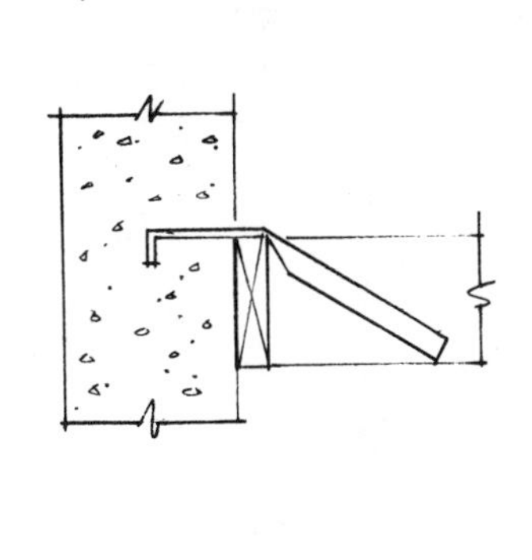

COLUMN BASE

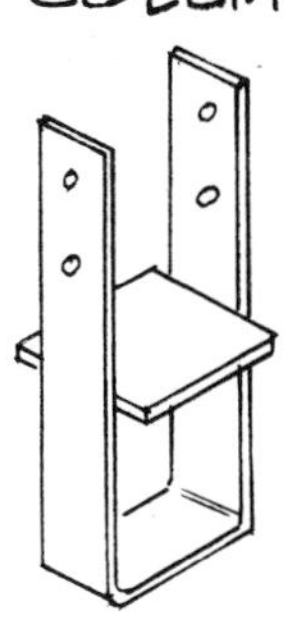

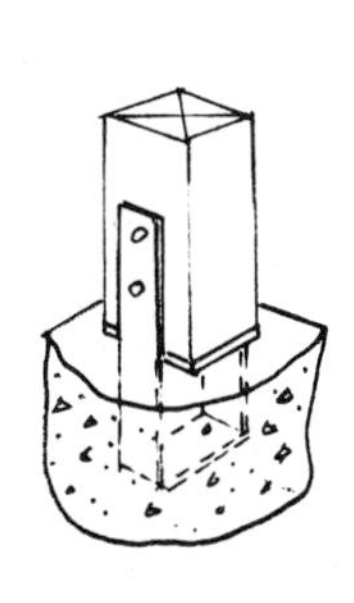

HOLDOWN

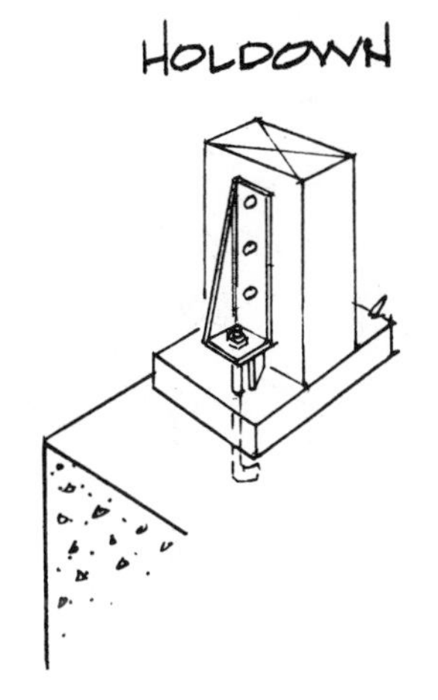

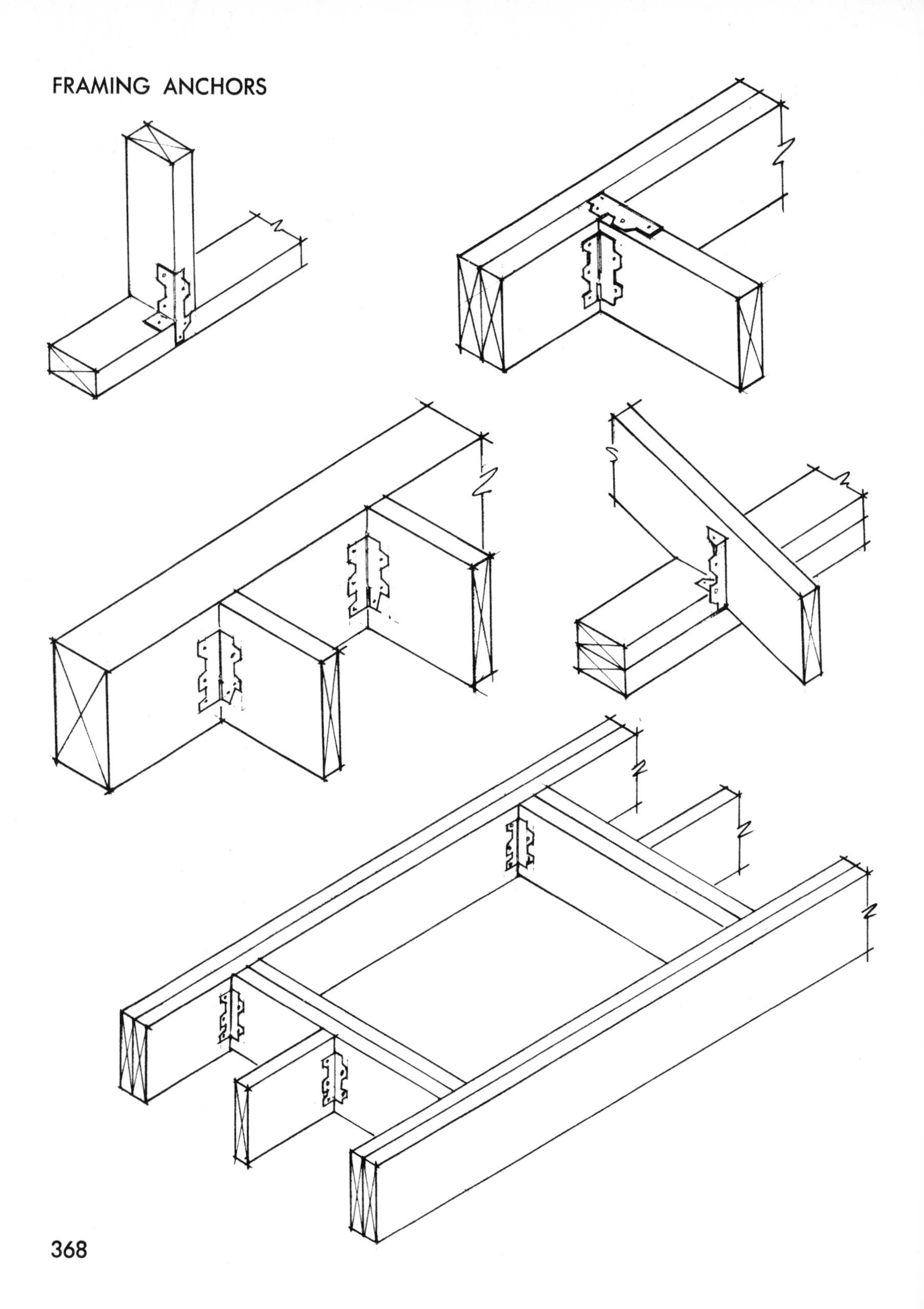

FRAMING CLIPS

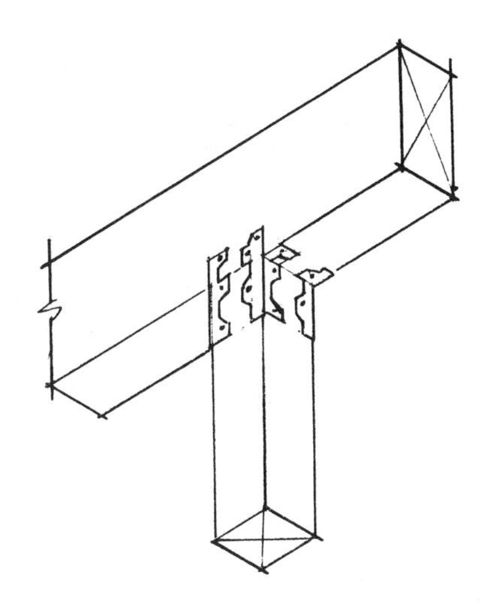

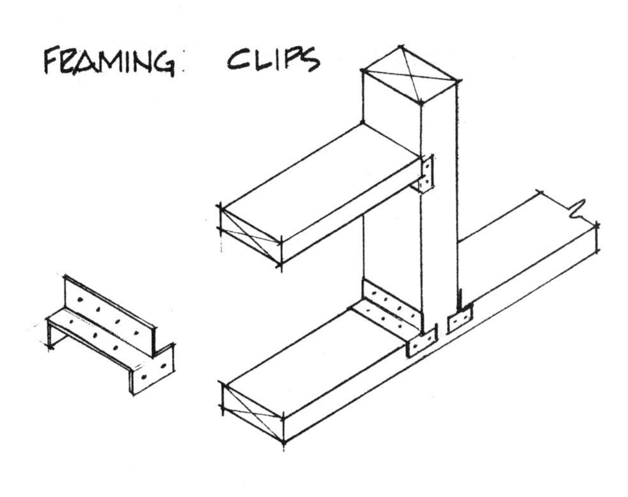

SEISMIC AND HURRICANE TIES

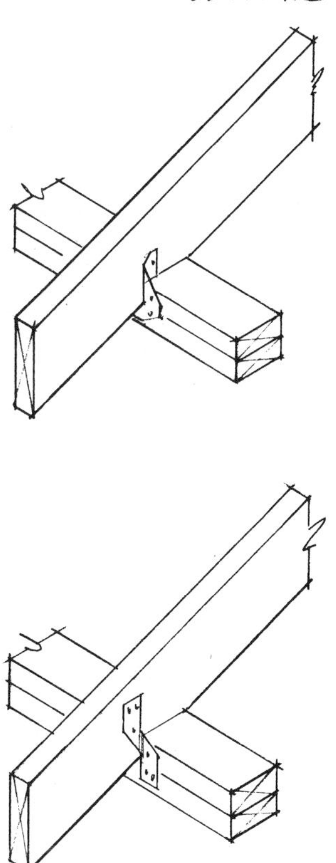

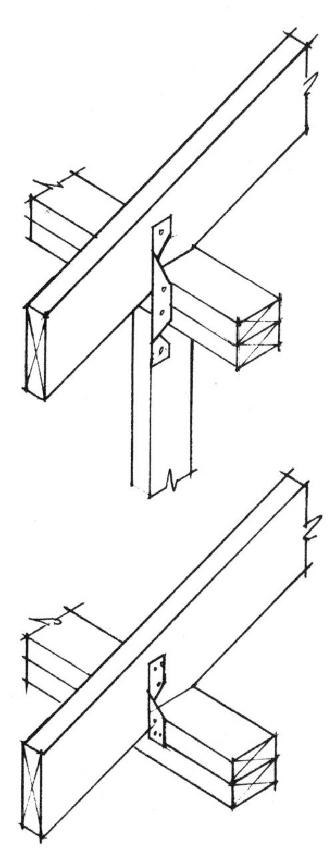

FRAMING TIE STRAP CONNECTORS

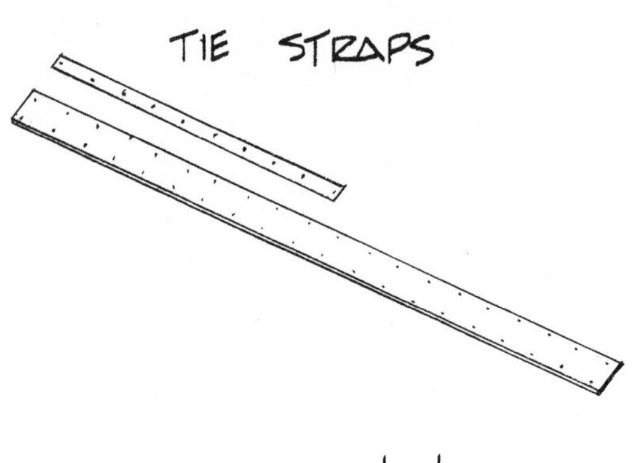

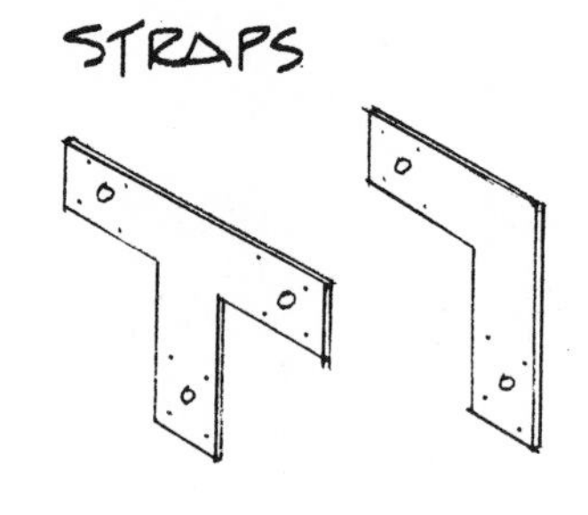

STRAP ANCHORS

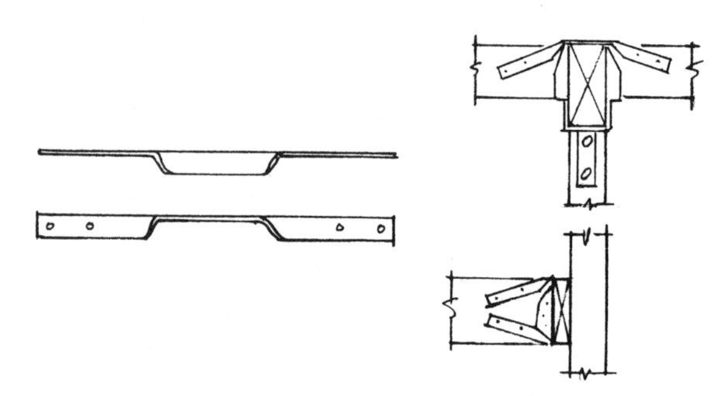

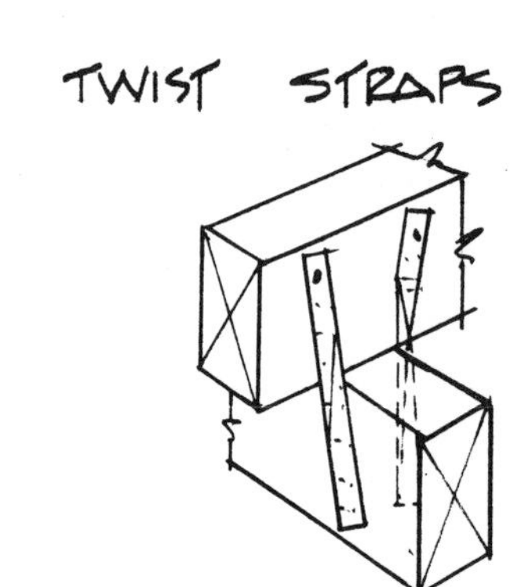

WALL STRAPS

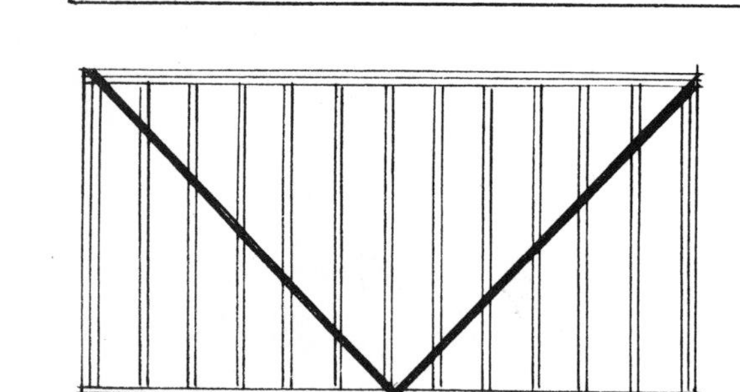

FRAMING HANGERS

HEADER HANGERS

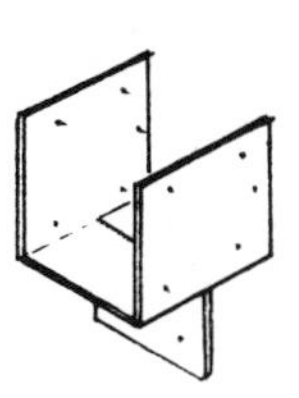

STANDARD JOIST HANGERS

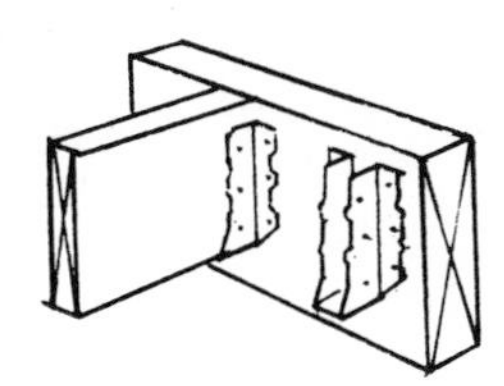

FORMED SEAT JOIST HANGERS

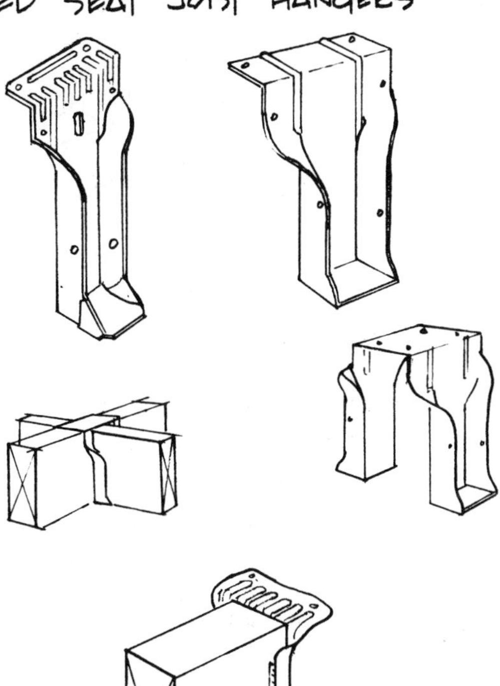

JOIST & PURLIN HANGERS

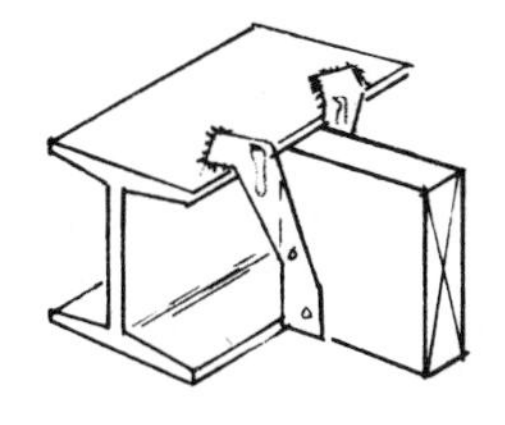

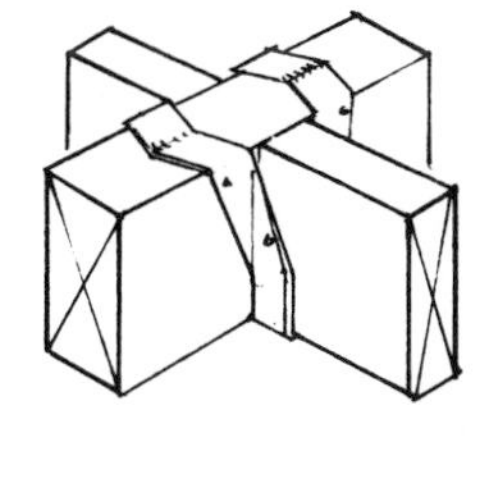

POST CAP CONNECTORS

POST CAP

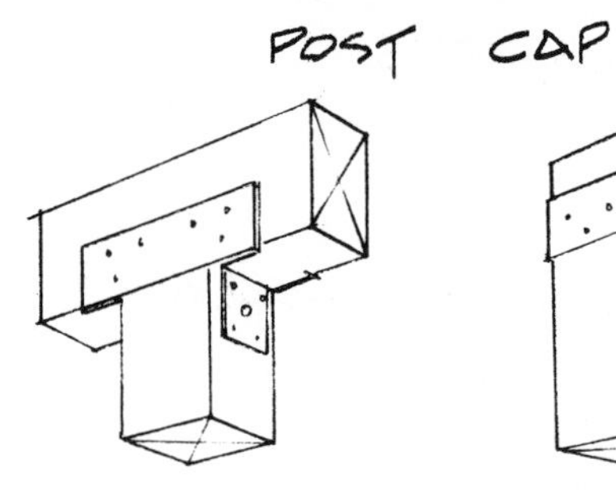

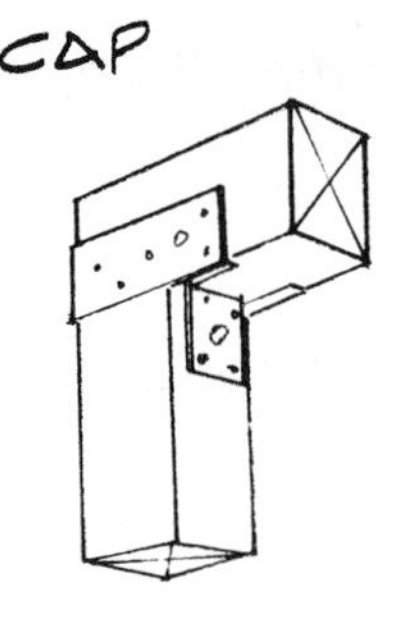

TWIN POST CAP

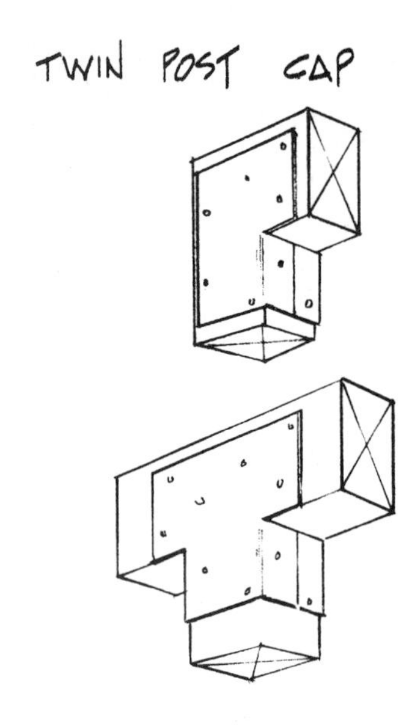

4-WAY POST CAP

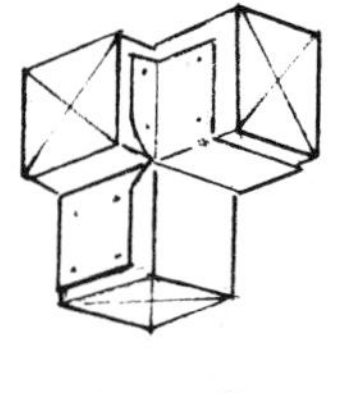

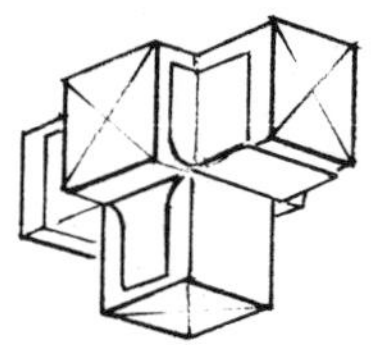

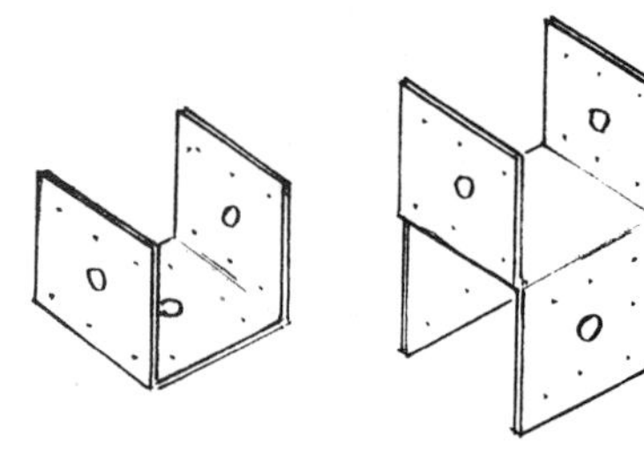

METAL BRIDGING

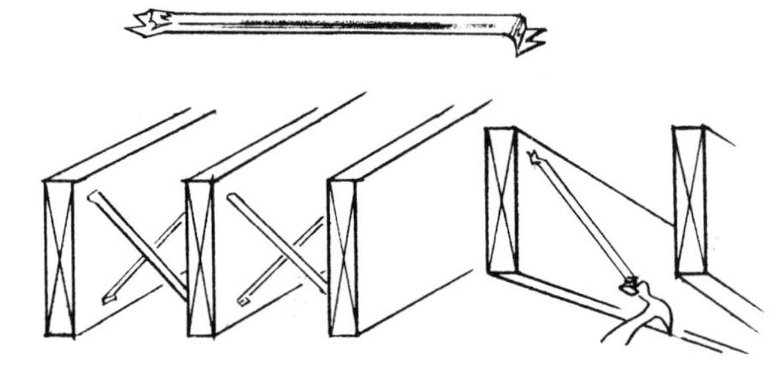

The estimator must carefully check the floor plan, roof plan, section views, and the structural details to make a list of needed framing hardware.

74

Rough Hardware

Rough hardware will usually include nails, lag bolts, bolts-nuts-washers, and screws. Various illustrations follow.

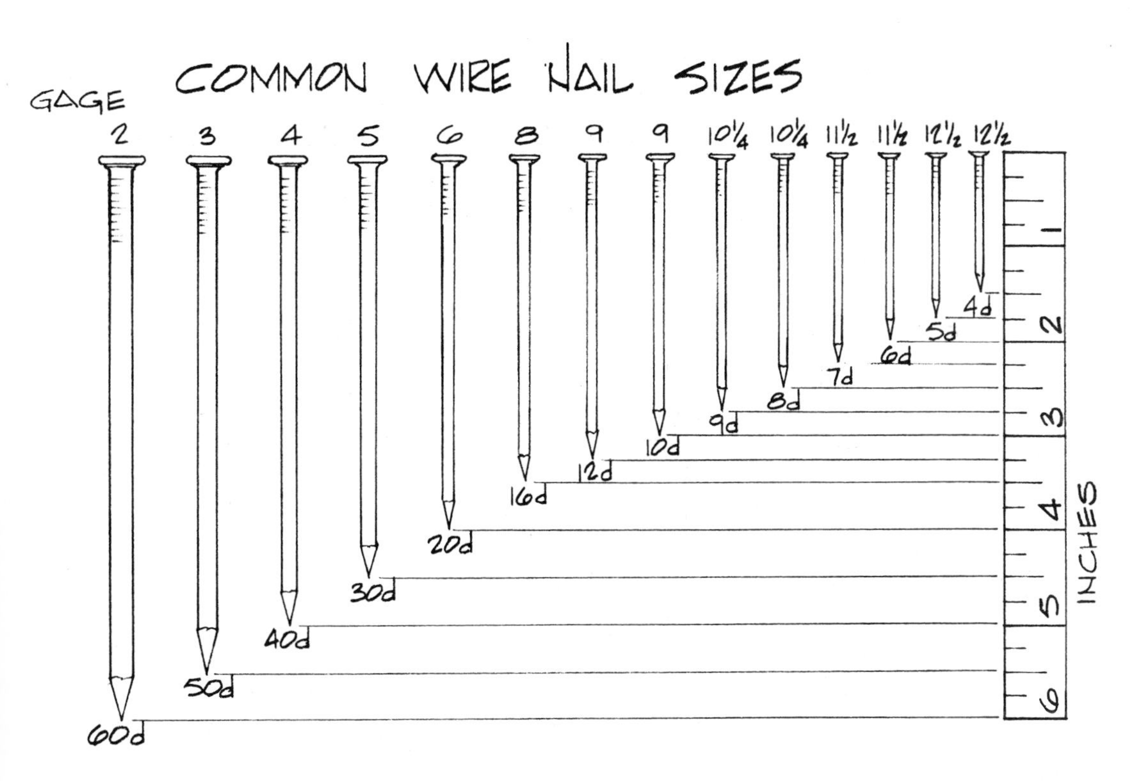

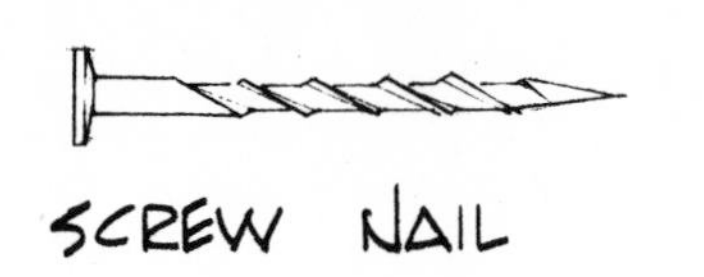

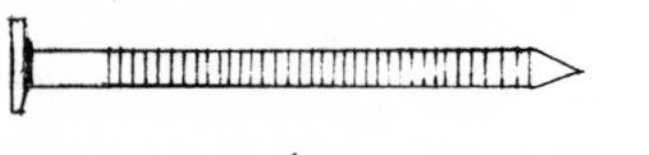

RING SHANK NAIL

FINISHING NAIL

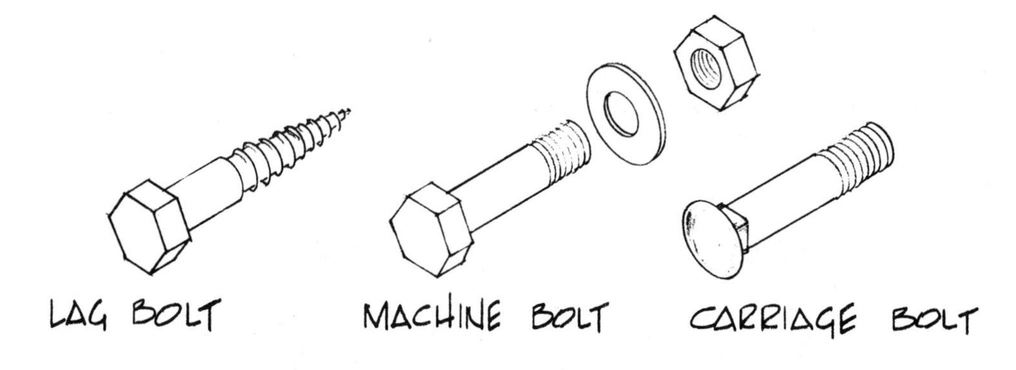

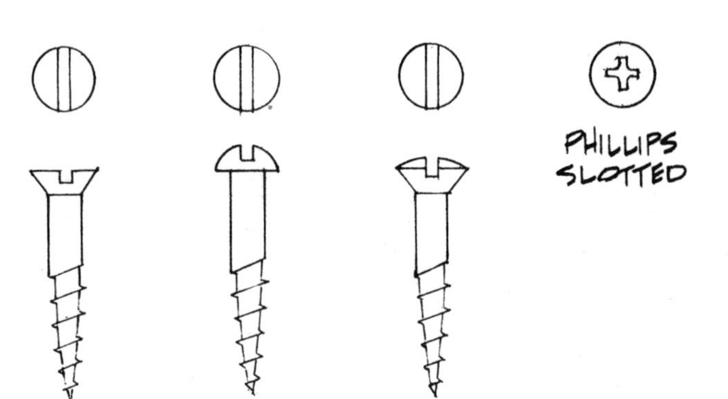

Bolts and screws will usually be designated on the framing sections and structural details of the plans, and should be taken off accordingly.

The nails used for rough framing will consist mainly of 6d, 8d, and 16d, common, box, or coated sinker nails. The 6d and 8d nails are used primarily for nailing plywood, board sheathing, and some framing toe-nailing. The 16d nails are used primarily for rough framing of wood members in a structure.

The least expensive way to buy nails is by the 50-pound keg or box. Most contractors buy them this way and carry the nails left over from one job to the next. This means the estimator usually approximates the nails needed for a job. The ability to do this is acquired by experience; it is not a simple task for the beginning estimator.

Many experienced estimators add off-the-head cost to their estimate for rough hardware. Others will add the cost for hardware by the square footage of the building, so many cents per square foot. This cost per square foot will be determined by the contractor's situation. For example, a contractor having full control of his materials on the job may add 10 to 12 cents per square foot for hardware. A contractor doing a large job, a tract or subdivision of homes, may not have full control of his hardware and may have to add 14 to 16 cents per square foot because of pilferage and theft.

When the estimator adds a lump-sum cost of this type to his bid, it will usually include framing hangers, anchors, and connectors, rough and finish nails, building paper, window and door flashing, caulking, bolts, and screws.

75

Finish Hardware

Finish hardware includes the hardware necessary for completing the doors, windows, and closets of a home. It will include the hinges, doorknobs, door bumpers, sliding or folding door tracks, and closet hardware that has previously been discussed in Unit 69 (Estimating Closet Trim).

Some of the details for finish hardware may be found on a finish schedule or in the specifications. In other cases there may be no mention of finish hardware at all, and the selection will be left up to the owner, contractor, or estimator.

Some of the finish hardware used in a typical home is illustrated below.

DOORKNOB SETS

PRIVACY

PASSAGE

EXIT

DUMMY KNOB

COMBINATION DEAD BOLT AND ENTRY LOCK WITH OPTIONAL TRIM

ENTRY HANDLE SETS

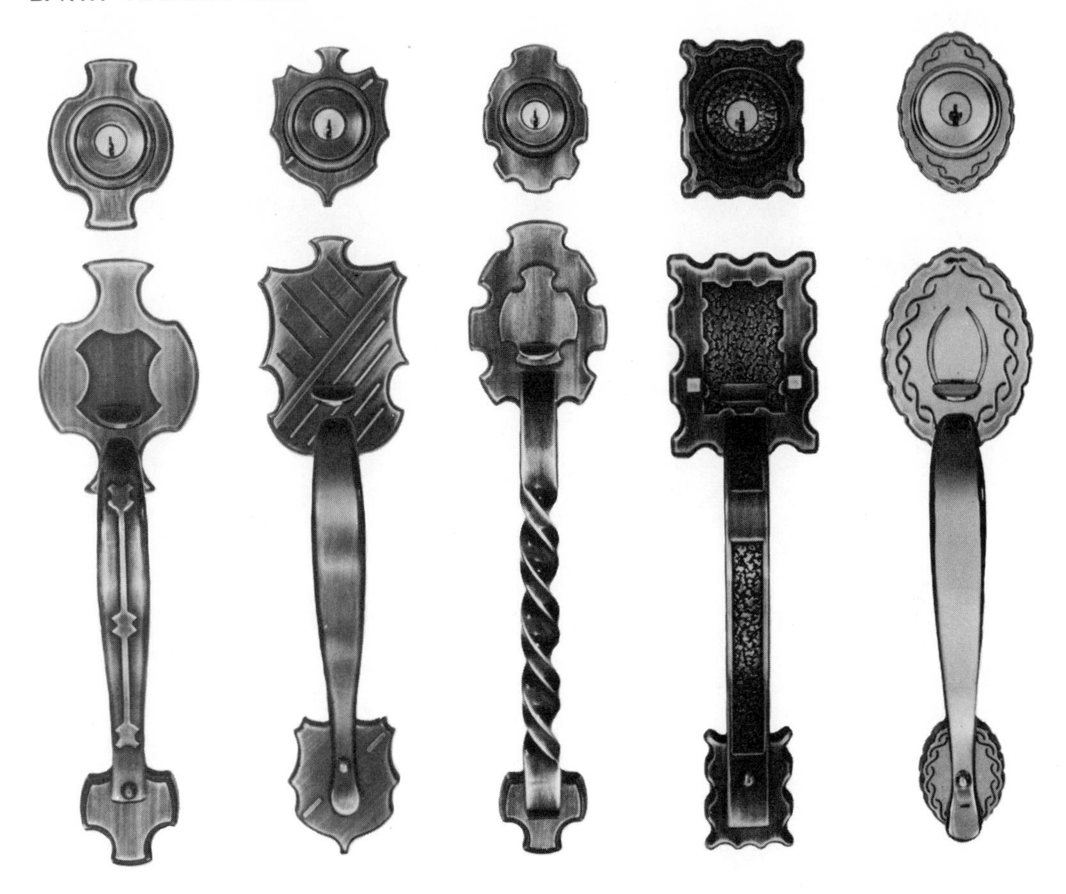

LEVER LOCK SETS

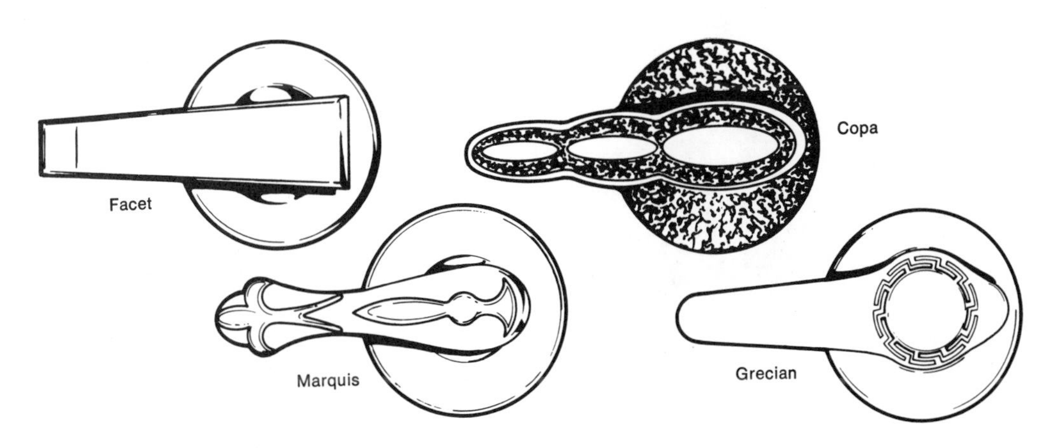

SLIDING POCKET DOOR HARDWARE

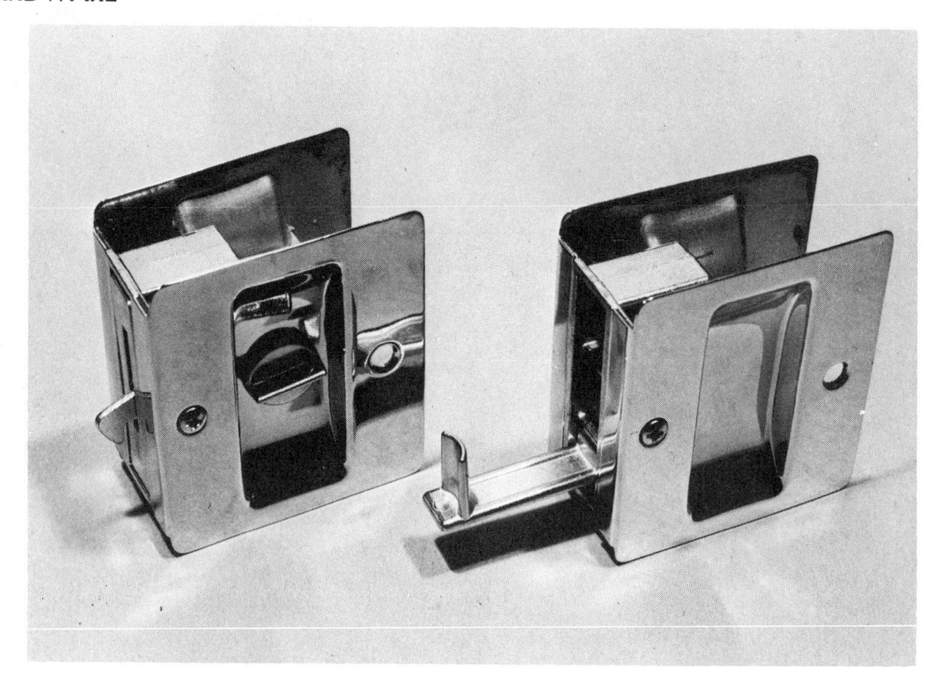

For passage type sliding doors not requiring a lock.

For bathroom or bedroom doors where privacy is desired. Locking turn button inside – emergency release outside.

DOOR TRIM ROSETTES

HARDWARE
BI-FOLD DOOR

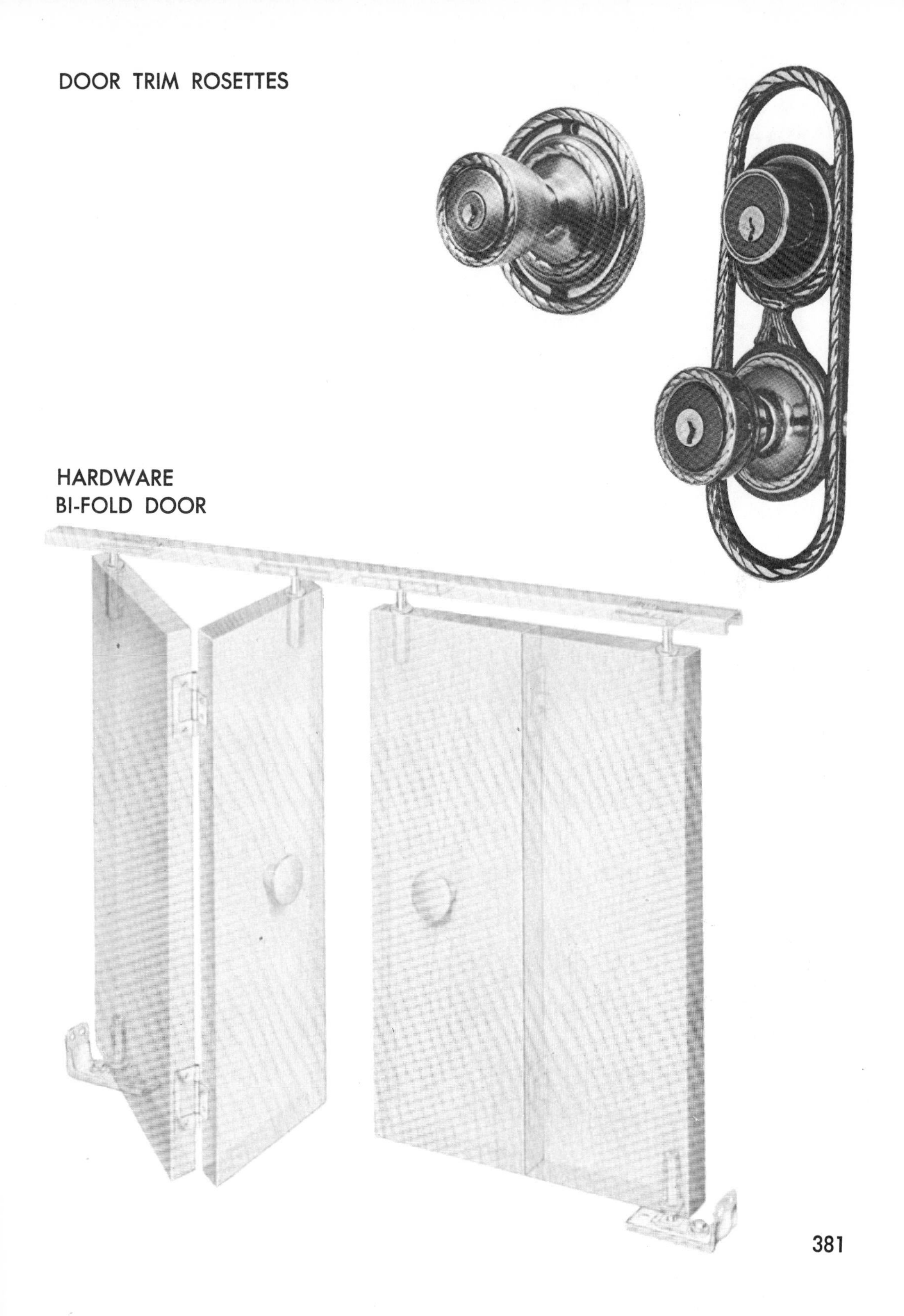

SLIDING DOOR HARDWARE

DOOR HINGES (BUTTS)

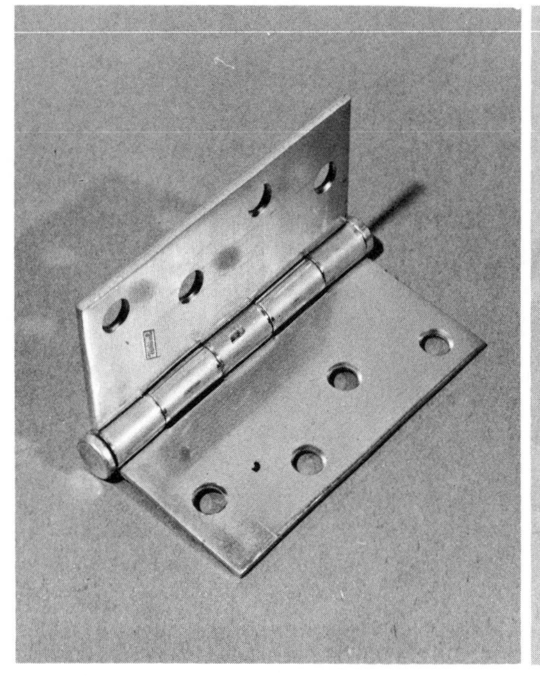

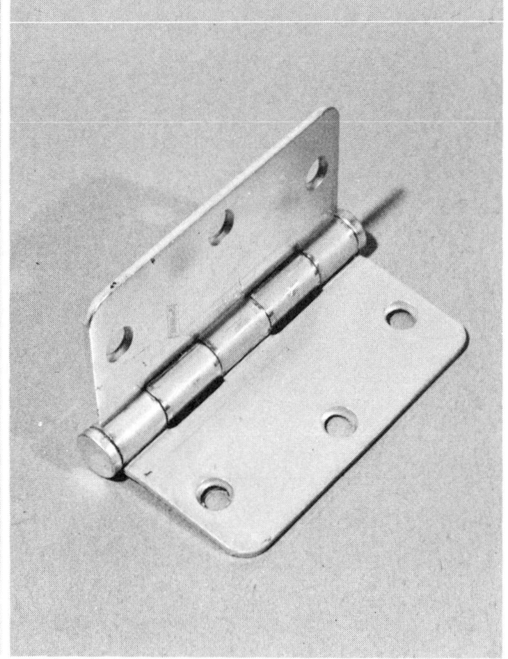

DOORSTOPS

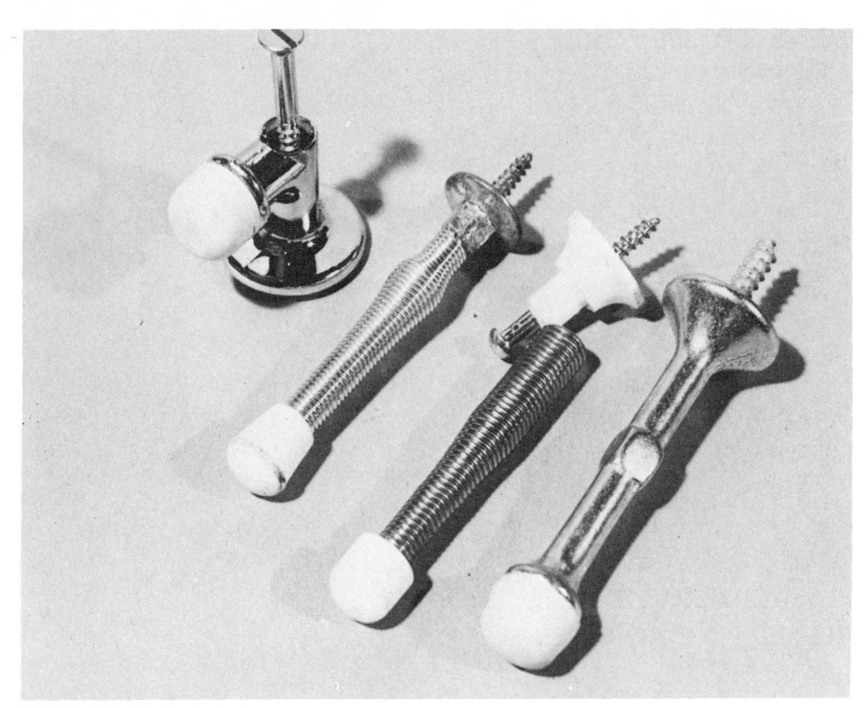

WINDOW HARDWARE

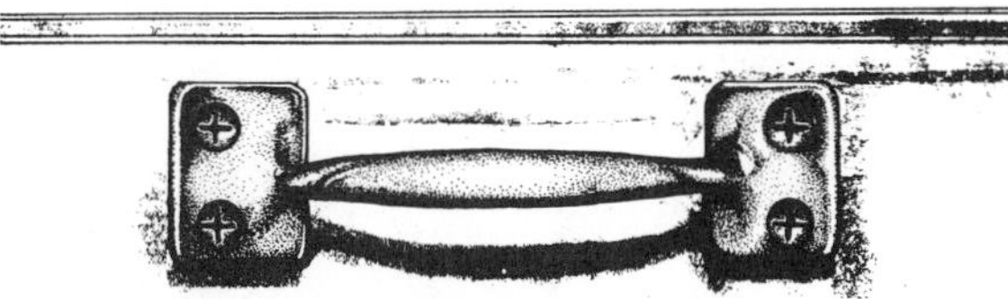

Sash Lift

Window Rail Lock

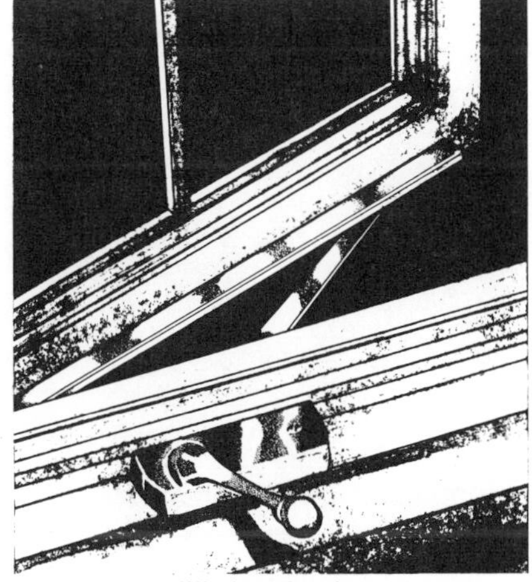

Window Crank

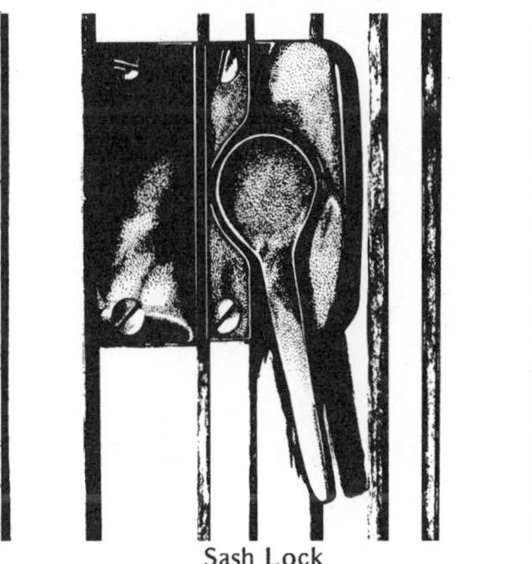

Sash Lock

METAL CLOSET POLE BRACKETS

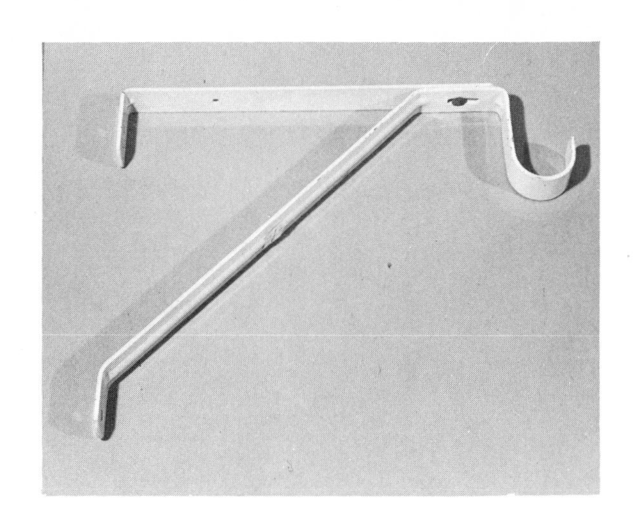

PLASTIC, METAL, OR WOOD CLOSET POLE ROSETTES

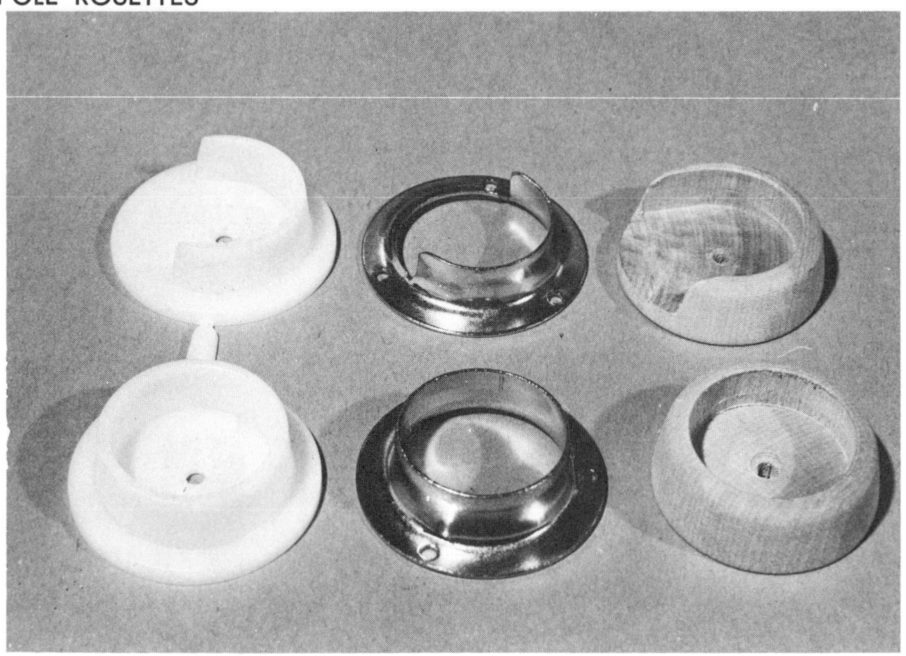

The best procedure for estimating finish hardware is to do a take-off room-by-room and then combine it into an organized hardware list.

76
Cabinet Hardware

Cabinet hardware, if there is any reference made to it, is usually found in the specifications. Often the specifications will state "Pulls to be selected by owner." In this case, the cabinet hardware is usually selected and prefit before the cabinets are finished.

Cabinet hardware consists of pull-type handles and knobs, hinges, latches, and catches. The contractor or estimator may state in the construction contract that the hardware is to be selected by the owner and is to be of builder's quality. Builder's quality refers to mass-produced cabinet hardware that may be found in most lumber and hardware stores. A certain amount has thus been set aside by the contractor for the hardware; but if the owner decides to upgrade the hardware, the contractor is entitled to the overage. Often arrangements can be made for the cabinet shop to provide the hardware after it has been selected by the owner. Or in other cases, the cabinet shop may include the hardware in its cabinet bid.

Some of the common types of hardware used today are illustrated here.

CABINET DOOR AND DRAWER PULLS

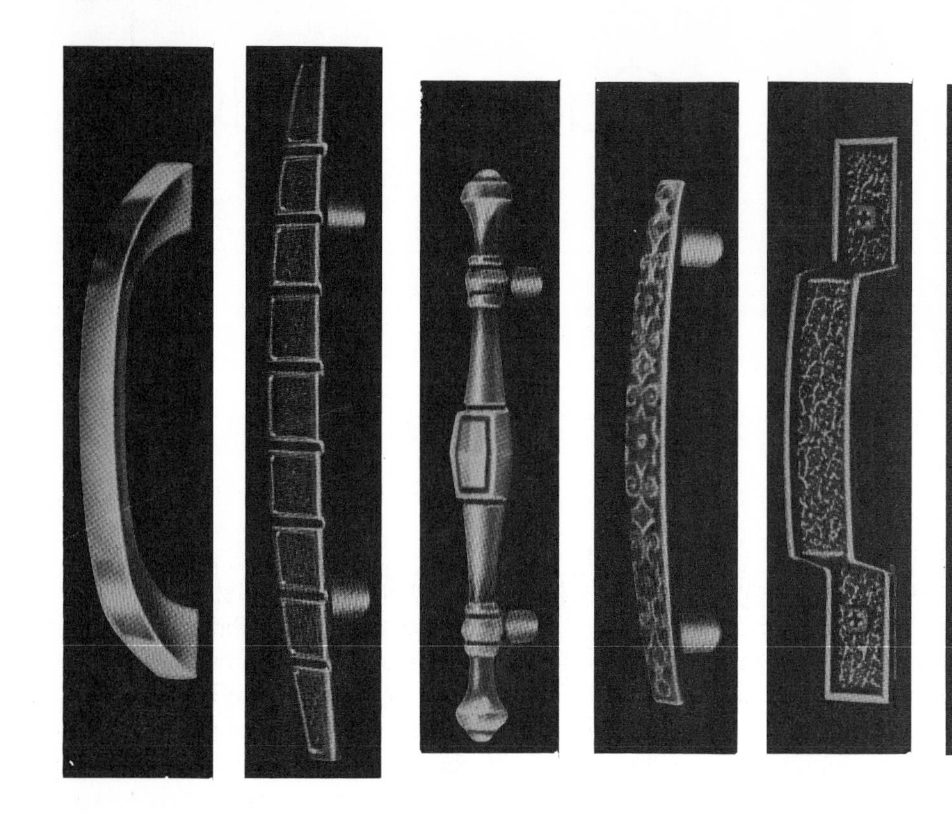

CABINET KNOBS

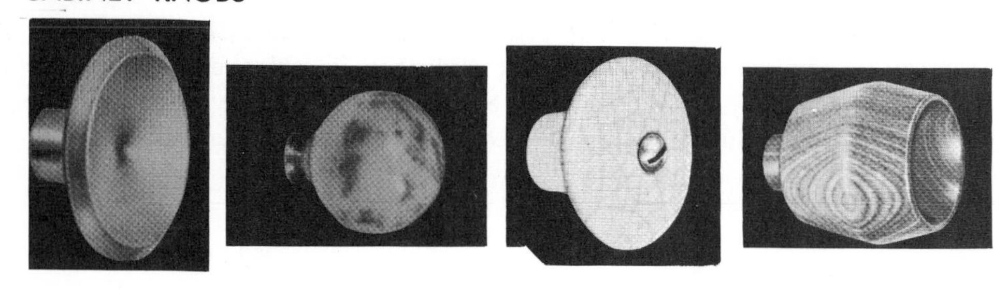

SELF-CLOSING
⅜″ INSET

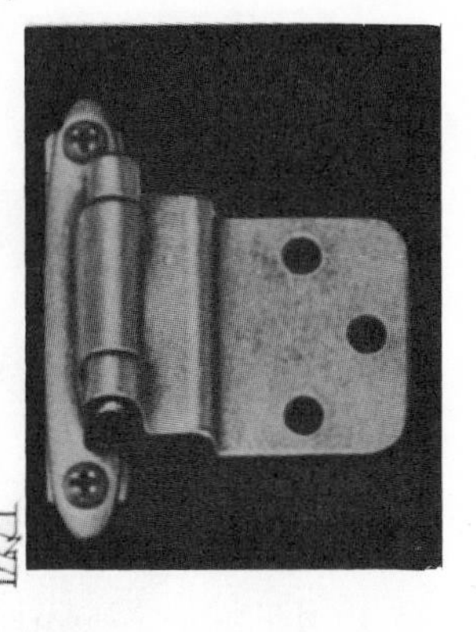

SURFACE HINGE FOR
⅜″ OFFSET DOORS

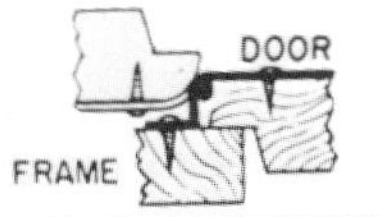

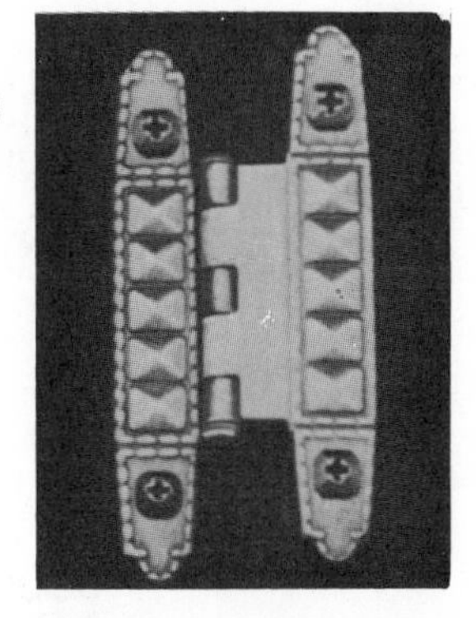

SELF-CLOSING
90° HI-KNUCKLE

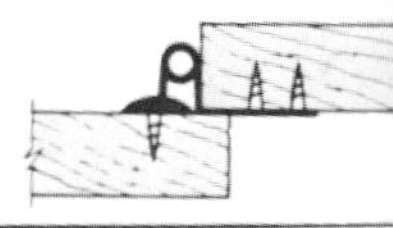

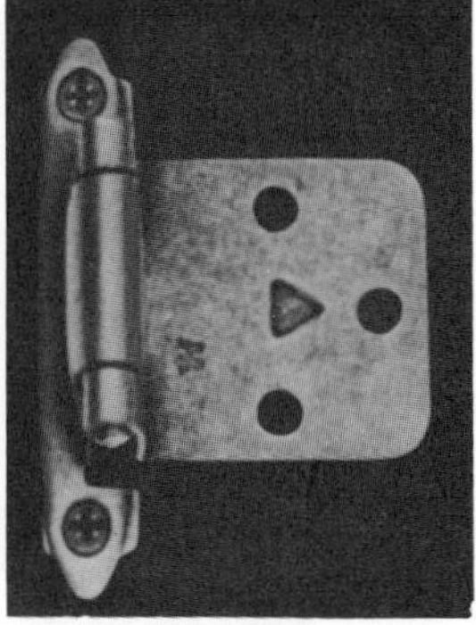

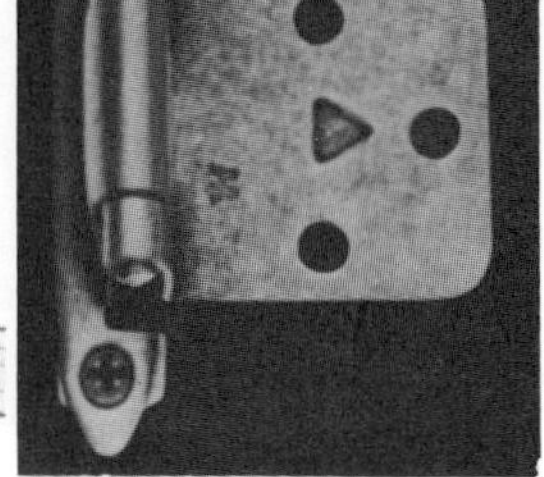

SELF-CLOSING
30° REVERSE BEVEL

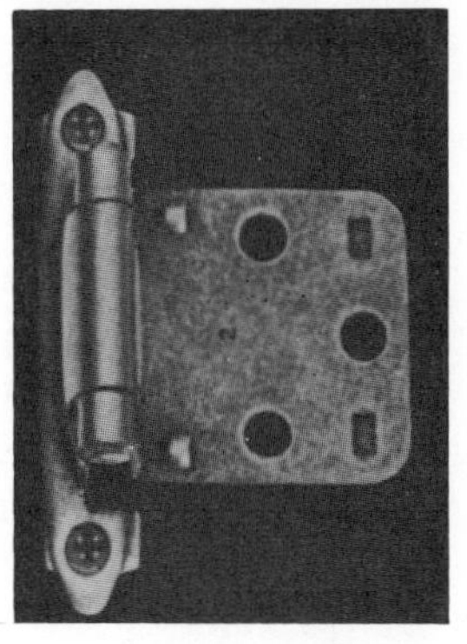

"H" SURFACE HINGE
FOR FLUSH DOORS

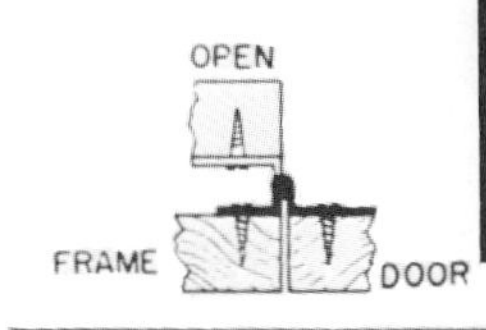

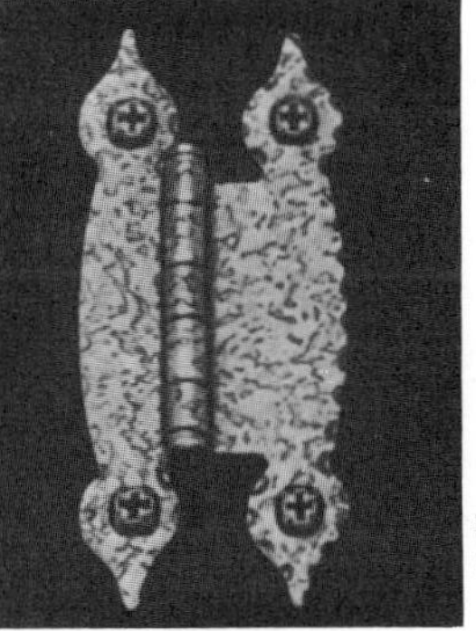

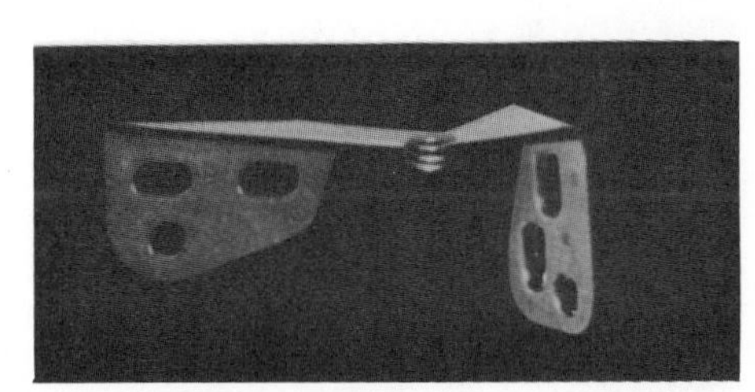

PIVOT HINGE—FRAME MOUNT FOR
¾″ FLUSH DOORS

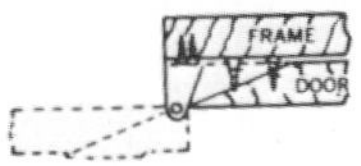

CABINET CATCHES

SINGLE ROLLER
CATCH W/2 STRIKES
FOR UNDER SHELF
STILE AND PARTITION
MOUNTING

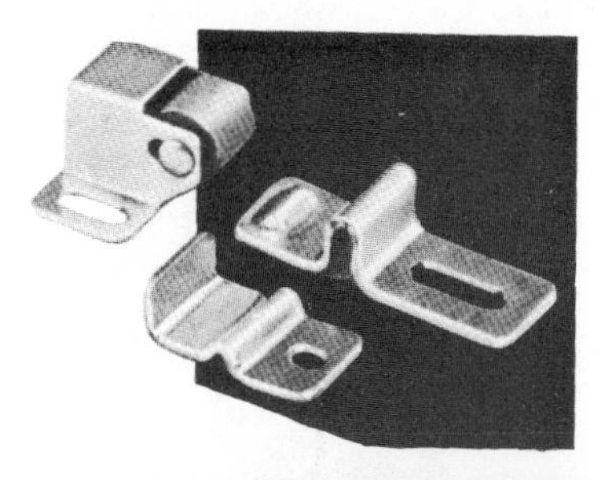

DOUBLE ROLLER
CATCH
SPRING-LOADED
W/WROUGHT STRIKE

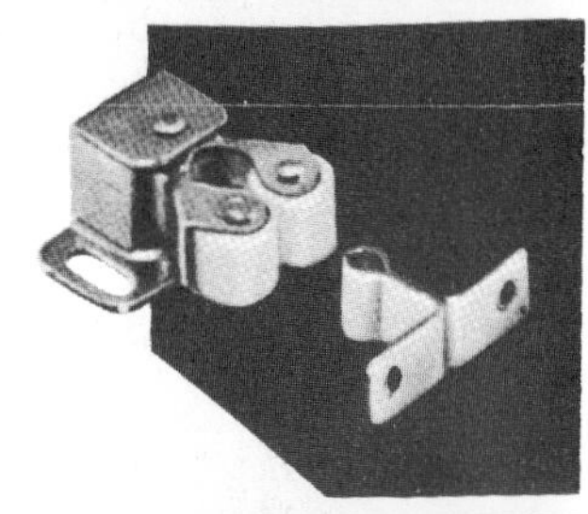

DOUBLE ROLLER
CATCH

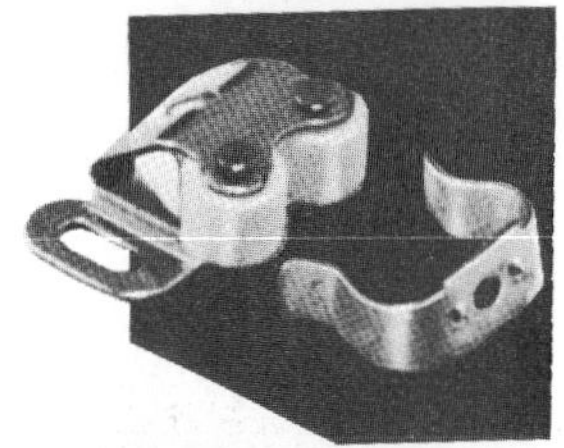

DOUBLE POST CATCH

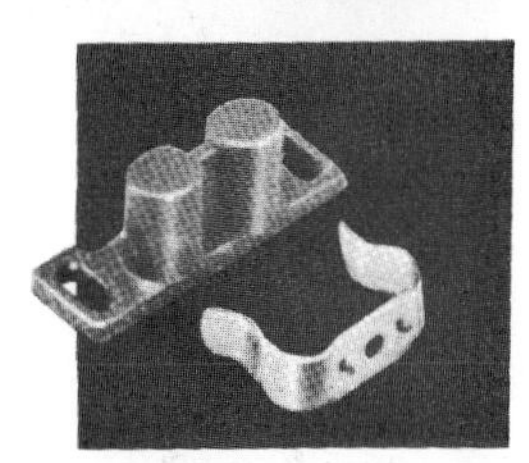

MAGNETIC CATCH
REVERSIBLE MOUNT

The number of hinges, pulls, and catches may be estimated from the cabinet details.

77

Miscellaneous Hardware

Miscellaneous hardware is the type of hardware that adds the extras. It is over and above the essential hardware necessary to turn out a house. These items are usually found in the specifications and include self-door closers, automatic garage door openers, dead bolts, special window and door locks, smoke detectors (sometimes required by the code), burglar alarms, mail slots, letter boxes, door knockers, street numbers, foot scrapers, etc., and small hardware fasteners such as eye bolts, hanger bolts, molly bolts, and toggle bolts.

Index